T0314976

Renewable Integrated Power System Stability and Control

IEEE Press
445 Hoes Lane
Piscataway, NJ 08854

Renewable Integrated Power System Stability and Control

Hêmin Golpîra
University of Kurdistan
Kurdistan, Iran

Arturo Román-Messina
The Center for Research and Advanced Studies of IPN
Guadalajara, Mexico

Hassan Bevrani
University of Kurdistan
Kurdistan, Iran

Registered Office
John Wiley & Sons, Inc., 111 River Street, Hoboken, NJ 07030, USA

Editorial Office
111 River Street, Hoboken, NJ 07030, USA

For details of our global editorial offices, customer services, and more information about Wiley products visit us at www.wiley.com.

Wiley also publishes its books in a variety of electronic formats and by print-on-demand. Some content that appears in standard print versions of this book may not be available in other formats.

Library of Congress Cataloging-in-Publication Data

Names: Golpîra, Hêmin, author. | Román-Messina, Arturo, author. | Bevrani, Hassan, author.
Title: Renewable integrated power system stability and control / Hêmin Golpîra, Arturo Román-Messina, Hassan Bevrani.
Description: Hoboken, NJ : Wiley-IEEE Press, 2021. | Includes index.
Identifiers: LCCN 2021003857 (print) | LCCN 2021003858 (ebook) | ISBN 9781119689799 (cloth) | ISBN 9781119689829 (adobe pdf) | ISBN 9781119689775 (epub)
Subjects: LCSH: Electric power systems. | Renewable energy sources.
Classification: LCC TK1001 .G65 2021 (print) | LCC TK1001 (ebook) | DDC 621.31/7–dc23
LC record available at https://lccn.loc.gov/2021003857
LC ebook record available at https://lccn.loc.gov/2021003858

Cover Design: Wiley
Cover Image: © kertlis/Getty Images

Set in 9.5/12.5pt STIXTwoText by SPi Global, Pondicherry, India

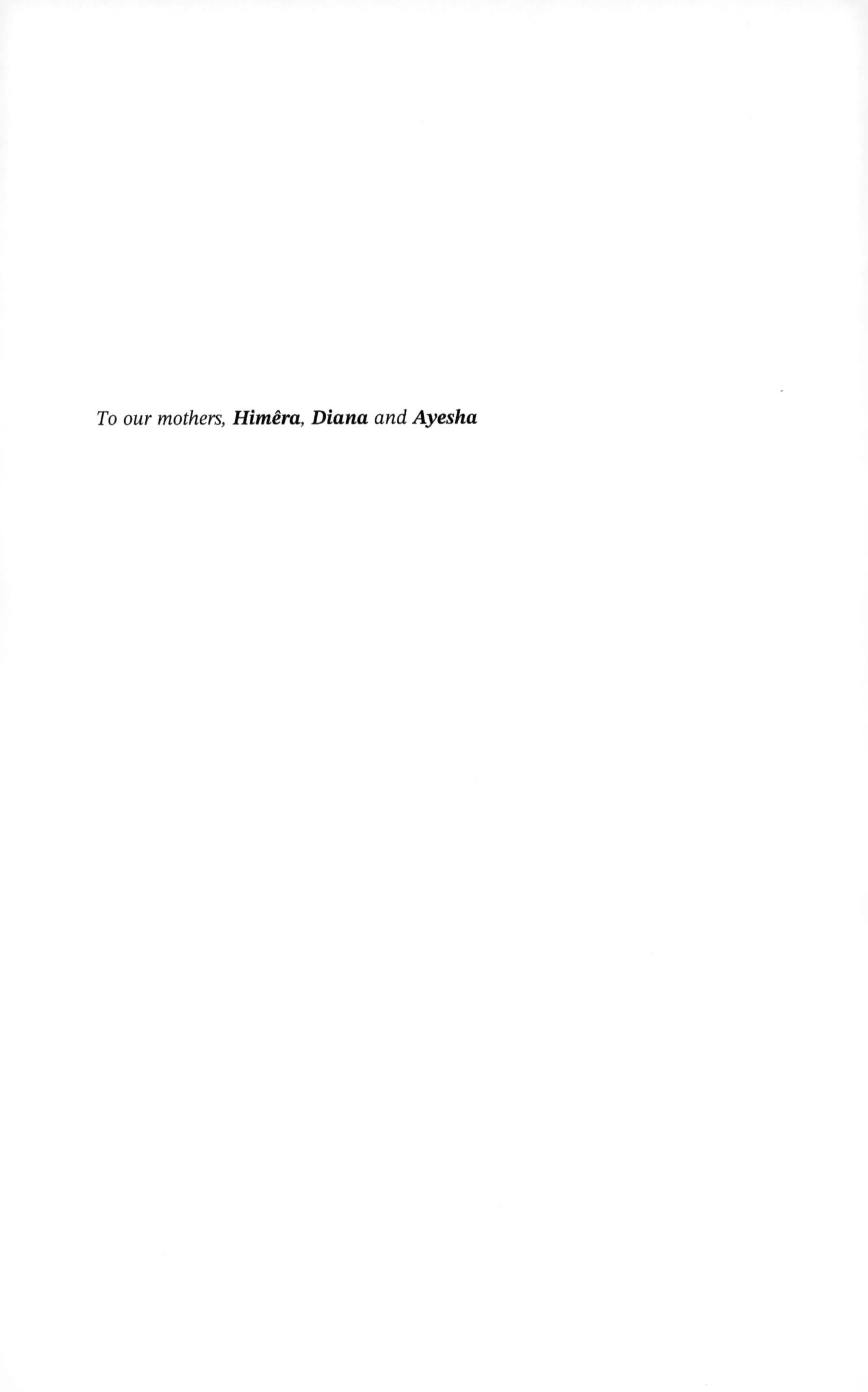

*To our mothers, **Himêra**, **Diana** and **Ayesha***

Contents

Preface

Increased needs for electrical energy as well as environmental concerns besides growing attempts to reduce dependency on fossil fuel resources have caused power system industries all around the world to set an ambitious target of renewable generation. Therefore, the capacity of installed inverter-based distributed generators (DGs) and renewable energy sources (RESs), individually or through the microgrids (MGs), in power systems is rapidly growing; and this increases the significance of *renewable integrated power system stability and control* as a challenging issue.

It is well known that low penetration of MGs/DGs has little influence on host grid stability and dynamics and thus the associated dynamics could be studied through simple power flow analysis. However, modern power grids face new technical challenges arising from the increasing penetration of power-electronic-interfaced MGs/DGs. Increasing renewable power penetration level may adversely affect frequency response, voltage and system control and lead to degraded performance of traditional control schemes. This, in turn, may result in large deviations and, potentially, system instability.

Moreover, the increasing penetration of inverter-interfaced DGs motivates the need to develop additional ancillary services to control undesired system dynamics. Advanced control of grid-connected MGs, however, has the potential to offset the intermittent nature of distributed energy resources and provide control support to the host utility during emergency conditions. To this end, new trends in power system modeling and dynamic equivalencing should be discussed.

In response to the above mentioned challenges, this book provides a comprehensive coverage of penetrated power grid dynamic analysis, data-driven estimation, modeling, and control synthesis. This book summarizes a long-term academic/industry research outcome and contributions and is influenced by the authors' practical experiences on power system dynamics and stability, MGs, electric network monitoring, and control and power electronic systems in several countries, universities, and power electric companies. The book also provides a thorough understanding of the basic principles of the penetrated power system modeling, model order reduction, and grid-connected MG equivalent model derivation.

One of the main parts of this book covers the modeling of power systems using PMU data for the purpose of renewable integrated dynamics identification and parameters estimation issues, as well as oscillation damping, voltage control, and frequency control design problems. For this purpose, in addition to real network data, several standard IEEE power system models are used as benchmarks for generating data that are used in system identification. Furthermore, in addition to dynamic stability analysis and controller synthesis, inertia challenge requirements and control levels are discussed, and recent advances in visualization of virtual synchronous generators (VSGs) and the associated effects on system performance are addressed.

This book could be useful for engineers and operators working on power systems dynamic, control, and operation, as well as postgraduate students and academic researchers. The book describes renewable integrated power system dynamics modeling and control issues from introductory to the advanced steps. This book is organized into eight chapters.

Chapter 1 discusses the term of power system stability and control with an updated brief review on the areas of frequency, voltage, and angle controls, concerning the penetration of RESs/DGs. In response to the existing challenges in penetration of more RESs/DGs to the grid, the necessity of using data-driven modeling, parameters estimation, and control synthesis in wide-area power systems is emphasized; a general scheme for wide-area measurement system and wide-area control is described.

Chapter 2 deals with dynamic equivalencing of penetrated power grid. Several methods are introduced to model the host grid as well as the distribution network. A center of gravity (COG)-based equivalent model is addressed to represent the power system dynamic behavior in terms of slow power and frequency dynamics. The relationship between the frequency of the COG and the motion of local centers of angle is analytically determined to compute local frequency deviations following major disturbances.

Chapter 3 addresses the power grid stability analysis from frequency, small signal and voltage points of view. Some analytical approaches have been discussed to determine maximum penetration level of MGs concerning the upward system stability. The given methods explicitly rely on the basic power system equations which, in turn, make the proposed indices completely independent of the test case; and this helps to solve the associated difficulties with the system dimensions.

Chapter 4 explains the VSG concept and its applications in renewable integrated power grids. The positive dynamic impacts of VSGs in a power system are discussed, and recent relevant achievements in the application of the advanced control methods in emulating virtual inertia are clarified. Afterwards, according to relevant dynamical metrics, dispatchable inertia is optimally placed in the system to enhance system stability and dynamics performance.

Chapter 5 examines the application of a measurement-based analysis technique to identify voltage control areas in renewable integrated power systems. The proposed technique combines the inherent abilities of graph theoretical techniques with spectral clustering and visualization methods to identify voltage control areas and reconstruct system behavior. The evaluation of voltage stability problems is done using both static and dynamic techniques. Numerical relevant issues are also discussed.

Chapter 6 proposes advanced control schemes rely on inertia manipulation in the system to improve frequency, voltage, and small signal stabilities. The proposed control approaches, which use stochastic equivalent model of power system (to enable high penetration levels of MGs), combine an adaptive dispatch strategy for energy storage systems with an MG-controlled islanding scheme to provide stability support for the host power grid.

Chapter 7 addresses some important issues in understanding the oscillatory performance of wind and solar PV penetrated power systems. A fundamental study of the characterization of power system dynamic behavior with increased RESs is presented. The study is motivated by the need to further clarify the participation of wind and PV farms in inter-area oscillations. The performance of data-driven model extraction techniques using two simple and complex power system examples is evaluated. Some analytical criteria to describe the energy relationships in the observed oscillations are derived, and a physical interpretation for the system modes is suggested.

Chapter 8 describes an experience in the analysis of wind and solar integration in a large-scale practical power system to examine the impact of high variable renewable generation on power system security. Both active and reactive power control strategies are considered. The study assesses the impact of large amount of wind and PV on system dynamics and identifies ways of improving system dynamic performance and stability through control and operating practices.

Hêmin Golpîra, *University of Kurdistan*
Arturo Román-Messina, *The Center for Research and Advanced Studies of IPN*
Hassan Bevrani, *University of Kurdistan*

January 2021

Acknowledgments

Most of the contributions, outcomes, and insight presented in this book were achieved through a long-term teaching and research cooperation on the renewable integrated power systems over the last 15 years. The materials given in the present book are mainly the research outcomes and original results of authors in Smart/Micro Grids Research Center-*SMGRC*, University of Kurdistan (Sanandaj, Iran) and in The Center for Research and Advanced Studies-*CINVESTAV* of the National Polytechnic Institute of Mexico (Guadalajara, Mexico). It is a pleasure to acknowledge the received supports from these sources, and the awards from Iran Grid Management Company (IGMC), Iran National Science Foundation (INSF), and Alexander von Humboldt (AvH) Foundation.

The authors would like to thank their colleagues Prof. J. Raisch and Prof. Bruno Francois for their kind support. Finally, the authors offer their deepest personal gratitude to their families for their patience during the preparation of this book.

Nomenclature

$\boldsymbol{X},\boldsymbol{Y}$	data matrices
ε	Kernel bandwidth
λ	eigenvalue
σ	singular value
ξ	damping
$\boldsymbol{\upsilon}$	vector of natural modes
$\boldsymbol{\psi},\boldsymbol{\varphi}$	eigenvectors
$\boldsymbol{\Psi\Phi}$	matrices of eigenvectors
$\boldsymbol{\Gamma}$	inflation operator
θ	phase angle
$\boldsymbol{\Lambda}$	diagonal matrix of eigenvalues
$\boldsymbol{\Sigma}$	diagonal matrix of singular values
δ^{s}	mechanical rotor angle (rad)
δ	rotor angle position (rad)
δ_{COI_i}	rotor angle position of the COI (rad)
ω	angular speed (rad/s)
ω^{s}	mechanical rotor angular speed (rad/s)
ω_0	rated angular speed (rad/s)
$T_m(t)$	mechanical input torque (*p.u.*)
$T_e(t)$	electrical output torque (*p.u.*)
M	inertia constant of the system (s)
M_i^{ESS}	inertia of ESS in area *i* (s)
$M'^{ESS}_{i,\min}$	minimum required ESS inertia, in compliance with RoCof, in area *i* (s)
$M''^{ESS}_{i,\min}$	minimum required ESS inertia, in compliance with frequency nadir, in area *i* (s)
D	damping coefficient
$I(t)$	impulse response of the system
$P(n)$	data sequence of interest

P_{in}	injected power of ESS to the host grid
K	number of sinusoidal components in noise
L	length of $P(n)$
L_x, L_y	latent variables
J	moment of inertia
a_k	magnitude
Φ_k	initial phase angle
ω_k	harmonic frequency in radius
A_k	complex magnitude of the kth-harmonic
s_i	eigenvectors associated with the noise subspace
e	signal eigenvector
e^U	complex-conjugate transpose of e
C_{cap}	capital costs ($/kW)
C_{PCS}	power conversion system costs ($/kW)
C_{stor}	storage section costs ($/kWh)
C_{BOP}	power balance costs ($/kW)
t_{ch}	charging/discharging time (h)
$C_{O\&M}$	operation and maintenance costs ($/kW-year)
$C_{R,a}$	annualized replacement costs ($/kW-year)
$C_{cap,a}$	annualized total capital costs ($/kW-year)
$C_{LCC,a}$	annualized life cycle costs ($/kW-year)
CRF	capital recovery factor
C_R	replacement costs ($/kWh)
$C_{FOM,a}$	fixed operation and maintenance costs ($/kW-year)
$C_{VOM,a}$	variable operation and maintenance costs ($/kWh)
n_{cycle}	number of discharge cycles per year
ζ_c	charging efficiency of the battery (%)
ζ_d	discharging efficiency of the battery (%)
η	power angle-based stability index
i (j)	area (bus) index
f	frequency (Hz)
$P_{\bullet}^{tie}$	virtual transferred power (pu)
$X_{\bullet}^{tie}$	fictitious reactance (pu)
$T^{tie}{}_{COIi,j}$	applied torques from bus j to COI
$T^{tie}{}_{COIi,COG}$	applied torques from COG to COI
A_i	area i
ΔP_i	size of disturbance in area i
ξ^-	deviations from the target value in negative direction

ξ^{+}	deviations from the target value in positive direction
ξ	target value
p_s	probability of each scenario
s	scenario counter
IC	internal combustion
SM	synchronous machine
M_P	slope of P-ω droop
K_{PI}	integral control gain
K_{PP}	proportional control gain
F_{CMD}	command fuel signal
E_{CMD}	exciter control signal
P_{meas}	measured value of real power
Q_{meas}	measured value of reactive power
$I.$	line current
K_{tf}	torque to fuel conversion ratio
η_{thr}	thermal constant
K_{cv}	calorific value
K_{fr}	fuel rate at rated speed
K_m	mechanic losses constant
τ_e	exciter machine time constant
P_{MG}	injected power of MG to the host grid
ω_{MGs}	angular speed at the point of common coupling
ζ	DGs re-dispatching time (s)
υ	DGs islanding time (s)
n	number of areas
β	frequency bias
P_{Genset}	generation of Genst
L	level arm length (m)
$M_{COI_i}^{Conv.}$	conventional synchronous inertia (s)
T_D	delivery time of primary frequency response (s)
$K_f(s)$	transfer function of the phase-locked loop
M^{MMG}	muti-micro-grid inerta constant (s)
V_0	initial values of terminal voltage
$\mathbf{Y}_{\mathrm{H}}$	Hankel matrix
$\lVert . \rVert$	norm

List of Abbreviations and Acronyms

A	amplitude
AGC	automatic generation control
AQR	automatic reactive power regulator
AVR	automatic voltage regulator
B	residue
BFV	best fitness value
COG	center of gravity
COI	center of inertia
DER	distributed energy resources
DFIG	doubly-fed induction generator
DG	distributed generation
DM	diffusion map
DMD	dynamic mode decomposition
E	energy
EMT	electromagnetic transient
ESS	energy storage system
EV	electrical vehicle
GA	genetic algorithm
HVDC	high-voltage direct current
KT	Kumaresan–Tuft
LCC	line commutated converter
MCL	Markov clustering
MG	microgrid
MMGs	multi-MGs
NERC	North American Electric Reliability Corporation
NYNE	New York New England
PC	principal component
PCTVAR	percentage of variation
PF	participation factor

PI	proportional-integral
PLL	phase-locked loop
PLS	partial least squares regression
PLSC	partial least squares correlation
PMSG	permanent magnet synchronous generators
PMU	phasor measurement unit
POIS	point of interconnection with the system
PS	pseudo spectrum
PSS	power system stabilizers
PV	photovoltaic
RES	renewable energy source
RoCoF	rate of change of frequency
SC	synchronous condenser
SCADA	supervisory control and data acquisition
SG	synchronous generator
SLB	static load bank
SOC	state of charge
SS	static switch
SVC	static VAR compensator
SVD	singular value decomposition
T-D	time domain
TSO	transmission system operator
UCTE	Union for the Coordination of the Transmission of Electricity
UFLS	underfrequency load shedding
ULTC	under load tap changer
VAR	volt–ampere reactive
VSC	voltage source converter
VSG	virtual synchronous generator
V2G	vehicle-to-grid
$\mathbf{V}_{vand}$	Vandermonde matrix
WAMS	wide-area measurement system
WF	wind farm
WT	wind turbine
z	complex amplitude

1 Introduction

The term *power system stability and control* is used to define the application of control theorems and relevant technologies to analyze and enhance the power system functions during normal and abnormal operations. Power system stability and control refers to keep desired performance and stabilizing power system following various disturbances, such as short circuits, loss of generation, and load.

The capacity of installed inverter-based distributed generators (DGs) and renewable energy sources (RESs) individually or through the microgrids (MGs) in power systems is rapidly growing, and a high penetration level is targeted for the next few decades. In most countries including developing countries, significant targets are considered for using the distributed microsources and MGs in their power systems for near future. The increase of DGs/RESs in power systems has a significant impact on CO_2 reduction; however, recent studies have shown that relatively high DGs/RESs integration will have some negative impacts on power system dynamics, frequency and voltage regulation, as well as other control and operational issues. Decreasing system inertia and highly variable dynamic nature of DGs/RESs/MGs are known as the main reasons. These impacts may increase for the dynamically weak power systems at the penetration rates that are expected over the next several years.

In this chapter, a brief discussion on the power system stability and control in modern renewable integrated power systems and the current state of this topic are given. Data-driven wide-area power system monitoring and control is emphasized, and the significance of measurement-based dynamic modeling and parameter estimation is shown.

1.1 Power System Stability and Control

Power system stability and control was first recognized as an important problem in 1920s [1]. Over the years, numerous modeling/simulation programs, synthesis/

Renewable Integrated Power System Stability and Control, First Edition.
Hêmin Golpîra, Arturo Román-Messina, and Hassan Bevrani.

analysis methodologies, and protection schemes have been developed. Power grid control must provide the ability of an electric power to regain a state of operating equilibrium after being subjected to a physical disturbance, with most system variables, i.e., frequency, voltage, and angle, bounded so that practically the entire system remains intact. Thus, the main control loops are known as frequency control, voltage control, and rotor angle (power oscillation damping) control [2].

In many power systems, advanced measurement devices such as phasor measurement units (PMUs) and modern communication devices are already being installed. Using these facilities, the parameters of existing power system controllers can be adjusted by an online data-driven control mechanism [3]. The PMU data after filtering are used to estimate some important parameters in the system (scheduling parameters). These parameters are then used in the control tuning algorithm that will adapt the controller parameters in frequency control, voltage control, and power oscillation control. Therefore, the controller's parameters are adapted according to the current status of the system.

One of the important steps of reliable and performant control system design is defining the performance specifications. It depends on the features of the controller design method, the constraints on the controller structure, the achievable performance that is limited by the physical constraints, the industrial standards on the limit of the variables, the limits of the actuators, etc. Finding the control specifications and making them compatible with the controller design approach require a deeper understanding of the physical system to be controlled.

The characteristics of three main control loops, i.e., frequency control, voltage control, and angle control, should be studied to enable the definition of achievable performance specifications and designing an effective control system.

- *Frequency control:* Since the frequency generated in an electric network is proportional to the rotation speed of the generator, the problem of frequency control may be directly translated into a speed control problem of the turbine generator unit. This is initially overcome by adding a governing mechanism that senses the machine speed and adjusts the input valve to change the mechanical power output to track the load change and to restore frequency to nominal value. Depending on the frequency deviation range, different frequency control loops, i.e., primary, secondary, and tertiary, may be required to maintain power system frequency stability [4].

  The secondary frequency control which is also known as load frequency control (LFC) initializes a centralized and automatic control task using the assigned spinning reserve. The LFC is the main component of an automatic generation control (AGC) system [5]. In large power systems, this control loop is activated in the time frame of few seconds to minutes after a disturbance. In a modern AGC system, based on the received area control error (ACE) signal, an online tuning algorithm must adjust the LFC parameters to restore the frequency and tie-line powers to the specified values.

- *Voltage control:* The generators are usually operated at a constant voltage by using an automatic voltage regulator (AVR) which controls the excitation of the machine via the electric field exciter system. The exciter system supplies the field winding of the synchronous machine with direct current to generate required flux in the rotor. A system enters a state of voltage instability when a disturbance changes the system condition to make a progressive fall or rise of voltages of some buses. Loss of load in an area, tripping transmission lines, and other protected equipment are possible results of voltage instability. Like frequency control, the voltage control is also characterized via several control loops in different system levels. The AVR loop which regulated the voltage of generator terminals is located on lower system levels and responds typically in a time scale of a second or less.
- *Angle control:* Rotor angle stability is the ability of the power system to maintain synchronization after being subjected to a disturbance. Angle stability refers to damping of power oscillations inside subsystems and between subsystems on an interconnected grid during variation beyond specified threshold levels. The risk of losing angle stability can be significantly reduced by using proper control devices inserted into the power grid to find a smooth shape for the system dynamic response.

  The power oscillation damping has been mainly guaranteed by power system stabilizers (PSSs). A PSS is a controller, which, beside the turbine-governing system, performs an additional supplementary control loop to the AVR system of a generating unit. Depending on the type of PSS, the input signal could be the rotor speed/frequency deviation, the generator active power deviation, or a combination feedback of rotor speed/frequency and active power changes. This signal to be passed through a combination of a lead-lag compensators. The PSS output signal is amplified to provide an effective output signal.

  In order to damp the inter-area oscillations, which have smaller oscillation frequency than the local oscillatory modes, a wide-area control (WAC) system is required. The WAC system is a centralized controller that uses the PMU signals and produces auxiliary control signals for the PSSs.
- *Virtual synchronous generator:* Additional flexibility may be required from various control levels so that the system operator can continue to balance supply and demand on the modern power grids in the presence of DGs/RESs/MGs. The contribution of DGs/RESs in regulation task refers to the ability of these grids to regulate their power output, by an appropriate control action. This can be regarded as adding virtual inertia to the grid and considered as a solution. Virtual inertia emulation requires the inverter to be able to store or release an amount of energy depending on the grid frequency's deviation from its nominal

value, analogous to the inertia of a conventional generator. This setup, which is known as virtual synchronous generator (VSG), will then operate to emulate desirable dynamics, such as inertia and damping properties, by flexible shaping of its output active and reactive powers as conceptually shown in Figure 1.1.

This VSG provides a promising solution to improve power grid stability and performance in the presence of a high penetration of DGs/RESs/MGs. The VSG is not only applicable for improving of frequency regulation and oscillations damping, particularly during the transient state following a disturbance, but also it is useful to support the voltage stability. The VSG system can use the available DGs/RESs, as primary sources to participate in power oscillation damping by adjusting their active and reactive power generations. The VSG is more discussed in Chapter 4.

1.2 Current State of Power System Stability and Control

Power system stability and control can take different forms, which are influenced by the type of instability phenomena. A survey on the basics of power system controls, literature, and achievements is given in [6, 7].

PMUs are sophisticated digital recording devices that communicate global positioning system (GPS) synchronized high sampling rate dynamic power system's data to the central control and monitoring stations. The recorded data by PMUs provide valuable information about the dynamic of the power system that can be used for data-driven modeling. An overview of system identification techniques for modeling of power systems using PMU data is given in [8]. In [9], a subspace identification method is used to identify a reduced order model for power oscillation control. The PMU data are used for the calibration of the parameters of the reduced-order model of a power generator in [10]. The feasibility of multi-input multi-output (MIMO) identification of power systems using low-level

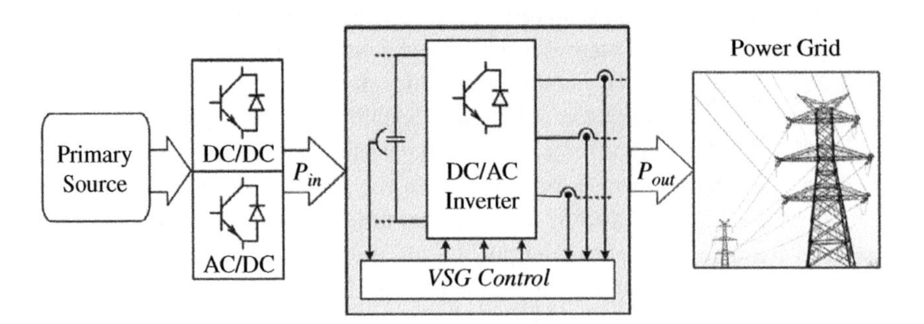

Figure 1.1 Conceptual structure of a virtual synchronous generator.

probing signal is shown in [11]. An online algorithm is used in [12] to identify the frequency response of power system dynamics, while it is combined with a selective modal analysis. The transfer function and state-space model identifications using PMU data are compared in [13] for electromechanical oscillation damping estimation. Several identification methods are compared for analysis of inter-area oscillatory modes of power systems [14].

The data from PMUs have already been used for estimation of some important power system parameters. The electromechanical modes of a power system and their confidence intervals are estimated using PMUs operational data in [15, 16]. Amplitude, frequency, and damping of power system oscillations are estimated using PMU measurements in [17, 18]. The PMU data are used in [19–21] to identify the topology (or change in topology) of a power system. Recently, some system identification methods have been employed to estimate the power system inertia using the operational PMU data (with no external excitation signal) [22, 23].

1.2.1 Frequency Control

Preliminary efforts in the field of power grid frequency regulation are reported in [24]. Subsequently, an IEEE working group prepared some standard definitions of significant terms and concepts on power system frequency control [25]. Considering the physical constraints and to cope with the advances in technologies and the changed system environment, dynamic modeling developments, security constraints, and communication delays, as well as modifications on the frequency control definitions, have been discussed over the years [26–30]. A comprehensive survey and exhaustive bibliography on frequency control up to 2014 are given in [31, 32].

Frequency control analysis, frequency response modeling, nonlinearity and uncertainty presentation, specific applications, frequency bias calculation, control performance standards, load characteristics impacts, and parameters identification are presented in several documents [29, 32–42]. A Considerable research on the time-delayed system is contained in [4, 30]. In addition, regarding parametric uncertainty, several self-tuning, adaptive, and robust control strategies are widely applied for power grid LFC system synthesis over the years [4, 43–51].

Dynamic impacts of intermittent DGs and high penetration of RESs on power grids frequency response are discussed in [32, 52–56]. A low inertia can negatively affect the grid frequency dynamic performance and stability. A number of recent works have suggested the application of inverter-based virtual inertia emulators to improve frequency stability and frequency response performance [57–61]. Furthermore, numerous research works have been recently focused on the use of DGs, RESs, MGs, electric vehicles, and storage devices to provide frequency control supports in the power grids [62–69]. Providing frequency control support via controllable loads and smart load technologies using the concepts of demand response (DR) is discussed in [41, 42, 70–76]. Two recent works in this area are [77, 78], that discuss the impact of a high integration of MGs on the frequency control of power systems, and propose a decentralized stochastic frequency control of MGs.

PMU-based/data-driven online tuning frequency control approach is not addressed in the abovementioned worldwide published works. In most cases, the secondary frequency control is designed using conventional frequency response model, which is very difficult to realize in a modern power grid with a highly variable structure and penetration of DGs/RESs.

1.2.2 Voltage Control

Since 1990s, supplementary control of generator excitation systems, static var compensator (SVC), and high voltage direct current (HVDC) converters is increasingly being used to solve power system oscillation problems [7]. There has also been a general interest in the application of power electronics-based controllers known as flexible alternating current transmission system (FACTS) controllers for the damping of system oscillations [79]. Following several power system collapses worldwide [80–82], in 1990s, voltage stability has attracted more research interests.

Recently, following the development of PMUs, communication channels, and digital processing, wide-area power system stabilization and control have become areas of interest [83, 84]. A typical generic of different voltage control levels is discussed in [85]. Optimal voltage control has long been successfully implemented in power systems, including the three-level hierarchical automatic voltage control in Europe [86–88], and the adaptive zone division method in China [89].

A supervisory voltage control strategy for large-scale solar photovoltaic (PV) integration in power network is proposed in [90, 91] to enhance the voltage stability. A survey of methods, mostly based on PMU data, for long-term voltage instability detection is given in [92]. In [93], a two-stage distributed voltage control scheme is proposed. The first stage is the local control of each DG based on sensitivity analysis, and the second stage acquires reactive power support from other DG units. In [94], a consensus-based cooperative control is proposed to regulate voltage by coordinating electric cars and active power curtailment of PVs. In [95], a distributed voltage stability assessment considering DG units is developed based on distributed continuation power flow. Coordinated voltage control is a technique which provides voltage control by means of adjusting, sequencing, and timing various kinds of controllers within a system. Some relevant works are reported in [96–98].

As mentioned above, several PMU-based voltage control methodologies have been reported worldwide; however, mostly presented a voltage recovery approach in an off-normal or emergency condition. Among existing three hierarchical levels of voltage control (primary, secondary, and tertiary controls), only few works are mainly focused on optimal supervisory on secondary voltage control, which is required to coordinate adjustment of the set-points of the existing voltage controllers. In this regard, the online adaptive tuning of available voltage control systems in a power grid with high integration of DGs/RESs is not well addressed. Furthermore, the overlap between voltage dynamics and frequency/active power as well as rotor angle dynamics in a modern power grid has not been highlighted in the published reports.

1.2.3 Oscillation Damping

Traditionally, the power system oscillations are damped through the generator local controllers, such as the exciter and governor, which are designed to ensure only the local stability of the generator (1–2 Hz). In order to increase the stability of the system, PSSs and power electronic converter-based FACTS are added into the grid [99–101]. In a broader context, the power system oscillation problem has also been related to voltage stability. The control interaction is discussed in [102, 103]. The exploitation of the wide-area measurements, provided by PMUs, for monitoring and controlling the power system led to the introduction of the wide-area monitoring and control (WAMC) systems [104]. The advent and application of synchronized measurement technology has enabled the detection and observation of poorly damped oscillations (such as the inter-area modes) and became the backbone for more development of the WAMC systems [105]. Inter-area oscillations are characterized by low frequency (0.2–1 Hz) and occur when generators of one group swing against generators of another group [106]. Integration of RESs into the WAC scheme for damping power oscillations is discussed in [107, 108]. The utilization of a networked control system model for the WAC design, according to linear matrix inequality techniques, is proposed in [111]. Furthermore, Ref. [110] presents a WAC design, based on particle swarm optimization, for improving the performance of the power system through the control of wind farms.

More specifically, WAC aims to utilize the synchronized phasor measurements in order to provide coordination signals to the local controllers, making them capable of damping effectively all the inter-area oscillations [100]. In the literature, various works deal with the development of a WAC system. The proposed WAC schemes are segregated mainly according to the components of the power system that the WAC is intended to coordinate [3, 110]. Multiple control methodologies have been developed for damping the inter-area oscillations deploying a WAMC. In [83], a decentralized/hierarchical architecture for wide-area damping control using PMU remote feedback signals was discussed. References [100, 111] proposed the design of wide-area damping controllers that provide supplementary damping control to synchronous generators (SGs). A networked control system model for wide-area closed-loop power systems is applied in [109]. A power oscillation damping controller is introduced in [112] based on a modal linear quadratic Gaussian methodology. A combination of controlling SGs and renewable sources in order to increase the overall damping capability of the system is shown in [101, 107, 108, 113]. Few LPV control solutions to power oscillation damping are proposed that use either a low-order first principle model of the system [114] or a reduced-order parametric LPV identified model [115].

In comparison of frequency and voltage control, a higher number of reports have been published in PMU-based oscillation damping (rotor angle control) field. However, most of the reported approaches require the detailed and accurate knowledge of the complete network model (both topology and parameter values), that is unavailable or corrupted in practice as a result of communication failures, bad data in state estimation etc. In addition, the impact of disturbances on the inter-area oscillations cannot be well captured by these methods.

1.3 Data-Driven Wide-Area Power System Monitoring and Control

Power grids modeling and control has become a more challenging issue due to the increasing penetration of RESs, changing system structure and the integration of new storage systems, controllable loads and power electronics technologies, and reduction of system inertia. Conventional modeling and control designs may not be any more effective to satisfy all specified objectives in various operation modes of modern power grids. These challenging issues set new demand for the development of more flexible, rapid, effective, precise, and adaptive approaches for power system dynamic monitoring, stability/security analysis, and control problems. Thanks to recent advances in control, communication, and computing technologies, it is possible to tackle mentioned challenges by implementing a data-driven-based modeling and control framework as shown in Figure 1.2.

The system data are collected from the distributed PMUs in the grid through a secure communication network. The development of information and communication technology (ICT) enables more flexibility in wide-area monitoring of power system with fast and large data transmission. Especially, the wide-area measurement system (WAMS) with PMUs is a promising technique as one of the smart grid technologies in the bulk power grid.

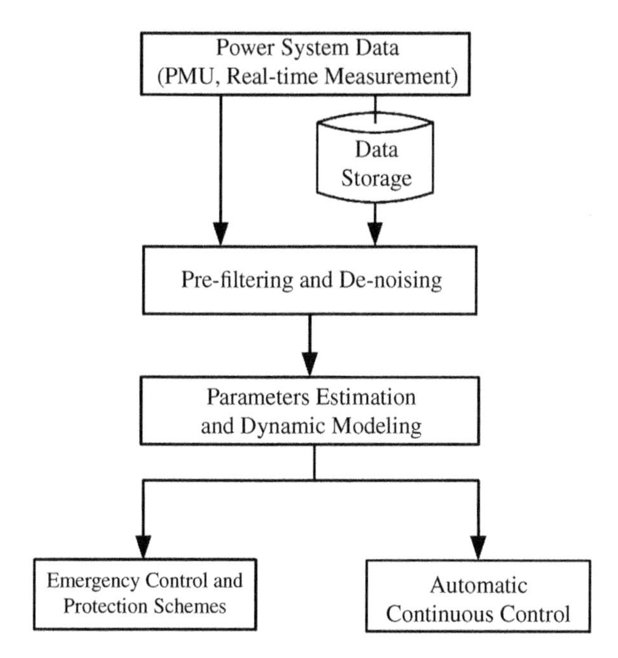

Figure 1.2 An overall data-driven control framework for renewable integrated power systems.

The measured data are locally saved and then collected by phasor data concentrators (PDCs) for the post analysis or sent to a remote location via a standard data format. These data with the time stamp of the synchronized GPS in real time may applied for parameter and state estimations and finally used for the system protection and/or real-time control. Figure 1.3 shows how a PMU-based WAMS can provide data for the power system control center to generate continuous (in normal states) and discontinuous control (in off-normal states) commands.

Before any application, the collected PMU data need to be cleaned and de-noised and employed by the data processors for estimation, modeling, and control purposes. The proposed de-noising method may use a rolling-averaging window with pre-specified length to remove noise from the recorded data. The block of parameters estimation algorithms contains high fast and precise algorithms for estimation of some important parameters and transient characteristics that are required to use in control tuning algorithm or to detect a contingency and triggering the emergency control and protection schemes. In case of crossing the assigned thresholds showing an off-normal and emergency condition, the recorded data and some estimated parameters are used to detect the amount of mismatch (size of disturbance) for the emergency control and protection schemes such as load shedding

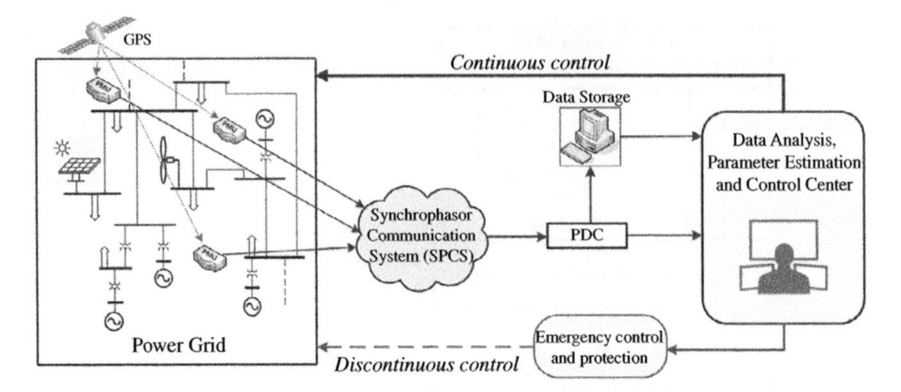

Figure 1.3 PMU-based wide-area measurement system and control.

algorithms. Otherwise, the estimated parameters such as scheduling parameters are employed by the continuous control systems.

As mentioned, using significant number of distributed micro-sources into power systems adds new technical challenges. As the electric industry seeks to reliably integrate large amounts of DGs/RESs into the power system in regulated environment, considerable effort is needed to accommodate and effectively manage the installed micro-sources. A key aspect is how to handle changes in topology and dynamics caused by penetration of numerous DGs/RESs in the network and how to make the power grid robust and able to take advantage of the potential flexibility of distributed micro-sources. In a modern control framework, a part of power produced by available DGs/RESs in the grid are used as a primary energy source of inertia emulator to provide virtual inertia as a supporting control for abovementioned controllers (like a fine tuner) to improve power grid stability.

1.4 Dynamics Modeling and Parameters Estimation

From a system dynamic point of view, the bulk generating units, due to their high inertia, provide a long time constant; such that the rotor speed and thus the grid frequency cannot alter suddenly, while the load changes. Hence, the total rotating mass enhances the dynamic stability. In future, a significant share of DGs/RESs/MGs in the electric power grids is expected. This increases the total system generation power, while does not contribute to the system rotational inertia. System dynamics are faster in power systems with low rotational inertia, making control and power system operation more challenging [32].

A complete understanding of reliability considerations via effective modeling/aggregation techniques is vital to identify a variety of ways that power grids can accommodate the large-scale integration of the distributed micro-sources in future. An accurate dynamic model is needed for the stability analysis and control synthesis in a grid with a high degree of DGs/RESs penetration. A proper dynamic modeling and aggregation of the DGs/RESs and MGs, for performance and stability studies, is a key issue to understand the dynamic impact of distributed micro-sources and simulate their functions in new environment.

The power system is a nonlinear multivariable time-varying system. It is represented by a nonlinear set of equations for the generators (swing equations), for the transmission lines and for the loads, which for a typical power system has a few hundreds of states. For the control design purpose, usually a reduced-order linearized model around an operating point is used and it is assumed that all system parameters are known and time-invariant. These assumptions, however, are not valid in a real power system with dominated DGs/RESs/MGs. The main dynamic modes of the system are varying stochastically during a day because of the variation of load and aggregated inertia. The dynamic modes will change more significantly by integration of new RESs into the power system (e.g. because of long-term variation of the mean value of the aggregated inertia). Therefore, a fixed linearized time-invariant model will not represent correctly the behavior of the power system.

The frequency response of the system can be identified offline/online using the data for different load and generation configurations (when the share of DGs/RESs is increased) and saved in a database for the models. The small variation of the system (originated from measurement noise, load variation, and system nonlinearity) will be modeled by frequency domain uncertainty. The long-term effect of change in system inertia can be considered by identifying several frequency-domain models for different levels of RES penetration. One can represent this model's database by an LPV model [116]. It should be mentioned that the model of the power system for the frequency, voltage, and rotor angle is different because they have different inputs and outputs and scheduling parameters.

1.4.1 Modeling of Frequency, Voltage, and Angle Controls

The participant bulk SGs with different participation factors are the main actuators for the frequency control system. Following a disturbance, the variation of frequency and tie-line power is applied to the LFC system via the ACE signal. Then, depending on the accessible amount of regulation power, the LFC system will be activated to compensate the power grid frequency and return it to the nominal value. The LFC system can attenuate the frequency and active power changes from tenth of seconds to few minutes. Therefore, the ACE signal may provide the output of system model for frequency control. Considering the frequency response dynamics [32], the candidate scheduling parameters are system inertia, aggregated generating time constant, droop and damping coefficient. The measurement-based dynamics identification and system modeling will be for adaptive control and online parameters tuning of the LFC system. The increasing size and diversification of demand/power sources magnify the importance of this issue in the modern power grids.

Unlike grid frequency, since the voltage is known as a local variable, a higher number of measured points are required. For proposing the data-driven models, several concepts like Thevenin equivalent system and oscillation model will be applied and the results will be used for the grid voltage analysis, and then optimal tuning of AVRs. The measurements of voltage, current, and phase deviations of existing nodes are considered as the most important inputs of model. The system output can be the terminal voltage change of the SGs. In order to construct an appropriate LPV model for voltage control, relevant scheduling parameters must be selected. There are several choices for the scheduling parameters (low-frequency resonance mode, reactive power of the system, Thevenin equivalent impedance/admittance, etc.) that must be compared and discussed using effective analytical and simulation-based studies. In choosing efficient scheduling parameters, a tradeoff between accuracy and simplicity of the resulting LPV model is needed. The measurements are also used to perform some important graphical tools and curves to evaluate the stress conditions and to analyze the voltage stability criteria. For instance, the data are fitted to the active power–voltage curve (PV curve) of the equivalent system by suitable fitting approaches such as least squares method.

The SGs equipped with a PSS and RESs participating in power oscillation damping are considered as the main actuators of the control system. It is assumed that some of the RESs are not operating at their maximum generated power, so that they can help for power oscillation damping by reducing/increasing a small percentage of their power generation. The system to be controlled is a multivariable system, where the inputs are the reference voltages for the AVR of the SGs as well as the auxiliary signals that will be added to the reference active and reactive powers of RESs. The system outputs can be the active power or the speed (or both, depending on the type of PSSs) of the SGs and the measured active and reactive powers of the RESs. In order to construct an appropriate model for power oscillation damping, we need to choose the scheduling parameters.

1.4.2 Parameters Estimation

For estimation of all required parameters, the recorded data from the installed PMUs can be used. As shown in Figure 1.2, firstly, a de-noising methodology, mostly based on rolling averaging windows, can be employed to prepare the received PMU data for further processing. Afterwards, some data-driven-based algorithms are used to estimate the most important parameters (e.g., system inertia, droop characteristic, and damping factor) required for building low order power system models such as frequency response model [32] and oscillation model [3]. In real-time operation, an accurate and fast estimation of parameters is required. The estimated parameters can be used in the auto-tuning algorithm of the controllers.

In a modern power system, in addition to scheduling parameters, several algorithms need to be developed for online estimating of other important system parameters such as synchronizing coefficient between various areas, rate of change of frequency/voltage (ROCOF/ROCOV), frequency/voltage nadir, and time occurrence of frequency/voltage nadir. The estimations must be fast enough and should cover the issues related to the existing time delay. These data-driven-based estimation algorithms can be analytically developed based on the concept of swing equation, base-case systems, regression, and curve fitting. These measurement-based dynamics identification and system modeling can be used for adaptive online parameters tuning of the targeted controllers. The increasing size and diversification of demand/power sources magnify the importance of this issue in the modern power grids. The estimation approaches may be applicable for both on-line and off-line methods. However, for on-line applications, shorter data windows must be used.

PMU data-based power-load imbalance estimation is a key estimation to successfully handle the emergency control strategies, e.g. load shedding, and protection plans. In case of detecting a contingency or an emergency condition, following comparing of frequency, voltage, and their rate of changes with the specified threshold values, an estimation algorithm must estimate the size of disturbance to use in the available emergency control systems and special protection schemes. This online estimation is an important issue to realize a successful load-shedding scheme with minimum amount of shed load.

Conventionally, for the estimation of the size of load-power mismatch, the swing equation is used. The estimated imbalance based on this method may far from real power mismatch as it relies on three worst assumptions: (i) there is no additional active power variation except for that of disturbance, (ii) there is a negligible reactive-power imbalance in response to sudden active power imbalance, and (iii) the inertia constant assumed to be known. This under/over-estimation causes inaccurate calculation of total amount of load to be shed.

In [117], for estimating the size of disturbance, some appropriate base-case features are selected from a set of pre-defined base cases. Moreover, an efficient yet simple logic is defined to select appropriate base-case for the received data. This approach benefits from the use of PMU data to precise calculation of the required amount of load to be shed, in one-step and in a short time in comparison with the actual time. The proposed scheme relies on fast, yet iterative, estimation of frequency nadir, and time of minimum frequency occurrence. Accordingly, the inertia constant as well as the size of power mismatch are estimated which, in turn, compares with the maximum size of imbalance, satisfying the pre-specified thresholds, to determine the amount of shed load.

1.5 Summary

Modern power grids face new technical challenges arising from the increasing penetration of power-electronic-connected RESs/DGs. Increasing MGs/DGs penetration level may adversely affect frequency response and voltage and system control and lead to degraded performance of traditional control schemes. This, in turn, may result in large deviations and, potentially, system instability.

This chapter provides the pre-requirement terminology and general background for the next chapters of this book. The term power system stability and control with an updated brief review on the areas of frequency, voltage, and angle controls, concerning the penetration of RESs/DGs, is discussed. In response to the existing challenges in penetration of more RESs/DGs to the grid, the necessity of using data-driven modeling, parameters estimation, and control synthesis in wide-area power systems is emphasized.

References

1. Steinmetz, C.P. (1920). Power control and stability of electric generating stations. *Transactions of the American Institute of Electrical Engineers* **XXXIX** (2): 1215–1287.
2. Smith, J.R., Andersson, G., and Taylor, C.W. (1996). Annotated bibliography on power system stability controls: 1986–1994. *IEEE Transactions on Power Systems* **11** (2): 794–800.
3. Bevrani, H., Watanabe, M., and Mitani, Y. (2014). *Power System Monitoring and Control*. Wiley.
4. Bevrani, H. and Hiyama, T. (2009). On load-frequency regulation with time delays: design and real time implementation, *IEEE Transactions on Energy Conversion* **24** (1): 292–300.
5. Bevrani, H. and Hiyama, T. (2011). *Intelligent Automatic Generation Control*. CRC Press.
6. Kundur, P. (1994). *Power System Stability and Control*, vol. **7**. New York: McGraw-Hill.

7. Kundur, P., Paserba, J., Ajjarapu, V. et al. (2004). Definition and classification of power system stability ieee/cigre joint task force on stability terms and definitions. *IEEE Transactions on Power Systems* **19** (3): 1387–1401.
8. Pierre, J.W., Trudnowski, D., Donnelly, M. et al. (2012). Overview of system identification for power systems from measured responses. *IFAC Proceedings Volumes* **45** (16): 989–1000.
9. Eriksson, R. and Soder, L. (2011). Wide-area measurement system-based subspace identification for obtaining linear models to centrally coordinate controllable devices. *IEEE Transactions on Power Delivery* **26** (2): 988–997.
10. Zhou, N., Lu, S., Singh, R., and Elizondo, M.A. (2011). Calibration of reduced dynamic models of power systems using phasor measurement unit (PMU) data. *2011 North American Power Symposium*, Boston, MA (2011), pp. 1–7. doi: https://doi.org/10.1109/NAPS.2011.6024873.
11. Zhang, J., Lu, C., and Han, Y. (2013). MIMO identification of power system with low level probing tests: applicability comparison of subspace methods. *IEEE Transactions on Power Systems* **28** (3): 2907–2917.
12. Wiseman, B.P., Chen, Y., Xie, L., and Kumar, P. (2016). PMU-based reduced-order modeling of power system dynamics via selective modal analysis. In: *2016 IEEE/PES Transmission and Distribution Conference and Exposition (T&D)*, 1–5. IEEE.
13. Liu, H., Zhu, L., Pan, Z. et al. (2016). Comparison of MIMO system identification methods for electromechanical oscillation damping estimation. *2016 IEEE Power and Energy Society General Meeting (PESGM)*, Boston, MA (2016), pp. 1–5. doi: https://doi.org/10.1109/PESGM.2016.7741834.
14. Tuttelberg, K., Kilter, J., and Uhlen, K. (2017). Comparison of system identification methods applied to analysis of inter-area modes. *Proceedings of International Power Systems Transients Conference 2017*, Seoul, South Korea (26–29 June 2017).
15. Ghasemi, H. and Canizares, C.A. (2008). Confidence intervals estimation in the identification of electromechanical modes from ambient noise. *IEEE Transactions on Power Systems* **23** (2): 641–648.
16. Dosiek, L., Pierre, J.W., and Follum, J. (2013). A recursive maximum likelihood estimator for the online estimation of electromechanical modes with error bounds. *IEEE Transactions on Power Systems* **28** (1): 441–451.
17. Uhlen, K., Warland, L., Gjerde, J.O. et al. (2008). Monitoring amplitude, frequency and damping of power system oscillations with PMU measurements. *2008 IEEE Power and Energy Society General Meeting – Conversion and Delivery of Electrical Energy in the 21st Century*, Pittsburgh, PA (2008), pp. 1–7, doi: https://doi.org/10.1109/PES.2008.4596661.
18. Tripathy, P., Srivastava, S.C., and Singh, S.N. (2011). A modified TLS-ESPRIT-based method for low frequency mode identification in power systems utilizing

synchrophasor measurements. *IEEE Transactions on Power Systems* **26** (2): 719–727.

19. Rogers, K.M., Spadoni, R.D., and Overbye, T.J. (2011). Identification of power system topology from synchrophasor data. *2011 IEEE/PES Power Systems Conference and Exposition*, Phoenix, AZ (2011), pp. 1–8, doi: https://doi.org/10.1109/PSCE.2011.5772462.
20. Nabavi, S. and Chakrabortty, A. (2013). Topology identification for dynamic equivalent models of large power system networks. *2013 American Control Conference*, Washington, DC, (2013), pp. 1138–1143. doi: https://doi.org/10.1109/ACC.2013.6579989.
21. Wang, X., Bialek, J.W., and Turitsyn, K. (2018). PMU-based estimation of dynamic state jacobian matrix and dynamic system state matrix in ambient conditions. *IEEE Transactions on Power Systems* **33** (1): 681–690.
22. Tuttelberg, K., Kilter, J., Wilson, D., and Uhlen, K. (2018). Estimation of power system inertia from ambient wide area measurements. *IEEE Transactions on Power Systems* **33** (6): 7249–7257.
23. Zeng, F., Zhang, J., Zhou, Y., and Qu, S. (2020). Online identification of inertia distribution in normal operating power system. *IEEE Transactions on Power Systems* **35** (4): 3301–3304. https://doi.org/10.1109/TPWRS.2020.2986721.
24. Concordia, C. and Kirchmayer, L. (1953). Tie-line power and frequency control of electric power systems [includes discussion]. *Transactions of the American Institute of Electrical Engineers. Part III: Power Apparatus and Systems* **72** (3): 562–572.
25. System Controls Subcommittee of the Power System Engineering Committee of the IEEE Power Group (1970). IEEE standard definitions of terms for automatic generation control on electric power systems. *IEEE Transactions on Power Apparatus and Systems* **PAS-89** (6): 1356–1364.
26. I. C. Report (1973). Dynamic models for steam and hydro turbines in power system studies. *IEEE Transactions on Power Apparatus and Systems* **PAS-92** (6): 1904–1915.
27. Jaleeli, N., VanSlyck, L.S., Ewart, D.N. et al. (1992). Understanding automatic generation control. *IEEE Transactions on Power Systems* **7** (3): 1106–1122.
28. Pathak, N., Bhatti, T.S., and Verma, A. (2017). Accurate modelling of discrete AGC controllers for interconnected power systems. *IET Generation, Transmission & Distribution* **11** (8): 2102–2114.
29. Moawwad, A., El-Saadany, E.F., and El Moursi, M.S. (2018). Dynamic security-constrained automatic generation control (AGC) of integrated ac/dc power networks. *IEEE Transactions on Power Systems* **33** (4): 3875–3885.
30. Ledva, G.S., Vrettos, E., Mastellone, S. et al. (2018). Managing communication delays and model error in demand response for frequency regulation. *IEEE Transactions on Power Systems* **33** (2): 1299–1308.

31. Ibraheem, P., Kumar, and Kothari, D.P. (2005). Recent philosophies of automatic generation control strategies in power systems. *IEEE Transactions on Power Systems* **20** (1): 346–357.

32. Bevrani, H. (2014). *Robust Power System Frequency Control*, 2e. Gewerbestrasse, Switzerland: Springer.

33. Ulbig, A., Borsche, T.S., and Andersson, G. (2014). Impact of low rotational inertia on power system stability and operation. *IFAC Proceedings Volumes* **47** (3): 7290–7297.

34. Jaleeli, N. and VanSlyck, L.S. (1999). NERC's new control performance standards. *IEEE Transactions on Power Systems* **14** (3): 1092–1099.

35. Hain, Y., Kulessky, R., and Nudelman, G. (2000). Identification-based power unit model for load-frequency control purposes. *IEEE Transactions on Power Systems* **15** (4): 1313–1321.

36. Chang-Chien, L.R., Hoonchareon, N.-B., Ong, C.-M., and Kramer, R.A. (2003). Estimation of /spl beta/ for adaptive frequency bias setting in load frequency control. *IEEE Transactions on Power Systems* **18** (2): 904–911.

37. Wilches-Bernal, F., Concepcion, R., Neely, J.C. et al. (2018). Communication enabled fast acting imbalance reserve (CE-FAIR). *IEEE Transactions on Power Systems* **33** (1): 1101–1103.

38. Zhang, G. and McCalley, J.D. (2018). Estimation of regulation reserve requirement based on control performance standard. *IEEE Transactions on Power Systems* **33** (2): 1173–1183.

39. Polajzer, B., Brezovnik, R., and Ritonja, J. (2017). Evaluation of load frequency control performance based on standard deviational ellipses. *IEEE Transactions on Power Systems* **32** (3): 2296–2304.

40. Avila, T., Gutierrez, E., and Chavez, H. (2017). Performance standard-compliant secondary control: the case of Chile. *IEEE Latin America Transactions* **15** (7): 1257–1262.

41. Douglas, L.D., Green, T.A., and Kramer, R.A. (1994). New approaches to the AGC nonconforming load problem. *IEEE Transactions on Power Systems* **9** (2): 619–628. https://doi.org/10.1109/59.317682.

42. Trovato, V., Sanz, I.M., Chaudhuri, B., and Strbac, G. (2017). Advanced control of thermostatic loads for rapid frequency response in Great Britain. *IEEE Transactions on Power Systems* **32** (3): 2106–2117.

43. Delavari, A. and Kamwa, I. (2018). Improved optimal decentralized load modulation for power system primary frequency regulation. *IEEE Transactions on Power Systems* **33** (1): 1013–1025.

44. Pan, C. and Liaw, C. (1989). An adaptive controller for power system load-frequency control. *IEEE Transactions on Power Systems* **4** (1): 122–128.

45. Vajk, I., Vajta, M., Keviczky, L. et al. (1985). Adaptive load-frequency control of the Hungarian power system. *Automatica* **21** (2): 129–137.

46. Wang, W., Li, Y., Cao, Y. et al. (2018). Adaptive droop control of VSC-MTDC system for frequency support and power sharing. *IEEE Transactions on Power Systems* **33** (2): 1264–1274.
47. Prostejovsky, A.M., Marinelli, M., Rezkalla, M. et al. (2018). Tuningless load frequency control through active engagement of distributed resources. *IEEE Transactions on Power Systems* **33** (3): 2929–2939.
48. Stankovic, A.M., Tadmor, G., and Sakharuk, T.A. (1998). On robust control analysis and design for load frequency regulation. *IEEE Transactions on Power Systems* **13** (2): 449–455.
49. Rerkpreedapong, D., Hasanovic, A., and Feliachi, A. (2003). Robust load frequency control using genetic algorithms and linear matrix inequalities. *IEEE Transactions on Power Systems* **18** (2): 855–861.
50. Ojaghi, P. and Rahmani, M. (2017). LMI-based robust predictive load frequency control for power systems with communication delays. *IEEE Transactions on Power Systems* **32** (5): 4091–4100.
51. Zhang, C., Jiang, L., Wu, Q.H. et al. (2013). Delay-dependent robust load frequency control for time delay power systems. *IEEE Transactions on Power Systems* **28** (3): 2192–2201.
52. Zhao, J., Mili, L., and Milano, F. (2018). Robust frequency divider for power system online monitoring and control. *IEEE Transactions on Power Systems* **33** (4): 4414–4423.
53. Aliabadi, S.F., Taher, S.A., and Shahidehpour, M. (2018). Smart deregulated grid frequency control in presence of renewable energy resources by EVs charging control. *IEEE Transactions on Smart Grid* **9** (2): 1073–1085.
54. Wang, D., Liang, L., Hu, J. et al. (2018). Analysis of low-frequency stability in grid tied DFIGs by non-minimum phase zero identification. *IEEE Transactions on Energy Conversion* **33** (2): 716–729.
55. Liu, Y., Jiang, L., Wu, Q.H., and Zhou, X. (2017). Frequency control of DFIG-based wind power penetrated power systems using switching angle controller and AGC. *IEEE Transactions on Power Systems* **32** (2): 1553–1567.
56. Pradhan, C. and Bhende, C.N. (2017). Frequency sensitivity analysis of load damping coefficient in wind farm-integrated power system. *IEEE Transactions on Power Systems* **32** (2): 1016–1029.
57. Golpira, H., Seifi, H., Messina, A.R., and Haghifam, M. (2016). Maximum penetration level of microgrids in large-scale power systems: frequency stability viewpoint. *IEEE Transactions on Power Systems* **31** (6): 5163–5171.
58. Leon, A.E. (2018). Short-term frequency regulation and inertia emulation using an MMC-based MTDC system. *IEEE Transactions on Power Systems* **33** (3): 2854–2863.

59. Rakhshani, E., Remon, D., Cantarellas, A.M. et al. (2017). Virtual synchronous power strategy for multiple HVDC interconnections of multi-area AGC power systems. *IEEE Transactions on Power Systems* **32** (3): 1665–1677.

60. Li, D., Zhu, Q., Lin, S., and Bian, X.Y. (2017). A self-adaptive inertia and damping combination control of VSG to support frequency stability. *IEEE Transactions on Energy Conversion* **32** (1): 397–398.

61. Wu, Y., Yang, W., Hu, Y., and Dzung, P.Q. (2019). Frequency regulation at a wind farm using time varying inertia and droop controls. *IEEE Transactions on Industry Applications* **55** (1): 213–224.

62. Fang, J., Li, H., Tang, Y., and Blaabjerg, F. (2018). Distributed power system virtual inertia implemented by grid-connected power converters. *IEEE Transactions on Power Electronics* **33** (10): 8488–8499.

63. Li, Y., Xu, Z., Ostergaard, J., and Hill, D.J. (2017). Coordinated control strategies for offshore wind farm integration via VSC-HVDC for system frequency support. *IEEE Transactions on Energy Conversion* **32** (3): 843–856.

64. Ahmadyar, A.S. and Verbic, G. (2017). Coordinated operation strategy of wind farms for frequency control by exploring wake interaction. *IEEE Transactions on Sustainable Energy* **8** (1): 230–238.

65. Izadkhast, S., Garcia-Gonzalez, P., Frias, P., and Bauer, P. (2017). Design of plug-in electric vehicle's frequency-droop controller for primary frequency control and performance assessment. *IEEE Transactions on Power Systems* **32** (6): 4241–4254.

66. Hwang, M., Muljadi, E., Jang, G., and Kang, Y.C. (2017). Disturbance-adaptive short-term frequency support of a DFIG associated with the variable gain based on the ROCOF and rotor speed. *IEEE Transactions on Power Systems* **32** (3): 1873–1881.

67. Attya, A.B.T. and Dominguez-Garcia, J.L. (2018). Insights on the provision of frequency support by wind power and the impact on energy systems. *IEEE Transactions on Sustainable Energy* **9** (2): 719–728.

68. Tielens, P. and Van Hertem, D. (2017). Receding horizon control of wind power to provide frequency regulation. *IEEE Transactions on Power Systems* **32** (4): 2663–2672.

69. Garmroodi, M., Verbic, G., and Hill, D.J. (2018). Frequency support from wind turbine generators with a time-variable droop characteristic. *IEEE Transactions on Sustainable Energy* **9** (2): 676–684.

70. Khooban, M., Dragicevic, T., Blaabjerg, F., and Delimar, M. (2018). Shipboard microgrids: a novel approach to load frequency control. *IEEE Transactions on Sustainable Energy* **9** (2): 843–852.

71. Benysek, G., Bojarski, J., Smolenski, R. et al. (2018). Application of stochastic decentralized active demand response (DADR) system for load frequency control. *IEEE Transactions on Smart Grid* **9** (2): 1055–1062.

72. Vrettos, E., Ziras, C., and Andersson, G. (2017). Fast and reliable primary frequency reserves from refrigerators with decentralized stochastic control. *IEEE Transactions on Power Systems* **32** (4): 2924–2941.
73. Short, J.A., Infield, D.G., and Freris, L.L. (2007). Stabilization of grid frequency through dynamic demand control. *IEEE Transactions on Power Systems* **22** (3): 1284–1293.
74. Molina-Garcia, A., Bouffard, F., and Kirschen, D.S. (2011). Decentralized demand-side contribution to primary frequency control. *IEEE Transactions on Power Systems* **26** (1): 411–419.
75. Zhao, H., Wu, Q., Huang, S. et al. (2018). Hierarchical control of thermostatically controlled loads for primary frequency support. *IEEE Transactions on Smart Grid* **9** (4): 2986–2998.
76. Yao, E., Wong, V.W.S., and Schober, R. (2017). Robust frequency regulation capacity scheduling algorithm for electric vehicles. *IEEE Transactions on Smart Grid* **8** (2): 984–997.
77. Ferraro, P., Crisostomi, E., Raugi, M., and Milano, F. (2017). Analysis of the impact of microgrid penetration on power system dynamics. *IEEE Transactions on Power Systems* **32** (5): 4101–4109.
78. Ferraro, P., Crisostomi, E., Shorten, R., and Milano, F. (2018). Stochastic frequency control of grid connected microgrids. *IEEE Transactions on Power Systems* **33** (5): 5704–5713.
79. Larsen, E. and Sener, F. (1996). Facts Applications. Catalogue No. 96TP116-0.
80. IEEE (1990). Voltage Stability of Power Systems: Concepts, Analytical Tools and Industry Experience. *IEEE Technical Report 90YH0358-2-PWR*. IEEE/PES.
81. Balu, C. and Maratukulam, D. (1994). *Power System Voltage Stability*. McGraw-Hill.
82. Van Cutsem, T. and Vournas, C. (2007). *Voltage Stability of Electric Power Systems*. Springer Science & Business Media.
83. Kamwa, I., Grondin, R., and Hebert, Y. (2001). Wide-area measurement based stabilizing control of large power systems – a decentralized/hierarchical approach. *IEEE Transactions on Power Systems* **16** (1): 136–153.
84. Taylor, C.W., Erickson, D.C., Martin, K.E. et al. (2005). WACS wide-area stability and voltage control system: R & D and online demonstration. *Proceedings of the IEEE* **93** (5): 892–906.
85. Andersson, G., Bel, C.A., and Canizares, C. (2009). Frequency and voltage control. In: *Electric Energy Systems: Analysis and Operation*. CRC Press.
86. Ilic, M.D., Liu, X., Leung, G. et al. (1995). Improved secondary and new tertiary voltage control. *IEEE Transactions on Power Systems* **10** (4): 1851–1862.
87. Corsi, S., Pozzi, M., Sabelli, C., and Serrani, A. (2004). The coordinated automatic voltage control of the Italian transmission grid-Part I: reasons of the choice and

overview of the consolidated hierarchical system. *IEEE Transactions on Power Systems* **19** (4): 1723–1732.

88. Corsi, S., Pozzi, M., Sforna, M., and Dell'Olio, G. (2004). The coordinated automatic voltage control of the italian transmission grid-Part II: control apparatuses and field performance of the consolidated hierarchical system. *IEEE Transactions on Power Systems* **19** (4): 1733–1741.

89. Guo, Q., Sun, H., Zhang, M. et al. (2013). Optimal voltage control of PJM smart transmission grid: study, implementation, and evaluation. *IEEE Transactions on Smart Grid* **4** (3): 1665–1674.

90. Xiao, W., Torchyan, K., El Moursi, M.S., and Kirtley, J.L. (2014). Online supervisory voltage control for grid interface of utility-level PV plants. *IEEE Transactions on Sustainable Energy* **5** (3): 843–853.

91. A. Awadhi, N. and Moursi, M.S.E. (2017). A novel centralized PV power plant controller for reducing the voltage unbalance factor at transmission level interconnection. *IEEE Transactions on Energy Conversion* **32** (1): 233–243. https://doi.org/10.1109/TEC.2016.2620477.

92. Glavic, M. and Van Cutsem, T. (2011). A short survey of methods for voltage instability detection. *2011 IEEE Power and Energy Society General Meeting*, Detroit, MI (2011), pp. 1–8, doi: https://doi.org/10.1109/PES.2011.6039311.

93. Robbins, B.A., Hadjicostis, C.N., and Dominguez-Garcia, A.D. (2013). A two-stage distributed architecture for voltage control in power distribution systems. *IEEE Transactions on Power Systems* **28** (2): 1470–1482.

94. Zeraati, M., Hamedani Golshan, M.E., and Guerrero, J.M. (2019). A consensus-based cooperative control of PEV battery and PV active power curtailment for voltage regulation in distribution networks. *IEEE Transactions on Smart Grid* **10** (1): 670–680.

95. Li, Z., Guo, Q., Sun, H. et al. (2018). A distributed transmission-distribution coupled static voltage stability assessment method considering distributed generation. *IEEE Transactions on Power Systems* **33** (3): 2621–2632.

96. Popovic, D.H., Hill, D.J., and Wu, Q. (2002). Optimal voltage security control of power systems. *International Journal of Electrical Power & Energy Systems* **24** (4): 305–320.

97. Larsson, M. and Karlsson, D. (2003). Coordinated system protection scheme against voltage collapse using heuristic search and predictive control. *IEEE Transactions on Power Systems* **18** (3): 1001–1006.

98. Ma, H. and Hill, D.J. (2018). A fast local search scheme for adaptive coordinated voltage control. *IEEE Transactions on Power Systems* **33** (3): 2321–2330.

99. Ghahremani, E. and Kamwa, I. (2016). Local and wide-area PMU-based decentralized dynamic state estimation in multi-machine power systems. *IEEE Transactions on Power Systems* **31** (1): 547–562.

100. Raoufat, M.E., Tomsovic, K., and Djouadi, S.M. (2016). Virtual actuators for wide-area damping control of power systems. *IEEE Transactions on Power Systems* **31** (6): 4703–4711.

101. Mohagheghi, S., Venayagamoorthy, G.K., and Harley, R.G. (2007). Optimal wide area controller and state predictor for a power system. *IEEE Transactions on Power Systems* **22** (2): 693–705.

102. Mithulananthan, N., Canizares, C.A., Reeve, J., and Rogers, G.J. (2003). Comparison of PSS, SVC, and STATCOM controllers for damping power system oscillations. *IEEE Transactions on Power Systems* **18** (2): 786–792.

103. Bian, X.Y., Geng, Y., Lo, K.L. et al. (2016). Coordination of PSSs and SVC damping controller to improve probabilistic small-signal stability \\of power system with wind farm integration. *IEEE Transactions on Power Systems* **31** (3): 2371–2382.

104. Padhy, B.P., Srivastava, S.C., and Verma, N.K. (2017). A wide-area damping controller considering network input and output delays and packet drop. *IEEE Transactions on Power Systems* **32** (1): 166–176.

105. Giri, J. (2015). Proactive management of the future grid. *IEEE Power and Energy Technology Systems Journal* **2** (2): 43–52.

106. Wu, X., Dorer, F., and Jovanovic, M.R. (2016). Input-output analysis and decentralized optimal control of inter-area oscillations in power systems. *IEEE Transactions on Power Systems* **31** (3): 2434–2444.

107. Zacharia, L., Hadjidemetriou, L., and Kyriakides, E. (2018). Integration of renewables into the wide area control scheme for damping power oscillations. *IEEE Transactions on Power Systems* **33** (5): 5778–5786.

108. Surinkaew, T. and Ngamroo, I. (2016). Hierarchical coordinated wide area and local controls of DFIG wind turbine and PSS for robust power oscillation damping. *IEEE Transactions on Sustainable Energy* **7** (3): 943–955.

109. Wang, S., Meng, X., and Chen, T. (2012). Wide-area control of power systems through delayed network communication. *IEEE Transactions on Control Systems Technology* **20** (2): 495–503.

110. Mokhtari, M. and Aminifar, F. (2014). Toward wide-area oscillation control through doubly-fed induction generator wind farms. *IEEE Transactions on Power Systems* **29** (6): 2985–2992.

111. Zhang, Y. and Bose, A. (2008). Design of wide-area damping controllers for inter-area oscillations. *IEEE Transactions on Power Systems* **23** (3): 1136–1143.

112. Zenelis, I. and Wang, X. (2018). Wide-area damping control for interarea oscillations in power grids based on PMU measurements. *IEEE Control Systems Letters* **2** (4): 719–724.

113. Youseian, R., Bhattarai, R., and Kamalasadan, S. (2017). Transient stability enhancement of power grid with integrated wide area control of wind farms and synchronous generators. *IEEE Transactions on Power Systems* **32** (6): 4818–4831.

114. El-Guindy, A., Schaab, K., Schurmann, B. et al. (2017). Formal lpv control for transient stability of power systems. In: *2017 IEEE Power & Energy Society General Meeting*, 1–5. IEEE.

115. Nogueira, F.G., Junior, W.B., da Costa Junior, C.T., and Lana, J.J. (2018). LPV-based power system stabilizer: identification, control and field tests. *Control Engineering Practice* **72**: 53–67.

116. Toth, R. (2010). *Modeling and Identification of Linear Parameter-Varying Systems*, vol. **403**. Springer.

117. Golpira, H. and Bevrani, H. (2020). Frequency Analysis Based Centralized Load Shedding and Island Detection Using PMU Data. *Technical Report, Iran Grid Management Company, IGMC1927*, (In Persian).

2

MG Penetrated Power Grid Modeling

Power grids worldwide have experienced a significant transformation, which has been characterized by increased penetrations of distributed renewable generation. With advances in communication, measurement, and control technologies, the technical and economical merits of utility-scale microgrids (MGs), as a group of distributed energy resources (DERs), interconnected loads, and energy storage systems (ESSs), make these technologies attractive to enhance the dynamic performance of future power grids.

Because of uncertainty in future power system projections, together with different load scenarios, and the uncertain behavior of inverter-based distributed generations (DGs), MG-integrated power system modeling is a critical aspect in the development of control strategies and coordinated operation with other generation resources. Two specific aspects of interest for successful integration of MGs are (i) the modeling of the host power grid and (ii) the development of MG dynamic equivalents as seen from their point of interconnection with the systems.

This chapter addresses the problem of power system modeling, with emphasis on the development of aggregate MG models for electromechanical stability studies. Both deterministic and stochastic models for distribution and transmission applications are considered, and the concept of power system dynamic equivalencing is introduced.

2.1 Introduction

The increased demand for electrical power, environmental concerns, pressing need to reduce dependence on fossil fuel, and technological developments have caused many countries to set an ambitious target for deployment of DERs [1, 2]. From the network perspective, aggregate models are needed that circumvent the need for central dispatch of a massive number of DERs and the inherent

Renewable Integrated Power System Stability and Control, First Edition.
Hêmin Golpîra, Arturo Román-Messina, and Hassan Bevrani.

difficulties associated with the control and coordination of DERs [3, 4]. MGs, acting as single controlled entities on the grid, can overcome some of the aforementioned limitations through local control of DERs/DGs, separation of generation and corresponding loads from the distribution system in the presence of disturbances, and providing high local reliability for loads. In addition to the DERs/DGs' potential benefits, the conversion of only one-third of fuel energy into electrical power and the poor efficiency of bulk power generation and high voltage transmission systems make DERs/DGs in distribution network more pervasive [3]. Moreover, the falling investment cost of small-scale power plants, development of data communications and control technologies, the emerging potential of DERs/DGs, and short installation time are the main incentives toward integrating significant amounts of DERs' generation [5, 6]. It is envisaged that a large number of DERs/DGs will be connected to the host grid shortly, which will increase the power system dimension and complexity.

In this chapter, a systematic methodology for modeling distributed MGs for integration studies is presented. This framework can be used to investigate the integration of high penetrations of DERs/MGs energy into the system.

2.2 Basic Concepts

2.2.1 Dynamic Equivalencing

High penetration of MGs may seriously affect distribution system dynamics which, in turn, may affect overall power system stability and dynamics. To analyze the associated impacts, a simple representation of the distribution network as a constant power load (*PQ*) is not appropriate [4]. Moreover, conventional approaches to model the host power grid fail to handle the complexity and uncertainties of the MGs-penetrated grid.

To facilitate the dynamical studies of MGs-penetrated power grids, appropriate dynamical models should be derived. Power system *dynamic equivalencing* is a powerful means to study system dynamics when a significant integration of MGs operating in grid-connected mode is considered. In this type of modeling, the structure (and dynamics) of part of the system is simplified, while attempting to preserve the important characteristics of the underlying dynamical processes being investigated.

Generally, power system equivalent models can be classified into three main categories: high frequency, low frequency, and wideband models depending on the transient phenomena of interest [6–8]. The focus of this chapter is on the derivation of low-frequency models to be used in stability studies. Commonly, low-frequency models are formally utilized in simulating rotor-angle stability of synchronous machines as well as for frequency stability studies.

Most of the research efforts in this area started in the 1970s and 1980s when computing power was dramatically less than today. The technical subcommittee of IEEE Power and Energy Society (PES) successfully reviewed the advances and challenges in the state of the art [7]. While the current industry practice is to simulate the full model due to the availability of fast computers, this is not the case for MGs-penetrated power grids. In other words, despite the availability of fast computation power, the need for low-frequency equivalent models is still relevant, especially in cases where the influence of power-electronics devices on low-frequency oscillations is of interest.

2.2.2 Background on Study Zone and External System

In dynamic equivalencing studies, the power network of interest is commonly divided into the *study zone* and *external system*. When the transient phenomena of interest occur in a study zone, the external system is commonly replaced with an appropriate equivalent model that preserves the main dynamic of concern [7]. Accordingly two commonly used power system simulation tools are (i) electromagnetic transient (EMT) models and (ii) transient stability models or "phasor models".

In EMT models, the conventional synchronous generator is replaced with a voltage source. This causes the elimination of electromechanical low-frequency dynamics which is not the case for transient studies. The EMT representations are well suited for the simulation of lightning and switching over voltages that is beyond the scope of this book. On the other hand, in modern power grids where power-electronic-interfaced sources play an important role, the high-frequency equivalent network representation for the external system is inadequate. By contrast, quasi-steady-state-based approaches rely on the assumption that the required frequency that defines the phasors and system parameters is equal to the nominal value [8]. Such an assumption is acceptable as long as only the rotor speed variations of synchronous machines are considered to regulate the frequency. However, an increasing number of devices characterized with zero rotational inertia, including flexible loads providing load demand programs, DERs, ESSs, and high voltage direct current (HVDC) transmission systems, are expected to contribute to frequency regulation. These devices do not generally impose the frequency at their connection point with the grid [8, 9].

Thus, there is, from a modeling point of view, the need to define with accuracy the local frequency at crucial buses of the network. Therefore, the power system equivalencing process considering the high penetration of renewable sources should accurately incorporate the computation of the local frequency to the system model.

2.3 Power Grid Modeling

This section firstly discusses conventional power system modeling considering local frequency estimation using a center-of-gravity (COG) formulation in physics. To account for the uncertain system behavior, a stochastic approach is utilized to model MGs/DGs in integration studies. Based on these representations, a measurement-based approach is combined with the model to derive a simple, yet analytical equivalencing approach.

2.3.1 The Notion of Center-of Gravity (COG)

The same reasoning used in mechanics to introduce the concept of the COG can be extended to define the center of inertia (COI). In physics, the COG is a unique point in a body or group of particles, where the resultant torque due to gravity forces vanishes [10, 11]. This section introduces COG-based approaches for power system dynamic equivalencing as well as local frequency estimation.

2.3.1.1 Key Concept

To introduce the proposed formulation, assume that an interconnected power system is divided into n areas $\{A_i, i = 1, \ldots, n\}$, where a local COI is associated with each area. The areas may be determined using coherency identification techniques or be associated with the use of time-varying single (multi)-machine equivalent (SIME) methods [12].

The motion of the COI for area A_i is given by [13]

$$M_{COI_i} \frac{d^2 \delta_{COI_i}(t)}{dt^2} = T_{m_{COI_i}}(t) - T_{e_{COI_i}}(t), \quad i = 1, \ldots, n \tag{2.1}$$

where $\delta_{COI_i} = \sum_{j \in A_i} M_j \delta_j / \sum_{j \in A_i} M_j$ is the position of the COI, expressed in terms of the individual rotor angles δ_j, $M_{COI_i} = \sum_{j \in A_i} M_j / \omega_o$ is the equivalent inertia, and $T_{m_{COI_i}}(t)$, $T_{e_{COI_i}}(t)$ are the time-varying mechanical input power and electrical output power, respectively, which include the effects of turbine dynamics and other controllers [12].

With the COI description (2.1), used to represent the area response to external forces, valuable information about local (bus) frequency behavior and inter-unit synchronizing oscillations between generators are eliminated. Ideally, fast synchronizing oscillations are suppressed when areas are chosen based on the coherency identification techniques (uniform frequency for an area), such that $f_j \in f_{A_i} = f_{COI_i}$, where $f_{COI_i} = \sum_{j \in A_i} M_j f_j / \sum_{j \in A_i} M_j$.

It is convenient to define fictitious transmission lines between key generators (buses) and the COI to study the key generators' motion, as illustrated in

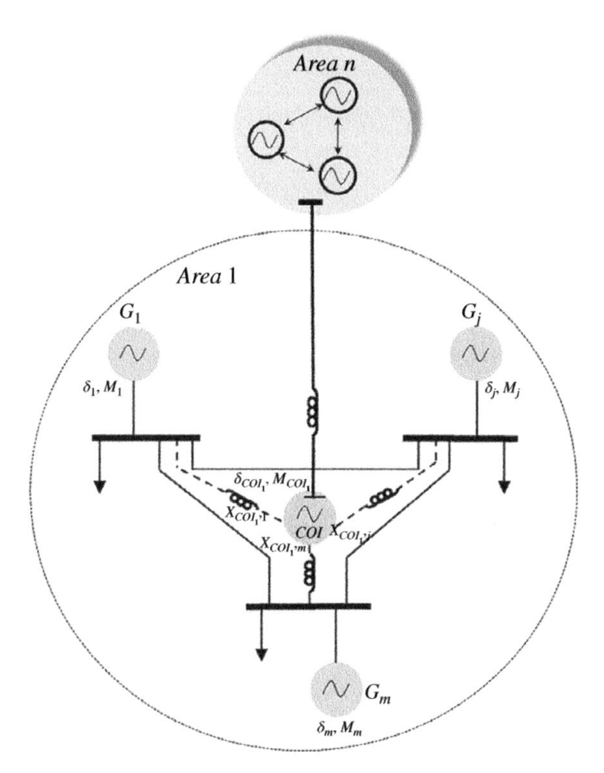

Figure 2.1 Area *i* illustrating the notion of the local COI. Dashed lines indicate virtual connections between the motion of the area *i* COI and key system buses.

Figure 2.1. Conceptually, this is equivalent to use weighted strings or plumb line techniques to find the COG in physical objects or systems.

Once the location of the COI is determined in Figure 2.1, expressions relating the local frequencies f_j, to the COI behavior can be obtained as

$$f_j = \frac{\partial f_j}{\partial f_{COI_i}} \Delta f_{COI_i}, \quad \forall j \in A_i \tag{2.2}$$

where the sensitivities are determined from the parameters of the fictitious tie-lines of Figure 2.1. By knowing these sensitivities, the propagation of slow frequency oscillations and tie-line power exchanges may be evaluated.

Drawing on the above framework, Figure 2.2 schematically represents the position of the COG relative to the local centers of inertia. Here, in analogy with the single-area representation, $P^{tie}_{COI_i,COG}$ and $P^{tie}_{COI_i,j}$ represent virtual power flows across the fictitious tie-lines interconnecting the ith local COI to the COG and the jth bus

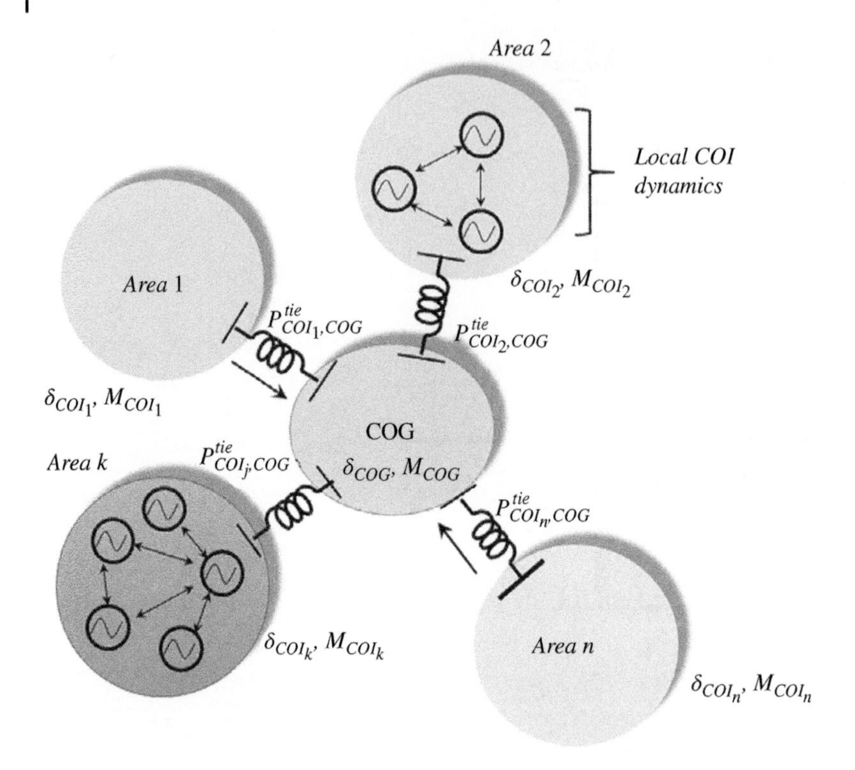

Figure 2.2 Interconnected power system divided into areas illustrating the notion of the COG. Dotted lines indicate virtual connections between the motions of centers of inertia and the system's COG.

to the associated local COI that are of interest for calculation. Two issues are of key interest here:

1) The determination of the COG and its associated parameters, and
2) The computation of physical relationships between the COG and the local centers of inertia variables. These issues are discussed in detail below.

The motion of the COG expresses in terms of the individual COIs as

$$M_{COG}\frac{d^2\delta_{COG}}{dt^2} = T_{m_{COG}} - T_{e_{COG}} = \sum_{i=1}^{n} M_{COI_i}\frac{d^2\delta_{COI_i}}{dt^2} = \sum_{i=1}^{n} T_{m_{COI_i}} - \sum_{i=1}^{n} T_{COI_i} \tag{2.3a}$$

subject to the boundary conditions

$$\sum_{i=1}^{n} T_{COI_i,COG}^{tie} = \sum_{i=1}^{n} \frac{P_{COI_i,COG}^{tie}}{2\pi f_{COG}} = 0 \tag{2.3b}$$

in which, power transfers from the ith local COI to the COG are assumed to be of the general form [14]:

$$P_{COI_i,COG}^{tie} = \frac{|V_{COI_i}||V_{COG}|}{X_{COI_i,COG}^{tie}} \sin\left(\delta_{COI_i} - \delta_{COG}\right) \tag{2.4}$$

where $X_{COI_i,COG}^{tie}$ is the virtual reactance between area i and the COG. On the other hand, the condition for the equilibrium of the COI relative to the COG and local buses requires that

$$\sum_{\substack{j=1 \\ i\in n}}^{n} T_{COI_i,j}^{tie} + T_{COI_i,COG}^{tie} = \sum_{j=1}^{n} \frac{P_{COI_i,j}^{tie}}{2\pi f_{COI_i}} + \frac{P_{COI_i,COG}^{tie}}{2\pi f_{COI_i}} = 0 \tag{2.5}$$

where $T_{COI_i,j}^{tie}$ $T_{COI_i,COG}^{tie}$ are the applied torques from bus j and COG, respectively.

Substituting (2.4) in (2.3b) results in

$$\sum_{i=1}^{n} T_{COI_i,COG}^{tie} = \sum_{i=1}^{n} \frac{P_{COI_i,COG}^{tie}}{2\pi f_{COG}} = \sum_{i=1}^{n} \frac{1}{2\pi f_{COG}} \frac{|V_{COI_i}||V_{COG}|}{X_{COI_i,COG}^{tie}} \sin\left(\delta_{COI_i} - \delta_{COG}\right) = 0 \tag{2.6}$$

and similarly, for area A_i,

$$\sum_{\substack{j=1 \\ i\in n}}^{n} T_{COI_i,j}^{tie} + T_{COI_i,COG}^{tie} = \sum_{j} \frac{1}{2\pi f_{COI_i}} \frac{|V_j||V_{COI_i}|}{X_{COI_i,j}^{tie}} \sin\left(\delta_j - \delta_{COI_i}\right) + \frac{P_{COI_i,COG}^{tie}}{2\pi f_{COI_i}} = 0 \tag{2.7}$$

In interpreting this model, note that the first term of (2.7) on the right-hand side explains the virtual transferred power between bus j and the local COI, while the second one describes the interactions between the local COI and the COG. The set of equations (2.1)–(2.7) describes the multi-area dynamic energy balance and is well suited for the efficient analysis of power and frequency transients of large interconnected power systems. In the new equivalent system of Figure 2.2, the equations are expressed in terms of variables that have the potential to approximate frequency behavior using a simplified model from which both local and global properties can be analyzed. In what follows, the relationship between the global frequency and the local frequencies is investigated. First, some basic assumptions are discussed to derive the equivalent model of Figure 2.2.

2.3.1.2 Basic Assumptions

Initially, several assumptions are introduced in the above framework to derive simple algebraic relationships between the frequency of the COG, the frequencies of the individual COIs, and the frequencies of key system buses, namely:

- The virtual impedances, i.e. $X^{tie}_{COI_i,COG}$ and $X^{tie}_{COI_i,j}$ in (2.6) and (2.7), are calculated based on the steady-state frequencies. This assumption can be easily understandable by considering that virtual impedances are calculated in relation to each other. In other words, since at the equilibrium condition, the resultant torque in the COG framework and *the summation of local COI torques are zero,* it seems reasonable to assume that the ratio of virtual impedances remains constant.
- Two different time horizons are assumed for voltage consideration in the procedure of calculation of virtual impedances: (i) the period from when the disturbance occurs to the time when the first under load tap changer (ULTC) tap movement takes place and (ii) the time in which the ULTCs operate and bus voltages increase to the reference values [15]. During the early period, bus voltage magnitudes gradually come down to values lower than the pre-disturbance values. The sensitivity of changes in the virtual impedances to the changes in bus voltage magnitude, in these two periods, is assumed to be negligible. This means that the ratio of impedances remains approximately constant over the time. Therefore, bus voltage magnitudes are set to 1 pu in the process of calculation of impedances.

2.3.1.3 Modeling Formulation

To determine the equivalent parameters of the COG, assume that COI angles and voltages are determined using the above procedures. It follows, therefore, that the problem of calculation of the equivalent reactances $X^{tie}_{COI_i,COG}$ and $X^{tie}_{COI_i,j}$ can be posed as the solution of an optimization problem of the form of

$$\min_{X^{tie}_{COI_i,COG}} \left| \sum_{i=1}^{n} \frac{1}{2\pi f_{COG}} \frac{|V_{COI_i}||V_{COG}|}{X^{tie}_{COI_i,COG}} \sin(\delta_{COI_i} - \delta_{COG}) \right| \tag{2.8}$$

and

$$\min_{X^{tie}_{COI_i,j}} \left| \sum_{j} \frac{1}{2\pi f_{COI_i}} \frac{|V_j||V_{COI_i}|}{X^{tie}_{COI_i,j}} \sin\left(\delta_j - \delta_{COI_i}\right) + T^{tie}_{COI_i,COG} \right| \tag{2.9}$$

Since the minimum value of an absolute function is zero, the solutions of (2.8) and (2.9) lead to the same results as solving (2.6) and (2.7). As the values of the virtual impedances in (2.8) and (2.9) are used to weight the participation of each area or bus in the multi-area dynamic energy balance, the optimization problem is enforced to satisfy the conditions $0 < X_{\bullet}^{tie} < 1$. The aim is to calculate the fictitious impedances which produce the local and global torques that add up to zero, and thus there is no need to guarantee the optimal solution. Genetic algorithms (GA) could be an efficient method to calculate the fictitious reactances.

The computational procedure involved in determining the fictitious reactances by the GA is given in the following steps:

- *Step 1.* Given rotor angle positions δ_j, $j = 1, \ldots, n_g$ and bus voltage magnitudes V_j, minimize the absolute resultant torque in (2.8) and (2.9) using a GA. An initial population of 200 chromosomes characterizes the GA. For a system including n areas, each chromosome consists of $10 \times n$ genes, representing the $X_{COI_i,COG}^{tie}$; each unknown variable, i.e. the fictitious reactance may be represented by 10 binary bits. Minimization of (2.9) requires defining $10 \times (m+1)$ genes, where m is the number of generator buses in the associated area. This means that $m+1$ fictitious reactances, including m $X_{COI_i,j}^{tie}$ and *one* $X_{COI_i,COG}^{tie}$, interact with each other to make that the resultant torque vanishes for the local COIs. As a result, only m $X_{COI_i,j}^{tie}$ are considered as unknown variables.
- *Step 2.* Set $X_{COI_i,COG}^{tie}$ to the value obtained from (2.8).
- *Step 3.* Compute frequency (and power) sensitivities in (2.2).
- *Step 4.* Calculate local and global frequency responses and tie-line net flows.

2.3.1.4 Local Frequency Estimation

From (2.3a), any instantaneous change in the COG frequency, caused by initial compensation of a disturbance by the kinetic energy of the equivalent rotating plant, would be represented by the swing equation as [13, 16]:

$$\frac{d(\Delta f_{COG})}{dt} = \frac{1}{2\pi M_{COG}}(\Delta P_{mech} - \Delta P_{elec}) \tag{2.10}$$

It then follows that the motion of the centers of inertia can be written as [17]:

$$\dot{f}_{COI_i} = \frac{1}{2\pi M_{COI_i}}\left[\Delta P_i - 2\pi D_i \Delta f_{COI_i} - P_{COI_i,COG}^{tie}\right], \quad i = 1, 2, \ldots, n \tag{2.11}$$

where $\dot{f}_{COI_i}$ and D_i are the COI frequency deviation for the area i following disturbance ΔP_i and load damping parameter, respectively. The equivalent damping coefficient D can be calculated using Prony analysis (refer to Section 7.2 for details of Prony analysis). The last term in (2.11) has a physical interpretation of interest. The net tie-line power exchange $P_{COI_i,COG}^{tie}$ can be rewritten in the alternative form of (2.4). By knowing the virtual impedances, the term $P_{COI_i,COG}^{tie}$ and (2.11) can be evaluated. Dividing (2.11) by (2.10) gives

$$\frac{df_{COI_i}}{df_{COG}} = \frac{M_{COG}}{M_{COI_i}} \frac{\left[\Delta P_i - 2\pi D \Delta f_{COI_i} - P^{tie}_{COI_i,COG}\right]}{(\Delta P_{mech} - \Delta P_{elec})} \tag{2.12}$$

then

$$\Delta f_{COI_i} = \underbrace{\frac{M_{COG}}{M_{COI_i}} \frac{\left[\Delta P_i - 2\pi D \Delta f_{COI_i} - P^{tie}_{COI_i,COG}\right]}{(\Delta P_{mech} - \Delta P_{elec})}}_{\frac{\partial f_{COI_i}}{\partial f_{COG}}} \Delta f_{COG} \tag{2.13}$$

ΔP_i in (2.13) is nonzero only for the disturbed area. Moreover, the term $\frac{df_{COI_i}}{df_{COG}}$ describes the sensitivity of the COI frequency to the COG frequency. Multiplying $\frac{df_{COI_i}}{df_{COG}}$ by the COG frequency variations for a given energy mismatch Δf_{COG} gives area's i COI frequency changes.

Rearranging terms in (2.13) results in:

$$\Delta f_{COI_i} + \frac{M_{COG}}{M_{COI_i}} \frac{2\pi D \Delta f_{COI_i}}{(\Delta P_{mech} - \Delta P_{elec})} \Delta f_{COG} = \frac{M_{COG}}{M_{COI_i}} \frac{\left[\Delta P_i - P^{tie}_{COI_i,COG}\right]}{(\Delta P_{mech} - \Delta P_{elec})} \Delta f_{COG} \tag{2.14}$$

or, equivalently,

$$\Delta f_{COI_i} = \underbrace{\frac{\frac{M_{COG}}{M_{COI_i}} \frac{\left[\Delta P_i - P^{tie}_{COI_i,COG}\right]}{(\Delta P_{mech} - \Delta P_{elec})}}{1 + \frac{M_{COG}}{M_{COI_i}} \frac{2\pi D}{(\Delta P_{mech} - \Delta P_{elec})} \Delta f_{COG}}}_{\frac{\partial f_{COI_i}}{\partial f_{COG}}} \Delta f_{COG} \tag{2.15}$$

Equation (2.15), which represents the area i-th COI frequency based on COG frequency (2.10), can be employed to analyze the propagation of disturbances to neighboring systems through tie-lines. It should be noted that as the overall rotating masses and loads are considered to be aggregated in the COG, the term $\Delta P_{mech} - \Delta P_{elec}$ in (2.15) represents the magnitude of load disturbance.

Following the same procedure as in (2.15), each bus frequency can be estimated based on the associated local COI frequency. For this purpose, the dynamic behavior of each local bus frequency f_j is expressed as

$$\dot{f}_j = \frac{1}{2\pi M_j} \left[\Delta P - 2\pi D_j \Delta f_j - P^{tie}_{COI_i,j}\right] \tag{2.16}$$

Dividing (2.16) by (2.11) and following the same procedure as in (2.15) gives:

$$\Delta f_j = \frac{\frac{M_{COI_i}}{M_j} \frac{\left[\Delta P_i - P^{tie}_{COI_i,COG}\right]}{\left(\Delta P_j - P^{tie}_{ij}\right)}}{1 + \frac{M_{COI_i}}{M_j} \frac{2\pi D}{\left(\Delta P_j - P^{tie}_{COI_i,j}\right)} \Delta f_{COI_i}} \Delta f_{COI_i} \tag{2.17}$$

While (2.15) ties each COI frequency to the COG frequency dynamics, (2.17) connects each bus frequency (local frequency) to the COI dynamics. This means that the frequency dynamics of each bus can be tied to the COG frequency using simple algebraic equations. In the other words, (2.15) and (2.17) illustrate how the frequency dynamics propagate in the system.

2.3.1.5 Simulation Results

Three test systems are employed for illustrating the capability of the modeling procedures in this chapter: (i) a simple two-area, four-machine test system, (ii) a 16-machine, five-area 68-bus test model of the New York New England (NYNE) test system, and (iii) the IEEE 50-machine test system.

Two-Area System

Figure 2.3 shows a single-line diagram of this system. All the generating units are modeled with sixth-order synchronous machine models with excitation systems [18]. Each generator is equipped with a simple turbine-governor model of Figure 2.4. The total system load is 2734 MW; the disturbance considered is the shedding of 1400 MW load, i.e. 14 pu at bus 14 in Area 2.

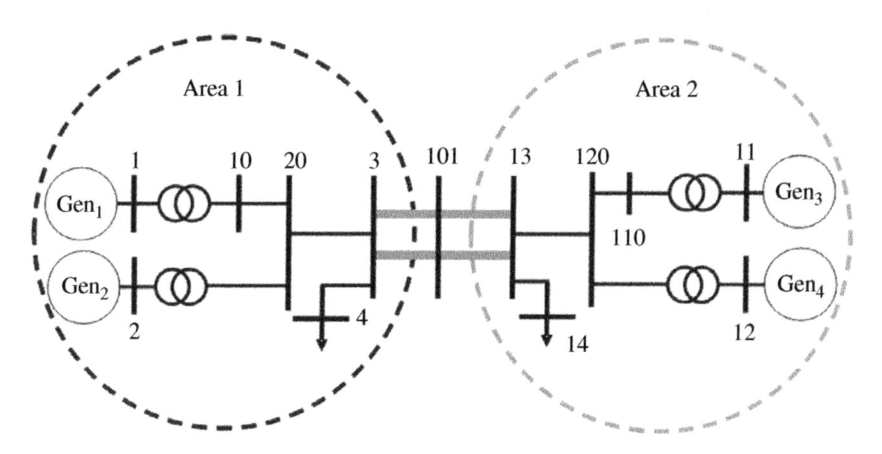

Figure 2.3 Single-line diagram of two-area system.

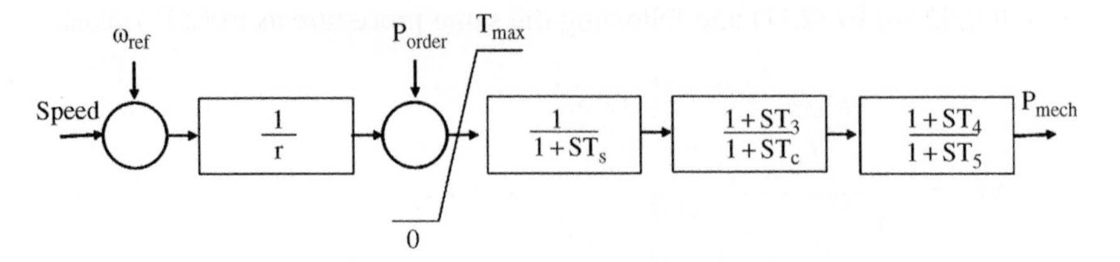

Figure 2.4 Simple Turbine Governor model ($r = 25$, $T_s = 0.1$, $T_3 = 0.0$, $T_4 = 1.25$, $T_5 = 5.0$, $T_c = 0.5$, $T_{max} = 1.0$).

Using (2.15) and (2.17), the frequency deviations at bus 1 in Area 1 can be expressed in terms of the COG frequency as

$$\Delta f_1 = \frac{1.06}{1 + 0.06\Delta f_{COI_1}} \Delta f_{COG} \tag{2.18}$$

Similarly, for bus 12 in Area 2, one can find that

$$\Delta f_{12} = \frac{1.78}{1 + 0.93\Delta f_{COI_2}} \Delta f_{COG} \tag{2.19}$$

The significant difference between (2.18) and (2.19) stems from the fact that the fault takes place in Area 1. Figure 2.5 shows the transient behavior of synchronous machines, obtained by a conventional washout filter, a frequency divider [8], and the frequency propagation approach in (2.15) and (2.17). For comparison purposes, the time constant of the washout filter T_f is set to the default value 0.01 [8].

Additional insight into the ability of the method to characterize the slow system dynamics can be obtained from the modal analysis of the reduced-order COG representation. Table 2.1 compares the frequency of the two slowest modes of the full

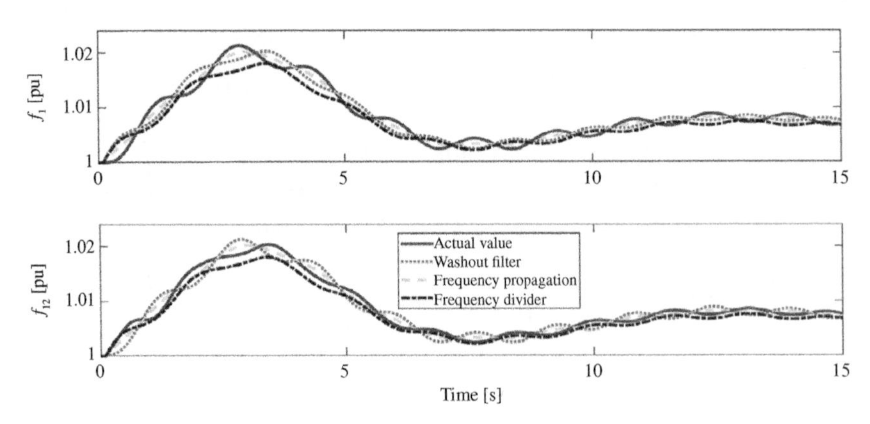

Figure 2.5 Frequency responses of two-area system following the loss of 14 pu of load at bus 14.

Table 2.1 Comparison of two slowest modes for original model and equivalent model.

Original model frequency	Equivalent model frequency	Error (%)
0.1936	0.1956	1.4
0.6802	0.7088	4.2

model with those of the COG-based equivalent model. As shown, the maximum error is less than 5%. This example suggests that simple algebraic equations of (2.15) and (2.17) can estimate local frequencies from the COG-based model of Figure 2.2 with high accuracy.

NYNE Test System

The NYNE test system, including five geographical regions, is used to test the ability of coherency-based frequency propagation analysis to characterize power and frequency deviations following large system perturbations. A diagram of the NYNE system, including areas and their interconnections, is shown in Figure 2.6. Detailed generating unit models and their controllers were included in the simulations. Generators 1–12 are equipped with a fourth-order type II power system stabilizer, tuned to provide sufficient damping [19, 20].

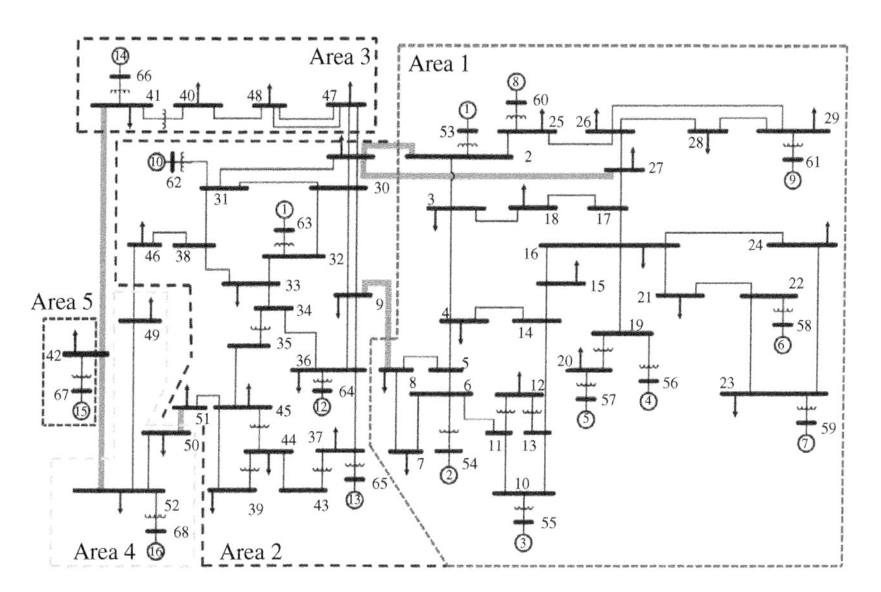

Figure 2.6 Single line diagram of the NYNE system showing coherent areas and their interconnections.

Numerical results comparing the estimated frequency are presented below. Four contingency scenarios (CSs) are considered in this analysis:

- CS1: Loss of 6000 MW at bus 37 in Area 2
- CS2: Three-phase fault at line 31–62, cleared in 12 cycles
- CS3: Three-phase fault at line 52–68 followed by the outage of generator G16
- CS4: Outage of line 50–52 resulting in system instability

Figures 2.7 and 2.8 illustrate the frequency and tie-line power dynamics at key system locations for contingency scenarios CS1 and CS2. The frequency of buses 59 (Area 1), 1, 63 (Area 2), and 47 (Area 3) and the net power flows to Areas 2 and 4 are selected for analysis.

In each case, the exact frequency behavior, obtained by a time-domain (T-D) simulation program, is used as a basis for evaluating the accuracy of the estimation methods. In the studies described below, (2.4) is employed to estimate the net power flow to areas in the equivalent system; the net power flow to Area 1 and Area 4 in the physical system is calculated by

$$P_{tie,1} = P^{tie}_{1,2} + P^{tie}_{1,27} + P^{tie}_{9,8} \tag{2.20}$$

and

$$P_{tie,4} = P^{tie}_{50,51} + P^{tie}_{52,41} \tag{2.21}$$

where $P^{tie}_{k,m}$ represents the power flow across the tie-line connecting bus k, located in the study area, to bus m, located in neighboring areas.

Careful analysis of the numerical results in Figure 2.8 reveals that the frequency dynamics of bus 63 in response to contingency CS2 shows some discrepancy with the actual frequency response. Such discrepancies may result from system topology changing conditions modifying the estimated virtual reactances. In the basic framework, the inputs to the COG system are selected as bus voltage magnitudes, angles, and frequencies, obtained using a T-D simulation program. As a result, the COG dynamics could be updated or renewed using various strategies. In the below given results, three main strategies (STs) to compute the virtual reactances in (2.8) and (2.9) are compared:

- *ST1*: Calculations based on pre-fault values
- *ST2*: Calculations based on post-fault steady-state values
- *ST3*: Calculations based on bus phase angles collected at a rate of 2 cycles/s

Table 2.2 compares the accuracy of the COG-based model in compliance with the aforementioned strategies for estimating first-swing oscillations. As shown in the table, by recalculating the virtual reactances, the accuracy of the proposed framework can be enhanced even for system topology changing disturbances.

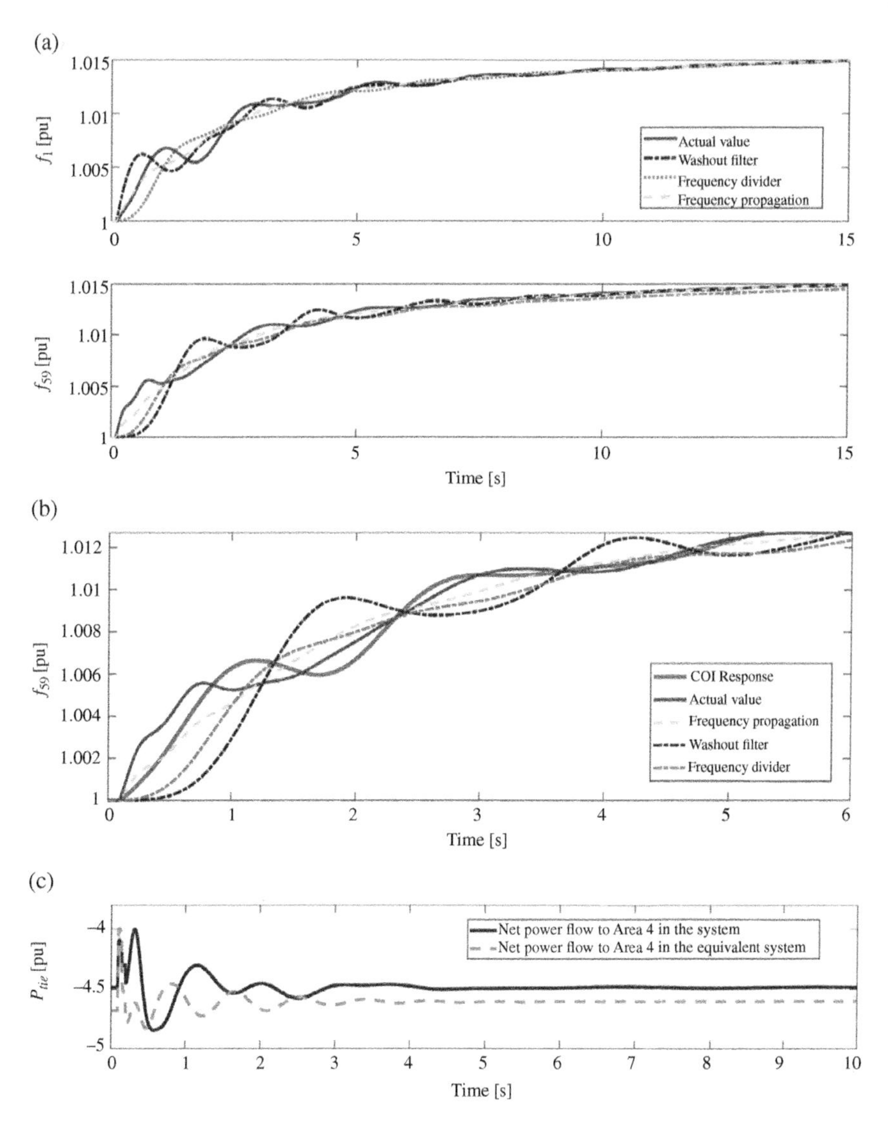

Figure 2.7 System response for contingency scenario CS1; (a) frequency responses of NYNE system, and (b) detail of frequency response showing the COI frequency; (c) net power flows to Area 4.

Figure 2.9 compares the frequency responses for the aforementioned strategies with those of actual response and frequency divider approach [8]. The results confirm the high accuracy of the frequency propagation approach to estimate both the transient and mid-term dynamics.

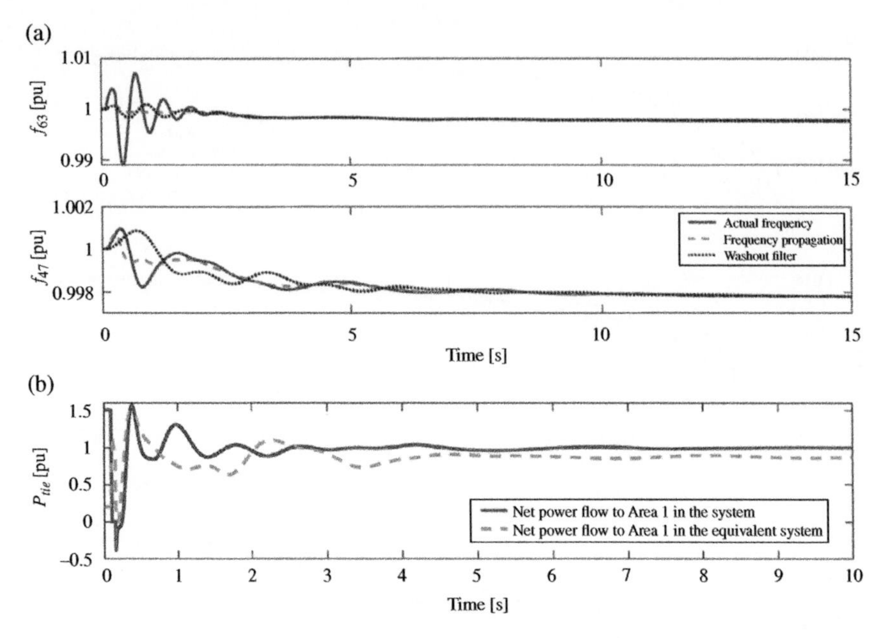

Figure 2.8 (a) Frequency responses of NYNE system for contingency scenario CS2, (b) net power flows to Area 3.

Table 2.2 Comparison of first swing amplitudes.

Strategy	Actual value	Frequency propagation	Error (%)
ST1	1.0483	1.0101	3.64
ST2	1.0483	1.0322	1.54
ST3	1.0483	1.0446	0.35

To further verify the theoretical basis behind the computation of fictitious reactances, Table 2.3 compares the ratio of the calculated impedances for different voltage magnitudes, including 1 pu voltage and the exact voltage magnitude after one, two, and three cycles, following the inception of fault. The results suggest that voltage magnitudes may not significantly affect the ratio of reactances.

Numerical experience shows that during a fault, when bus voltage magnitudes deviate from nominal values, the COG-based method can still exhibit satisfactory performance. This could be justified by the fact that the COG-based equivalent

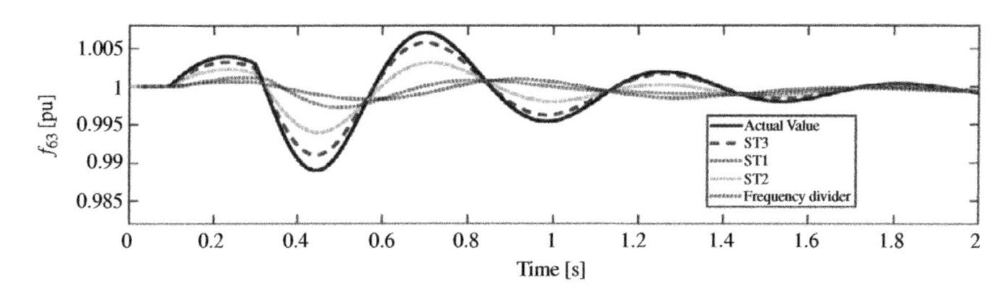

Figure 2.9 Frequency responses of different strategies for contingency scenario CS2.

Table 2.3 Sensitivity of virtual reactances to voltage variations.

	$\frac{X^{tie}_{COI_2,COG}}{X^{tie}_{COI_1,COG}}$	$\frac{X^{tie}_{COI_3,COG}}{X^{tie}_{COI_1,COG}}$	$\frac{X^{tie}_{COI_4,COG}}{X^{tie}_{COI_1,COG}}$	$\frac{X^{tie}_{COI_5,COG}}{X^{tie}_{COI_1,COG}}$
$v = 1$ pu	1.01	1.32	2.18	0.73
Snapshot 1	1.03	1.31	2.13	0.74
Snapshot 2	1.01	1.29	2.23	0.73
Snapshot 3	1.01	1.29	2.2	0.73

model mitigates the response of generators to voltage deviations, reflected as inter-unit synchronizing oscillations between generators.

Modal analysis was further conducted to evaluate the method's ability to assess the propagation of low-frequency oscillations. The NYNE has four inter-area modes of concern. Table 2.4 compares the eigenvalues of the full system and the COG representation in Figure 2.2. The reduced model includes five areas, five interconnecting lines, and their associated buses. As shown in Table 2.4,

Table 2.4 Comparison of modal analysis results.

	Original system model		**COG-based equivalent model**	
No.	**Eigenvalue**	**Frequency**	**Eigenvalue**	**Frequency**
1	$-0.7018 \pm 1.9710j$	0.3137	$-0.7356 \pm 1.7514j$	0.2787
2	$-0.1184 \pm 3.2666j$	0.5199	$-0.1004 \pm 3.0123j$	0.4794
3	$-0.3590 \pm 3.7108j$	0.5906	$-0.3913 \pm 4.0108j$	0.6383
4	$-0.1657 \pm 4.8915j$	0.7785	$-0.1535 \pm 4.6454j$	0.7393

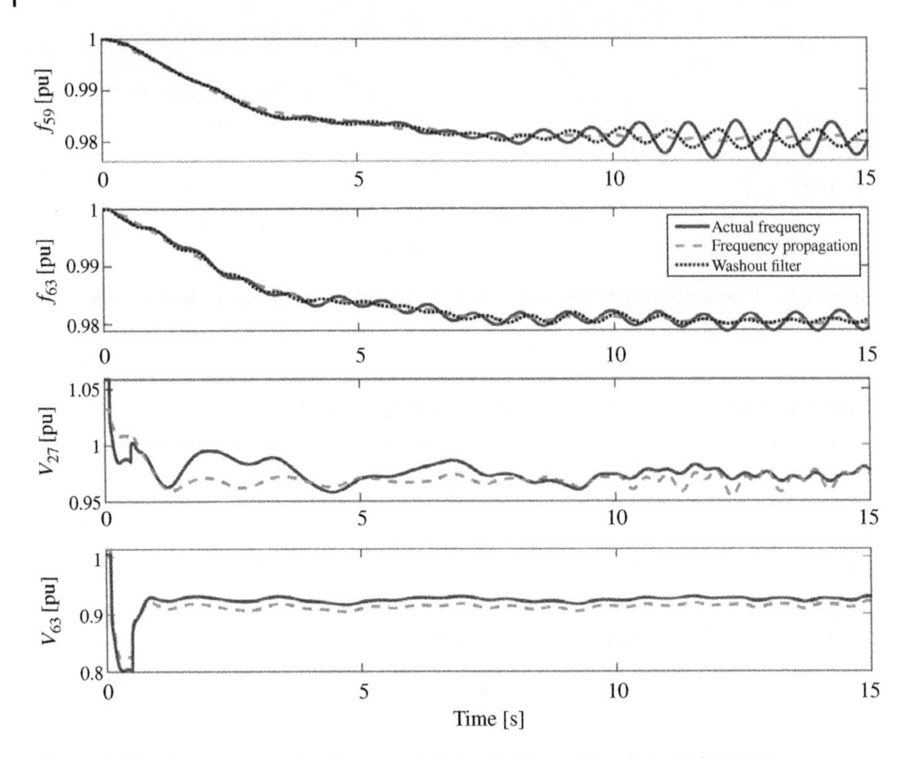

Figure 2.10 Frequency and voltage responses of 16-machine for contingency scenario CS3.

the COG-based equivalent model provides consistent results with eigen analysis results of the original (unreduced) system. This result also suggests that the approach can be effectively applied when characterizing long-term system behavior.

To further confirm the validity of the model, Figures 2.10 and 2.11 show the performance of the method for SC3 and SC4 resulting in system instability. The method is seen to accurately characterize system instability for both voltage and frequency signals.

During transients associated with cascading events, the stiffness of the COG-based method mitigates the variations of the center of angles and bus voltages. This is shown in Figure 2.11 which demonstrates that while such variations degrade the performance and accuracy of the proposed method to characterize short-term behavior, the overall trend of the unstable frequency oscillation is accurately visualized.

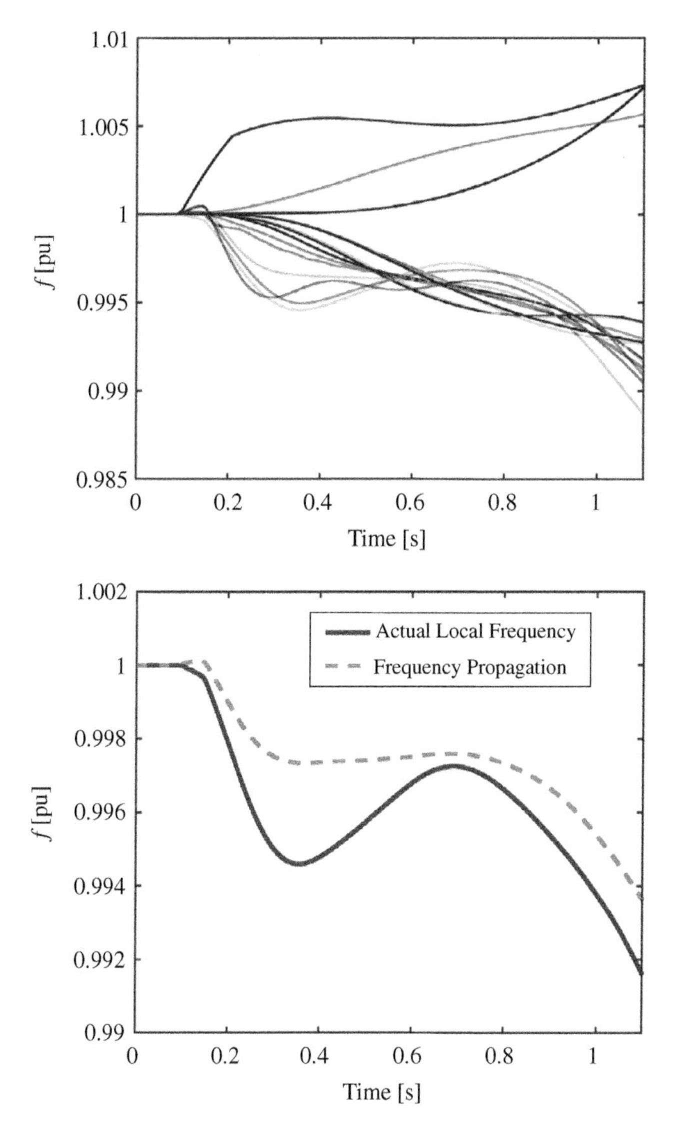

Figure 2.11 Frequency responses of 16-machine for contingency scenario CS4.

Finally, the effectiveness of the COG-based method is assessed to estimate frequency dynamics of interest in control and protection applications, including frequency nadir, and the rate of change of frequency (RoCoF) [17]. Table 2.5 compares the frequency nadir and RoCoF for the washout filter and the frequency propagation algorithm. This comparison is provided for two different scenarios:

Table 2.5 Comparison of frequency dynamics using the washout filter and proposed method.

	No.	Bus type	Nadir (FPA)	Nadir (WF)	Error (%)	RoCoF (FPA)	RoCoF (WF)	Error (%)
Loss of load	1	Dynamic load	1.022	1.020	0.2	0.022	0.0242	0.1
	8	Dynamic load	1.018	1.017	0.1	0.018	0.0234	0.3
	51	Dynamic load	1.025	1/023	0.2	0.025	0.0350	0.4
	53	Generator	1.017	1.016	0.1	0.017	0.0221	0.3
	68	Generator	1.026	1.026	0.0	0.026	0.0286	0.1
Loss of generation	1	Dynamic load	0.971	0.970	0.1	0.029	0.0290	0.0
	8	Dynamic load	0.968	0.968	0.0	0.032	0.0512	0.6
	51	Dynamic load	0.97	0.970	0.0	0.030	0.0450	0.5
	53	Generator	0.976	0.976	0.0	0.024	0.0312	0.3
	68	Generator	0.979	0.982	0.3	0.021	0.0252	0.2

FPA: Frequency propagation algorithm; WF: washout filter.

2470 MW load shedding at bus 52, and tripping of 4000 MW generation at bus 68. Numerical results demonstrate the accuracy of the COG-based method to capture slow system motion.

IEEE 50-Machine Test System

The 50-machine power network consists of 145 buses, 453 transmission lines, 52 transformers, and 60 loads; the total load is 2.83 GW and is used to investigate the performance of the method for more complex systems representations.

In the first scenario, the result obtained by applying a three-phase fault on bus 67 is considered. The fault is assumed to be cleared in 10 cycles by primary protection. Results obtained by the COG-based method are compared with those of using T-D simulation. A close examination of simulation results in Figure 2.12 shows that there is a good agreement with the results obtained by the frequency propagation-based method.

Further, the effectiveness of the COG-based method in the presence of inertia-less generating units is investigated. For this purpose, 10% of the system load is assumed to be supplied by wind farms. Figure 2.13 demonstrates the efficiency

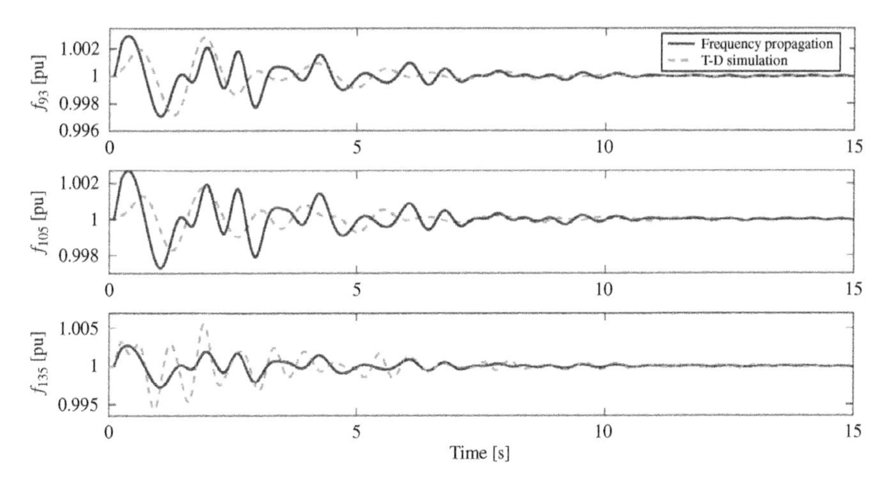

Figure 2.12 Frequency responses of 50-machine system following a three-phase fault on lines 67–119.

of the frequency propagation approach by comparing the estimated frequency dynamics and T-D simulation results.

Analysis of frequency studies in Figures 2.13 indicates that unrealistic frequency behavior associated with variations of the active power of variable generations is mitigated using the COG-based estimation method. To further investigate the accuracy of the modeling, the estimated frequency is numerically compared with

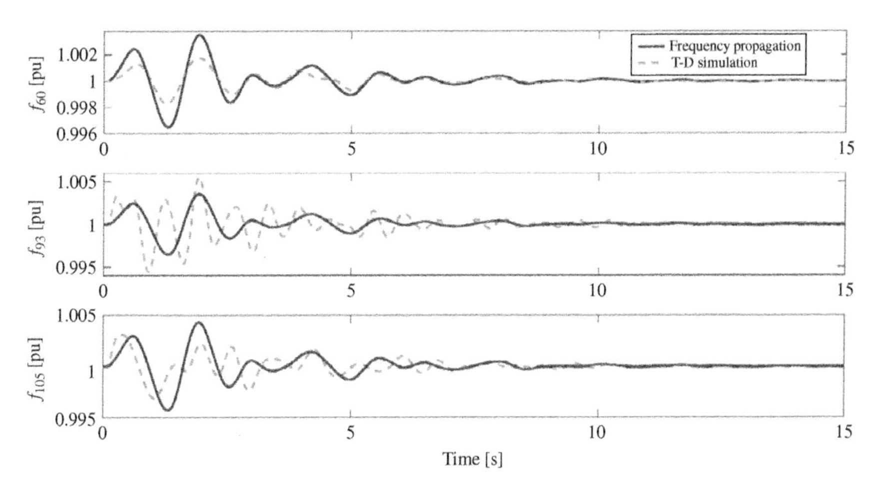

Figure 2.13 Frequency responses of 50-machine system including wind power facing three-phase fault.

Table 2.6 Computational burden of different bus frequency estimators.

Scenario	T-D simulation	Frequency propagation
With wind	535	29.34
Without wind	25	9

the actual frequency obtained using T-D simulation. In this analysis, the best fitness value (BFV), a measure of modeling accuracy, is calculated as

$$BFV = 1 - \frac{y - \hat{y}}{y - \bar{y}} \tag{2.22}$$

where y is the frequency response obtained from the T-D simulation, $\hat{y}$ is the frequency response obtained from equations (2.15) and (2.17), and $\bar{y}$ is the mean of y. Simulation results reveal that while the *BFV* for the COG-based equivalent model is more than 85% in Figures 2.7–2.12, it is less than 65% for the system considering wind farms. This illustrates the necessity of modifying the COG-based model through stochastic approaches which are discussed in the next section.

It may also be noted that the time required to estimate local frequencies in the frequency propagation algorithm in the presence of wind power is about 30 seconds, while it is more than 500 seconds for close T-D simulation. Table 2.6 compares the computational burden of frequency propagation paradigm and close T-D simulation. The significant difference between the frequency propagation technique and T-D simulation can be justified by noting that the number of state variables in the COG-model based estimation is much less than that for T-D simulation.

Finally it should be emphasized that the results of Figures 2.12 and 2.13 suggest that the tie-line power flows and the local frequencies converge to steady state in a short-time horizon. This finding justifies the fact that the COG-based method could be used to rapidly and accurately determine the post-fault stable equilibrium points following perturbations as well as to assess the effect of remedial control schemes on frequency dynamics.

2.3.2 An Enhanced COG-Based Model

2.3.2.1 Key Concept

The concept of COG was introduced in Section 2.3.1 to study long-term power-frequency transients following large perturbations. The COG model shows high efficiency to capture power and frequency oscillations, and thus it could be employed with high reliability in small-signal, voltage, and frequency stability studies. However, simulation results in the presence of variable generations suggest the need for modifying the deterministic COG-based model. The theory developed in this section extends the considerations of the effect of variable generations and the associated uncertainties on the COG model formulation. Without the loss of generality, DGs and ESSs included in a multi-MGs (MMGs) are assumed as variable generations to realize penetrated power grids.

Using the COG concept, the given MGs integrated-power grid can be represented by a simplified equivalent model of Figure 2.14, obtained by solving a simple minimization problem of (2.8) and (2.9). Two limitations are inherent for the representation of (2.8) and (2.9) to assess the impact of MMGs integration: (i) uncertainties in the representation of MMGs cannot be properly incorporated and (ii) different operating modes, i.e. grid-connected and islanded modes, cannot be properly represented [9, 21].

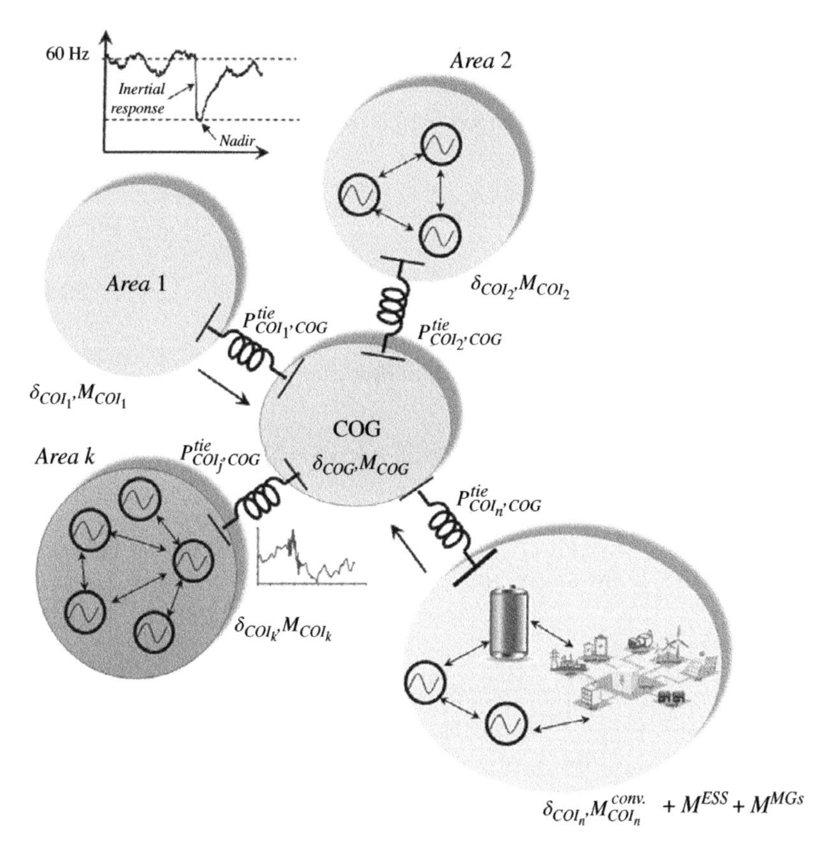

Figure 2.14 Interconnected power system divided into areas illustrating the notion of the COG considering MMGs and ESSs.

Incorporating the dynamics of MMGs into the equivalent model representations requires a transformation of the simple deterministic COG formulation of (2.8) and (2.9) to a complex uncertain optimization problem. In what follows, the use of the COG framework in the case of an MG-integrated power system with several MMGs is reviewed in the light of a robust approach to study transient power-frequency behavior.

To introduce the proposed formulation, assume that several MMGs are integrated with ESSs to provide frequency control during emergency conditions as well as to improve the grid's resiliency. Taking the contribution of MMGs into account, the balance constraint (2.3b) in the penetrated grid can be represented as

$$\sum_{i=1}^{n} T_{COI_i,COG}^{tie} = \sum_{i=1}^{n} \frac{P_{COI_i,COG}^{tie} - \Delta P_i^{MMG}}{\omega_{COG}} = 0 \tag{2.23}$$

where the virtual tie-line power flow, $P_{COI_i,COG}^{tie}$, has the form of (2.4). The problem of interest again is to determine the fictitious reactance $X_{COI_i,COG}^{tie}$ in (2.23). For this purpose, a four-step procedure based on a data-driven goal programming approach is proposed as follows:

Step 0: Insert (2.3b) into (2.23), as

$$\sum_{i=1}^{n} \frac{1}{\omega_{COG}} \left(\frac{|V_{COI_i}||V_{COG}|}{X_{COI_i,COG}^{tie}} \sin\left(\delta_{COI_i} - \delta_{COG}\right) - \Delta P_i^{MMG} \right) = 0 \tag{2.24}$$

Step 1: Reformulate (2.24) as a minimization problem

$$\begin{aligned} &\min \sum_{i=1} \frac{1}{\omega_{COG}} \left(\frac{|V_{COI_i}||V_{COG}|}{X_{COI_i,COG}^{tie}} \sin\left(\delta_{COI_i} - \delta_{COG}\right) - \Delta P_i^{MMG} \right) \\ &s.t. \qquad 0 < X_{COI_i,COG}^{tie} \leq 1 \end{aligned} \tag{2.25}$$

Step 2: Recast (2.25) in the linear form

$$\begin{aligned} &\min_{X_{COI_i,COG}^{tie}} \sum \xi^- - \xi^+ \\ &s.t. \quad \sum_{i=1}^{n} \frac{1}{\omega_{COG}} \frac{|V_{COI_i}||V_{COG}|}{X_{COI_i,COG}^{tie}} \sin\left(\delta_{COI_i} - \delta_{COG}\right) + \xi^- - \xi^+ = \sum_{i=1}^{n} \Delta P_i^{MMG} \\ &\xi^-, \xi^+ \geq 0 \\ &0 < X_{COI_i,COG}^{tie} \leq 1 \end{aligned} \tag{2.26}$$

where ξ^{-} and ξ^{+} define deviations from the target value, ξ, in the negative and positive directions. Goal programming assigns a goal value to each of the objective measures, i.e. zero to (2.25). Undesired deviations from the target value ξ are then minimized using (2.26). From linear programming theory, at least one of the ξ^{-} and ξ^{+} must be zero [22].

Step 3: Generate scenarios for ξ and ΔP_i^{MMG}
Step 4: Redefine (2.26), taking the expected value of ξ into account, as

$$\min_{X^{tie}_{COI_i,COG}} \sum_{s=1}^{S} p_s \sum \xi_s^{-} - \xi_s^{+} \tag{2.27}$$

s.t.

$$\begin{aligned} &\sum_{i=1}^{n} \frac{1}{\omega_{COG}} \frac{|V_{COI_i}||V_{COG}|}{X^{tie}_{COI_i,COG}} \sin\left(\delta_{COI_i} - \delta_{COG}\right) + \xi_s^{-} - \xi_s^{+} = \sum_{i=1}^{n} \Delta P_{i,s}^{MMG} \\ &\xi_s^{-}, \xi_s^{+} \geq 0 \\ &0 < X^{tie}_{COI_i,COG} \leq 1 \end{aligned} \tag{2.28}$$

where p_s denotes the probability of each scenario and s denotes the scenario.

2.3.2.2 Simulation Results

The efficiency of the goal programming-based equivalent model of (2.27) and (2.28) is investigated through a comparison of frequency response features as well as computational time with those of the GA-based equivalent model (2.8) and (2.9) and T-D simulation results for two area power system model of Figure 2.3. For this purpose, ξ^{-} and ξ^{+} are set to 0 and 0.1, respectively. Table 2.7 demonstrates the high efficiency of the equivalent model to estimate the frequency dynamics of interest.

Also of interest, a simple interval approach is utilized to assess the robustness of the method. The interval approach assumes that the uncertain parameter takes value in a range. The aim is to find the lower and upper bounds of the objective function in compliance with the interval [23]. It could be reinterpreted as the probabilistic modeling with a uniform probability density function (PDF). Table 2.8 reports the lower and upper bounds of the objective function for 5% uncertainty in the power of MMG. The results confirm the robustness feature of the goal programming-based model which is a key point when dealing with highly penetrated power grids.

To calculate the reported results of Tables 2.7 and 2.8, 10 random scenarios of equal probability, i.e. $P_s = 0.1$ in (2.27), are generated. Minimization of

Table 2.7 Computational burden of different bus frequency estimators.

Scenario	Frequency nadir	RoCoF	Computational time
T-D simulation	1.021	0.61	23
Equivalent model (2.8) and (2.9)	1.016	0.58	2.1
Stochastic model (2.27) and (2.28)	1.009	0.55	9.14

Table 2.8 Uncertainty analysis of the equivalencing.

Scenario	Upper bound	Lower bound
Equivalent model (2.8) and (2.9)	0.5712	0.112
Stochastic model (2.27) and (2.28)	0.461	0.432

(2.27) subject to (2.28) reveals that the reduced equivalent model can be characterized by

$$X^{tie}_{COI_1,COG} = 0.2134; \quad X^{tie}_{COI_2,COG} = 0.3172 \tag{2.29}$$

2.3.3 Generalized Equivalent Model

2.3.3.1 Basic Logic

Not only the deterministic COG-based model of (2.8) and (2.9) and stochastic COG-based model of (2.27) and (2.28) in the previous sections cause nonunique results but also the time consumption feature affects the effectiveness of the models. Therefore, in this section, the so-far models would be modified to have an efficient analytical model. Figure 2.15 schematically demonstrates the logic behind the modified equivalencing approach. The lines between areas in Figure 2.15 are considered as the auxiliary reactances which in turn simplify the calculation of fictitious reactance of interest in (2.6).

Indeed, Figure 2.15 is a modified version of Figure 2.2 where a combination of star and end to end topologies are considered in the process of calculation of the fictitious reactances. The logic behind the modeling procedure is simple: "*while the summation of applied torques to COG is zero, the summation of torques in each area is non-zero.*" According to (2.1) and considering $RoCoF = \frac{d(\Delta\omega)}{dt}$, the summation of torques in each area could be explained by

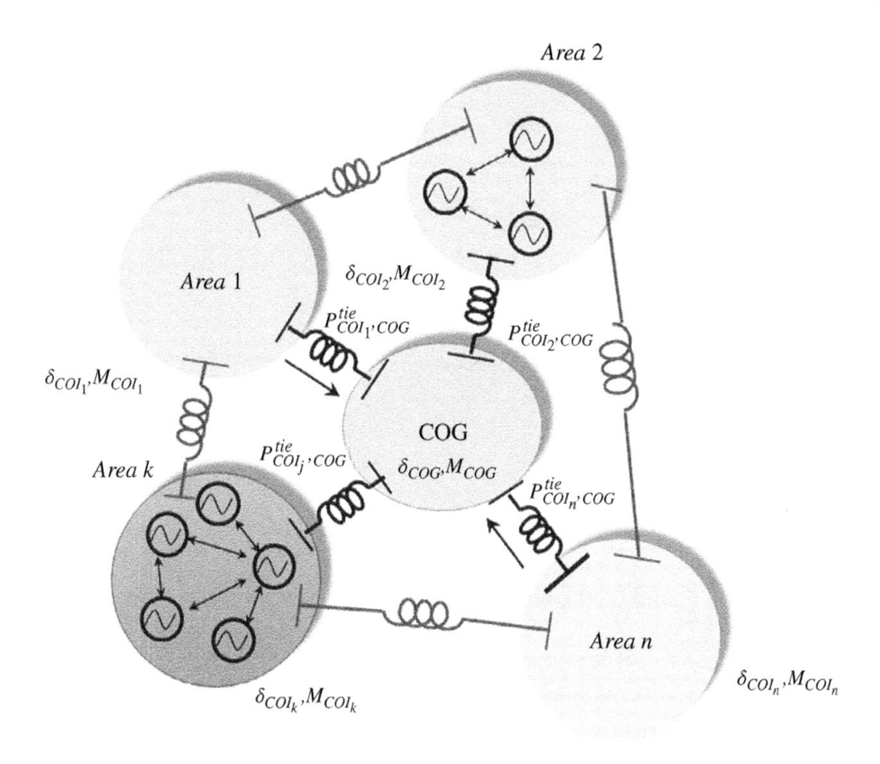

Figure 2.15 Interconnected power system divided into areas illustrating the notion of the modified COG concept.

$$M \times RoCoF = T_m - T_e \tag{2.30}$$

and hence, (2.6) would be rewritten for area i in Figure 2.15 as

$$T^{tie}_{COI_i,COG} + \sum_{j=1}^{k} T^{tie}_{COI_i,COI_j} = M \times RoCoF \tag{2.31}$$

Considering (2.6) and (2.31), a set of n equations with n undefined fictitious reactances would be derived.

2.3.3.2 Simulation and Results

The effectiveness of Figure 2.15 to represent the power grid equivalent model is investigated on two test systems. In the first scenario, the outage of generator number 4 is considered in the two-area power grid of Figure 2.3. Figure 2.16 shows the equivalent model of the system in the introduced framework.

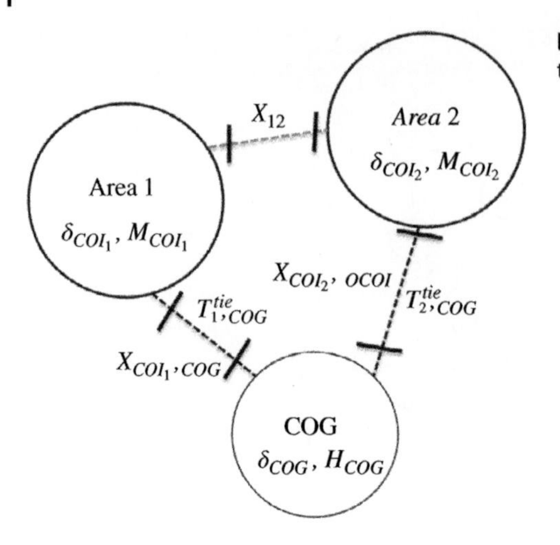

Figure 2.16 Equivalent model of two-area system.

According to (2.6) and (2.31), one could write:

$$\begin{aligned} &0.37X_{COI_1COG} - 0.58X_{COI_2COG} = 0 \\ &0.13X_{COI_1COG} + 0.73X_{12} = 0.45 \\ &0.51X_{COI_2COG} - 0.42X_{12} = 0.36 \end{aligned} \tag{2.32}$$

Figure 2.17 compares the results of T-D simulation with those of obtained using (2.32).

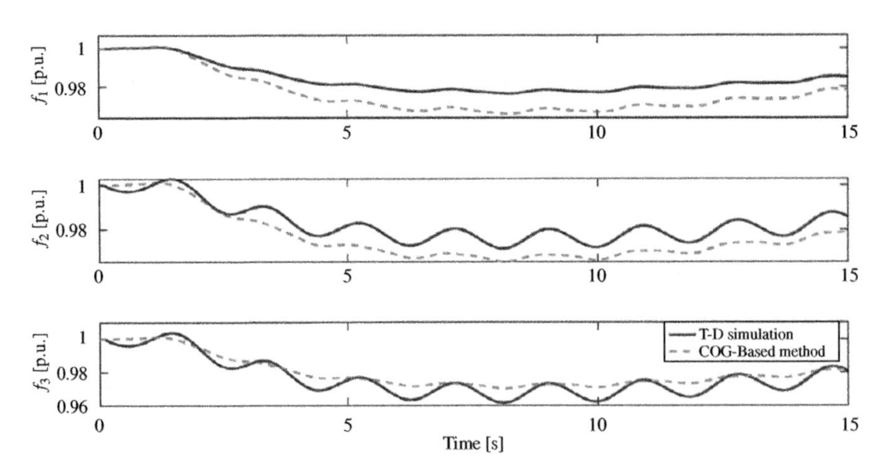

Figure 2.17 Frequency responses of two-area system for the outage of generator 4.

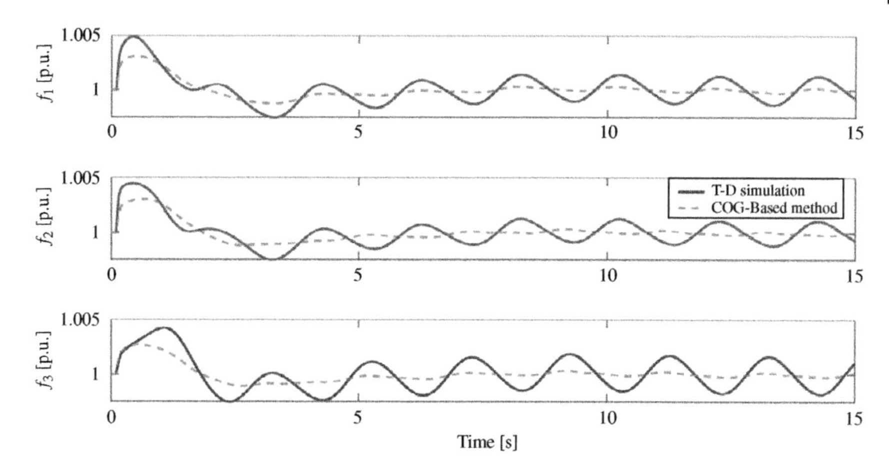

Figure 2.18 Frequency responses of two-area system for tripping of load.

In another attempt, the efficiency of the generalized COG-based method is investigated for the rejection of a load. Following the same procedure as (2.32) gives:

$$\begin{aligned}&0.18X_{COI_1COG} - 0.32X_{COI_2COG} = 0\\&0.72X_{COI_1COG} + 0.23X_{12} = 0.15\\&0.08X_{COI_2COG} - 0.12X_{12} = 0.07\end{aligned} \tag{2.33}$$

which in turn gives rise to the dynamic behavior of Figure 2.18.

Finally, the effectiveness of the model is further justified using NYNE test system of Figure 2.6 for four different scenarios:

- Scenario 1: Outage of generator 16 (Figure 2.19)
- Scenario 2: Three-phase fault on line 49–52 (Figure 2.20)
- Scenario 3: Outage of generator 15 (Figure 2.21)
- Scenario 4: Tripping of load at bus 37 (Figure 2.22)

2.4 MG Equivalent Model

Developing MG equivalent models is of high importance when dealing with MGs-integrated power system stability analysis. The equivalent model should consider uncertainties related to output power fluctuations and different operating modes of the sources. This section begins with a subsection devoted to the modeling of DGs in grid-forming mode (islanded mode) and will be continued by deriving

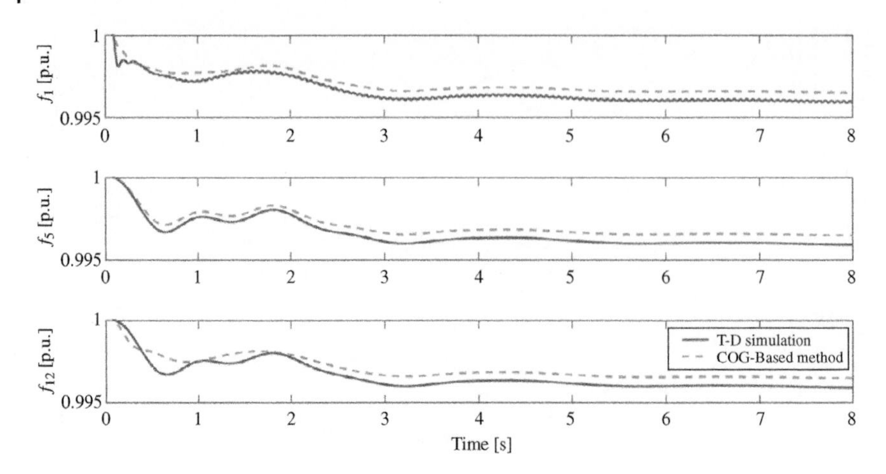

Figure 2.19 Frequency responses of NYNE system for scenario 1.

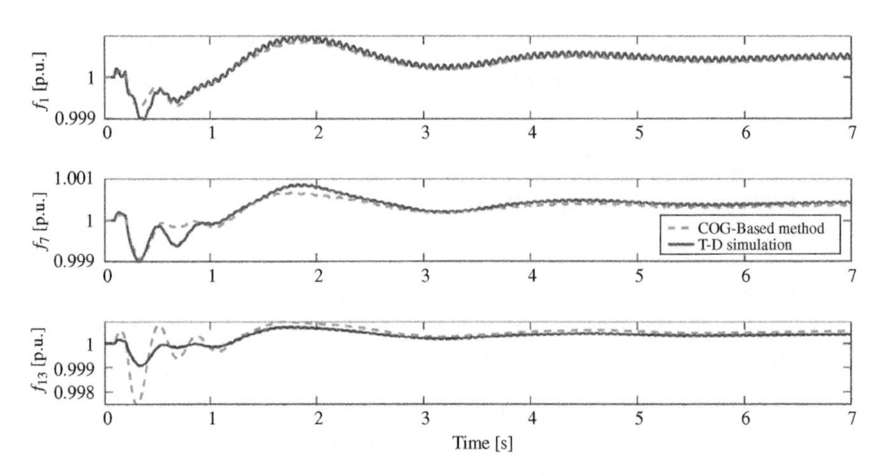

Figure 2.20 Frequency responses of NYNE system for scenario 2.

an experimental-based equivalent model of MG in grid-supporting mode (grid-connected mode). University of Kurdistan MG (UOK-MG), represented in Figure 2.23, is employed to validate the results in this section.

2.4.1 Islanded Mode

2.4.1.1 Synchronous-Based DG

Synchronous-based DG, i.e. Genset, includes an internal combustion (IC) engine driven by explosive combustion of gasoline and a wound field synchronous machine (SM) [24–26]. In the Genset modeling, there are two main components: the IC engine that converts the fuel to mechanical power and regulates the

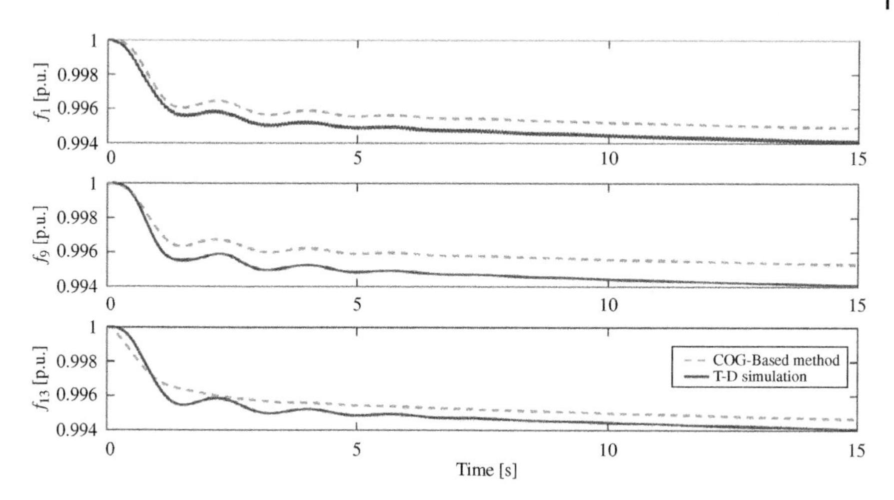

Figure 2.21 Frequency responses of NYNE system for scenario 3.

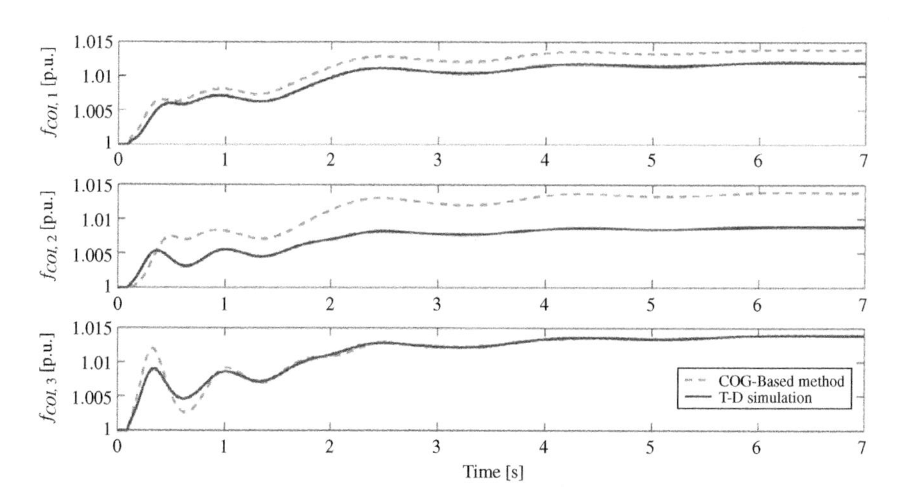

Figure 2.22 Frequency responses of NYNE system for scenario 4.

frequency of Genset and the exciter that regulates the terminal voltage of the Genset [27–30]. Details of the Genset controller are shown in Figure 2.24 [26].

Fuel command F_{CMD} is the output of the fuel controller Figure 2.25. Using this framework, the measured speed of the synchronous machine compares with the reference signal to produce an error signal. The error signal is fed into a proportional-integral (PI) controller to produce the torque signal [27, 28, 30].

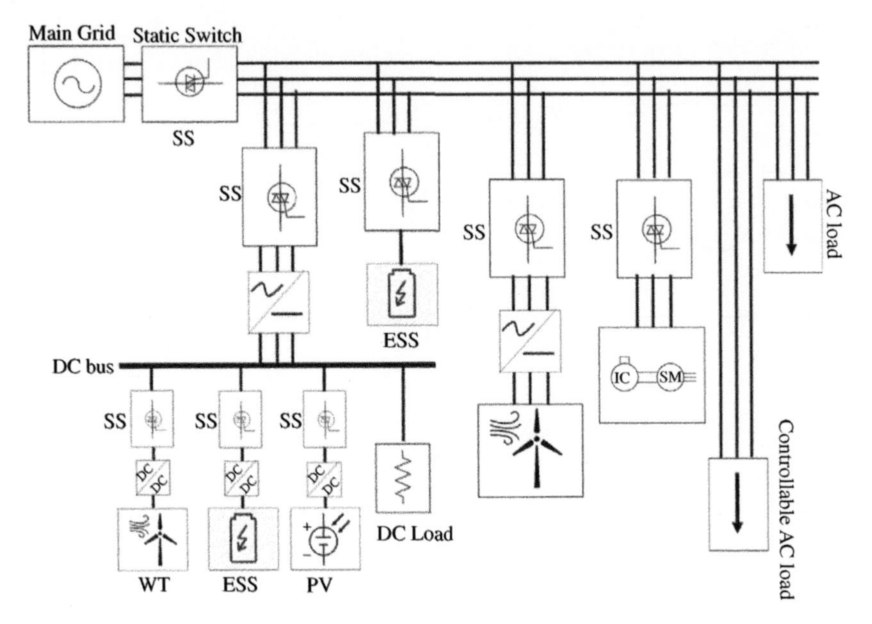

Figure 2.23 Three-phase schematic representation of the UOK-MG; WT, wind turbine; PV, photovoltaic; SS, static switch; IC, internal combustion; SM, synchronous machine.

The limiter often implements in the block diagram of Figure 2.25 to avoid unrealistic commands during large load transients. The resultant torque then converts to the fuel command signal using the torque to fuel conversion ratio K_{tf}. The output fuel command signal of Figure 2.25 then applies to the simplified IC engine model of Figure 2.26 to produce mechanical power. In this way, first, the fuel command signal converts into the torque signal. Furthermore, engine combustion delay affects the resultant torque. Finally, the torque converts to the mechanical output power while losses are removed from the resulting value [27, 28, 30].

The exciter mechanism of Figure 2.27 is simply represented by a first-order transfer function. Furthermore, the output DC field voltage multiplies by the electrical speed, producing the AC voltage magnitude. The implemented limiter in Figure 2.27 represents the saturation of the DC exciter field. The output is the field voltage of SM [27, 28, 30].

The input signal to the exciter of Figure 2.27 comes from the voltage regulator of Figure 2.28. It could be observed from Figure 2.28 that the error signal combines with the feedforward value, which is the expected value required for nominal voltage at the terminal. Feedforwarding this value allows for quicker initial convergence without the integrator having to wind up [27–31].

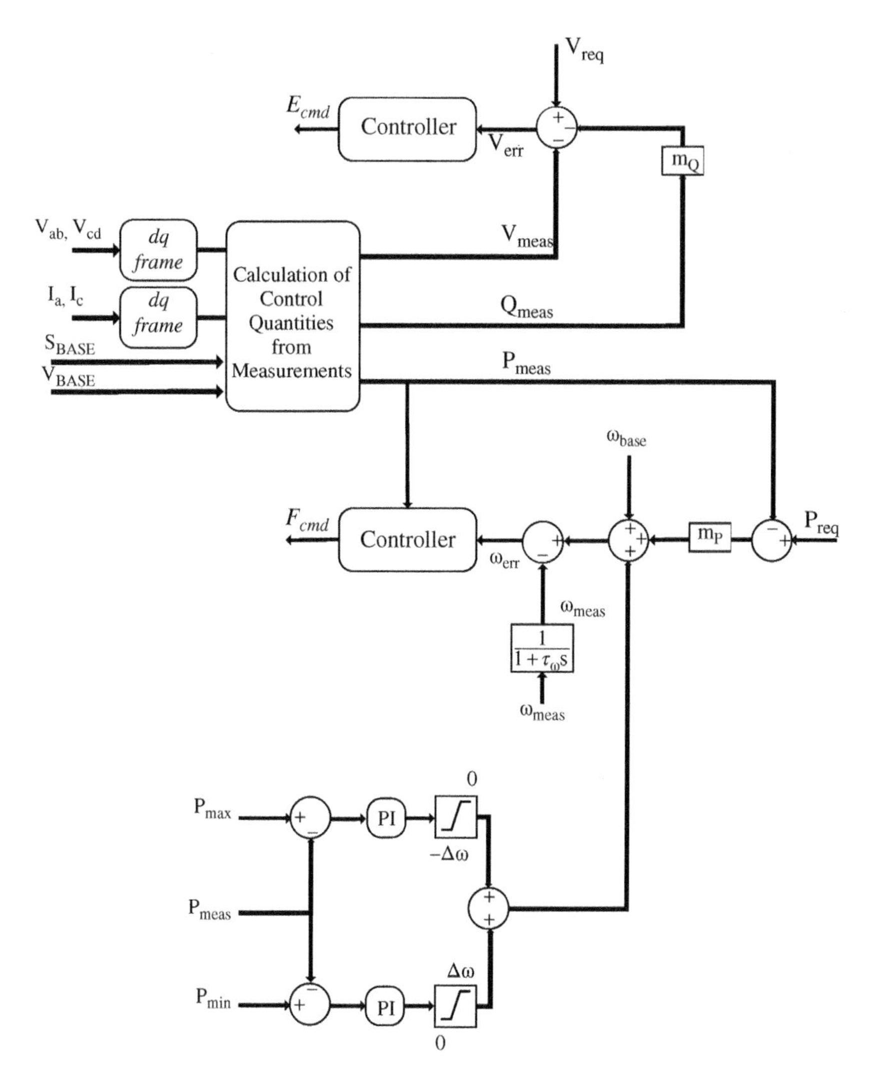

Figure 2.24 Genset controller scheme; M_Q, slope of $Q-V$ droop; M_P, slope of $P-\omega$ droop; K_{PI}, integral control gain; K_{PP}, proportional control gain; F_{CMD}, command fuel signal; E_{CMD}, exciter control signal; P_{meas}, real-time value of real power; Q_{meas}, real-time value of reactive power; $I_{.}$, line current; $\Delta\omega$, allowable frequency change.

2.4.1.2 Genset Model Validation

Figure 2.29 shows the frequency dynamics of the Genset in response to turn on a 4 kW Static Load Bank (SLB) at 2 seconds. A comparison of the experimental and simulation results reveals the high efficiency of the model.

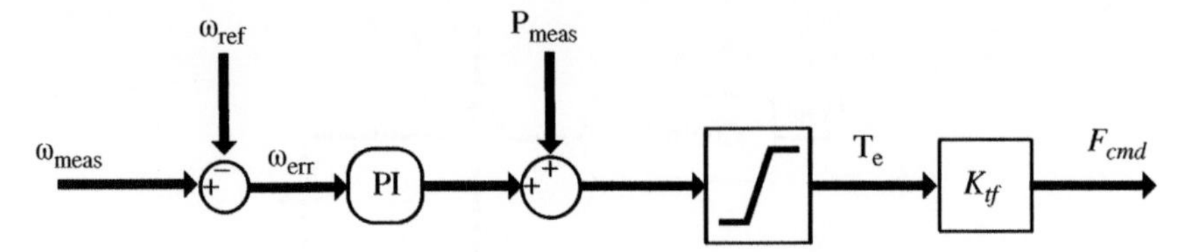

Figure 2.25 Fuel controller of Genset; K_{tf}, torque to fuel conversion ratio.

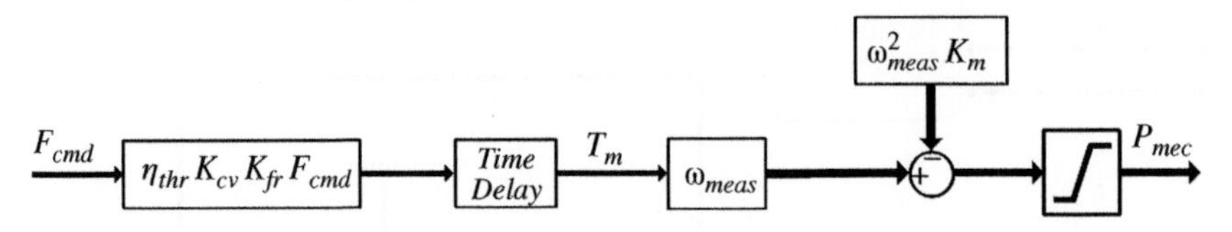

Figure 2.26 Simplified model of the IC engine; η_{thr}, thermal constant; K_{cv}, calorific value; K_{fr}, fuel rate at rated speed; K_m, mechanic losses constant.

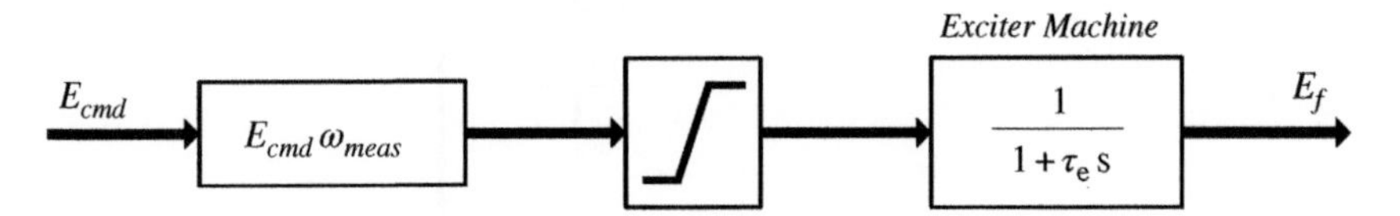

Figure 2.27 Exciter model; τ_e, exciter machine time constant.

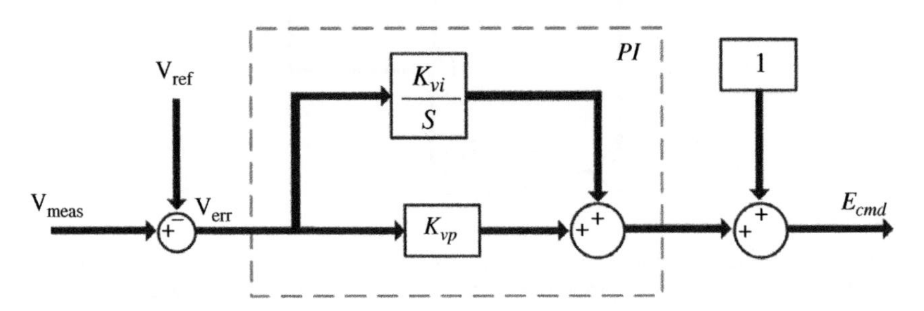

Figure 2.28 Voltage regulation diagram; K_{vi}, integral controller gain; K_{vp}, proportional controller gain.

2.4.1.3 Inverter-Based DG

Figure 2.30 shows the block diagram representation of an inverter-based DG [31]. Owing to the turn on/off effects of high-frequency switches, output signals have a fundamental component and higher harmonics. However, a high rate of

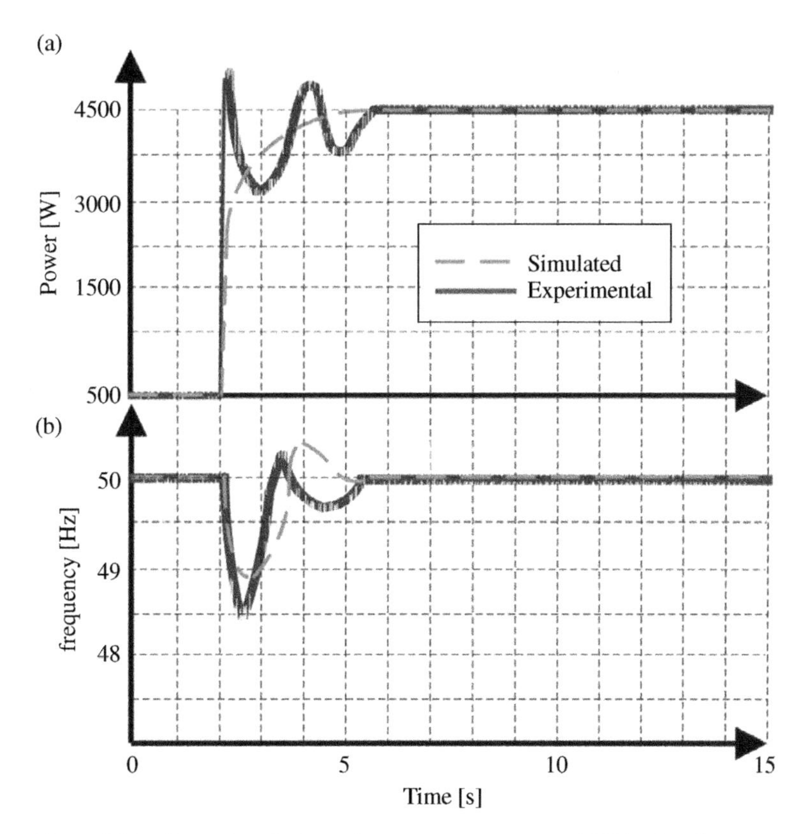

Figure 2.29 Dynamics of Genset in response to turn on 4 kW load at 2 seconds, (a) frequency response, (b) active power dynamics.

switching, in the range of 10 kHz, together with the implemented LC filters significantly mitigates harmonics [26, 29, 31]. Hence, the inverter, in the islanded operation mode, is represented as an ideal, balanced three-phase voltage source as shown in Figure 2.31 [31]. Accordingly, the instantaneous three-phase bus voltages can be expressed as

$$\begin{aligned} V_{ab}^{ESS} &= mV_{DC}\cos(\omega t + \theta) \\ V_{bc}^{ESS} &= mV_{DC}\cos(\omega t + \theta + 2\pi/3) \\ V_{ca}^{ESS} &= mV_{DC}\cos(\omega t + \theta - 2\pi/3) \end{aligned} \tag{2.34}$$

In this model, the modulation index m in (2.34) is a scalar coefficient that controls the voltage at the inverter terminals [26, 31]. Moreover, the same reasoning that uses to represent the Genset controller in the islanded mode can be extended to control inverter-based sources. However, the controller command signals are

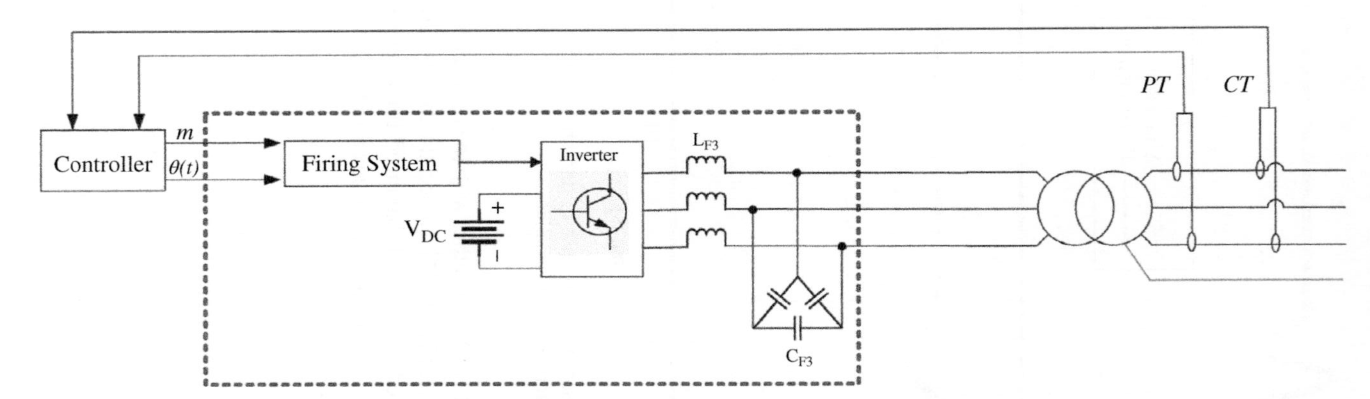

Figure 2.30 Inverter-based DG block diagram; m, modulating index; $\theta(t)$, angle for the voltage at the inverter terminals.

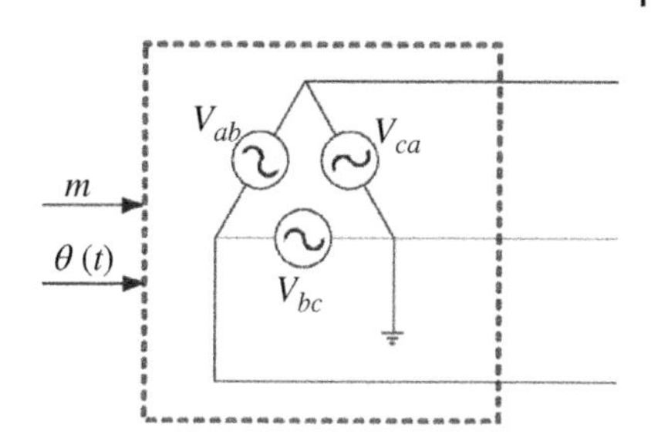

Figure 2.31 Ideal source model.

inverter frequency and voltage modulation index (Figure 2.32) [32–34]. Furthermore, in Figure 2.32, conventional droop characteristics implement using frequency and voltage droop coefficients [35].

2.4.1.4 Inverter-Based DG Model Validation

Figure 2.33 compares the ESS simulation results, obtained for switching on of 4 kW SLB at 1st second and switching off of 3 kW SLB at 5th seconds of the simulation, with those of experiments. Except for high-frequency noise, the simulated waveforms for real and reactive powers follow the experiments with high accuracy. The oscillation-free behavior of the simulated waveforms stems from the fact that the source of power models as an ideal voltage source with the only fundamental frequency. Furthermore, the offset between the simulated and experimental reactive powers is due to the neglecting of the line and transformer models.

2.4.2 Grid-Connected Mode

2.4.2.1 Basic Logic

After successful derivation of the models for DGs in islanded mode, this section discusses the dynamic equivalencing of a cluster of DGs and local loads, visualized in the form of grid-connected MG, for power-frequency transient studies. The analysis framework can be summarized in four steps:

- *Step 1:* Define an arbitrary number of operating points
- *Step 2:* Different commitment of power sources in the MG to realize operating points in *Step* 1
- *Step 3:* Metering of the injected power of MG to the host grid, i.e. P_{MG}, for each operating point in *Step 2*
- *Step 4:* Modal analysis (Prony analysis) of P_{MG}, metered in *Step 3*, to calculate inertia constant, i.e. M. Applying Prony analysis to P_{MG} gives:

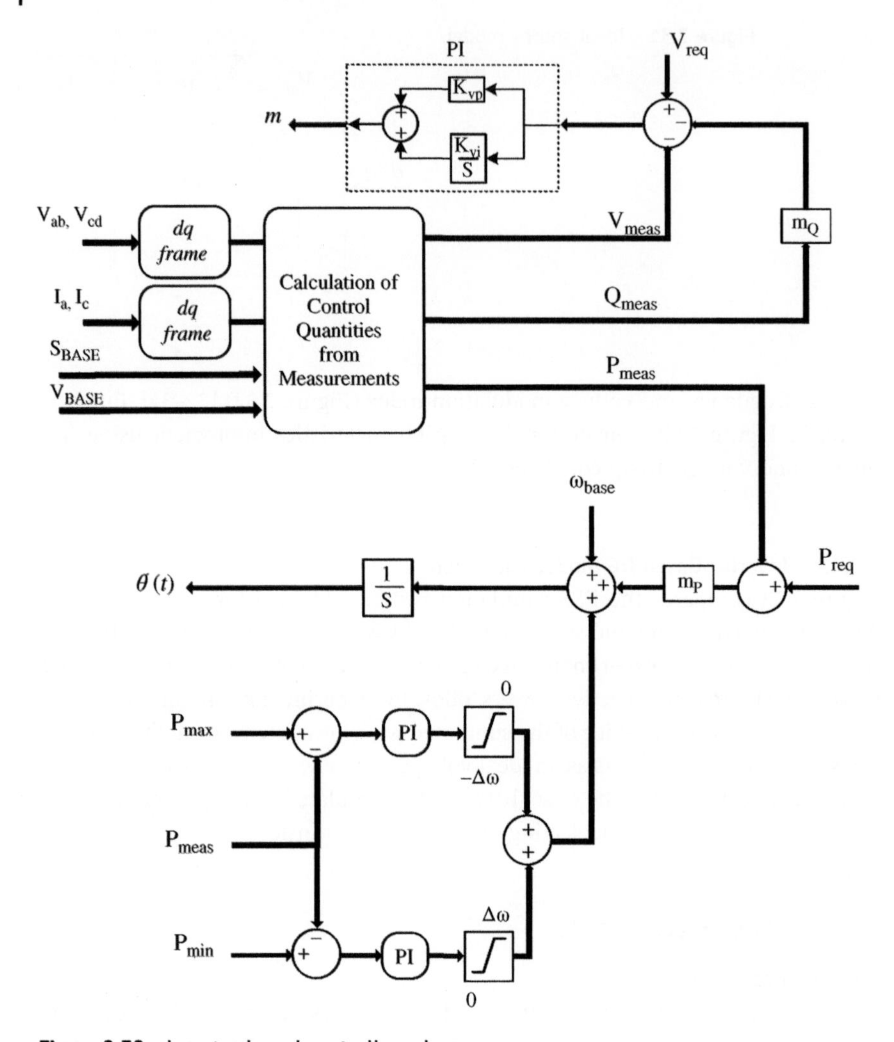

Figure 2.32 Inverter-based controller scheme.

$$\frac{f}{P_{MG}} \propto \frac{a_i}{s + c_i} \rightarrow \frac{f}{P_{MG}} \propto \frac{1}{\frac{1}{a_i}s + \frac{c_i}{a_i}} \tag{2.35}$$

where a, c are constant parameters which may be defined by Prony analysis; refer to Section 7.2 for details of Prony analysis. Equation (2.35) follows the same characteristics as the classical swing equation of the form

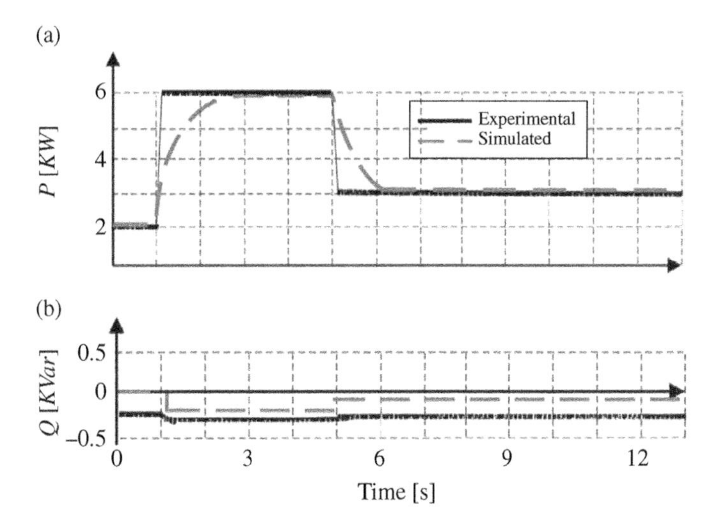

Figure 2.33 Experimental and simulated waveforms for real and reactive power output for inverter-based DG operation in the UOK-MG. (a) active power, (b) reactive power.

$$\frac{f}{P_{MG}} \propto \frac{1}{M_i s + D_i} \tag{2.36}$$

Hence, (2.35) with a reinterpretation of $\frac{1}{a_i} \triangleq M$; $\frac{c_i}{a_i} \triangleq D$ could be employed to calculate MG inertia constant. A conceptual scheme of the adopted model is shown in Figure 2.34.

Generally stated, Figure 2.34 suggests that MG response, from an upward point of view, would be mapped onto the conventional synchronous generator. Accordingly, the classical swing equation (2.36) represents the MG dynamics.

- *Step 5:* Return to *Step 2*, repeat until the prespecified number of operating points are considered
- *Step 6:* Extract relationship, using curve fitting tools, between inertia constant and committed DGs.

The above steps are summarized in the flowchart of Figure 2.35.

2.4.2.2 Model Validation

Experimental results of UOK-MG in grid-connected mode are utilized to demonstrate the efficiency of the modeling procedure. Furthermore, a real-time digital power system simulator is employed to visualize the host grid for testing purposes. Figure 2.36 shows the dynamic responses of the MG, host grid, and constituent DG for a sequence of events.

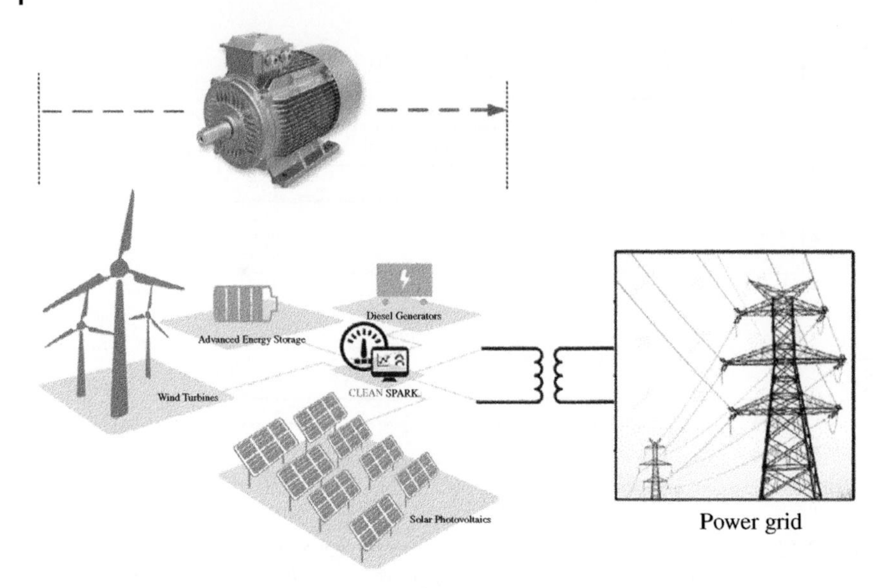

Figure 2.34 A conceptual scheme of the proposed MG dynamic equivalent model.

While event 1 refers to the importing of power to the MG, event 2 refers to the islanding of the MG from the grid. It is evident that, between events 1 and 2, the host grid provides the power of the MG with inertia constant. Following (2.36) for the injected power of the host grid to the MG in the transient period, in Figure 2.36, gives

$$\frac{\omega_{MG}}{\Delta P_{MG}} \cong \frac{0.37}{0.27 + 0.63s} \tag{2.37}$$

and in the steady state, (2.37) would be rewritten as

$$\frac{\omega_{MG}}{\Delta P_{MG}} = 1000 \tag{2.38}$$

where ω_{MGs} and ΔP_{MGs} are the frequency at the point of common coupling and the injected power of the MG, respectively. Equations (2.37) and (2.38) suggest that the MG mimics the behavior of a synchronous generator and a constant power source in the transient and steady-state periods, respectively. This brings a first-order circuit response to a pulse function in mind. Therefore, the same reasoning that uses to assess the first-order circuit in response to a pulse function could be adopted to approximate the effects of MG. In this way, one could represent the classical swing equation of MG as

Figure 2.35 Flowchart representation of the equivalencing approach.

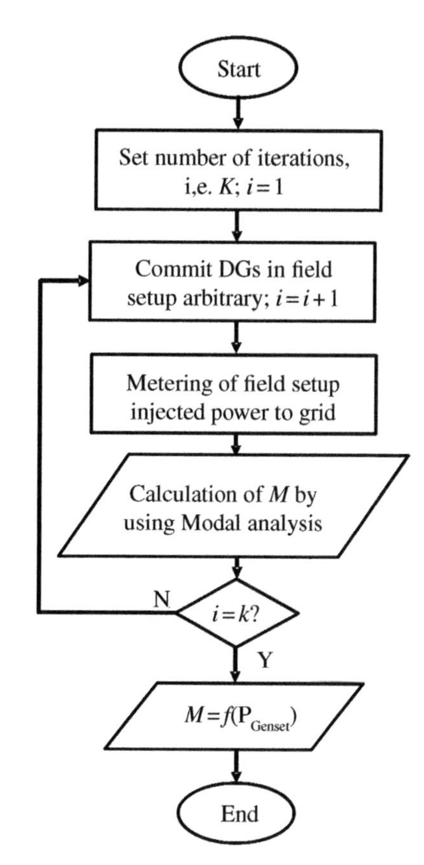

$$M_{MG}\frac{df(t)}{dt} = [T_m(t) - T_e(t)][u(\zeta) - u(\upsilon)] \tag{2.39}$$

where ζ and υ ($\zeta < \upsilon$) are DGs redispatching and islanding times, respectively.

Also of interest, event 2 refers to the islanding of the MG where droop characteristic affects frequency response. The ratio of the grid power variation to the steady-state frequency deviation of the islanded MG would be defined as the droop characteristic. One could write this for Figure 2.36 as

$$R = \frac{\Delta P}{\Delta f} = \frac{\frac{0 - 1000}{5000}}{49.83 - 50} = 1.17\left[\frac{\text{pu}}{\text{Hz}}\right] \tag{2.40}$$

Finally, a relationship between inertia constant of (2.37) and the MG capacity should be derived to facilitate assessing of impact of penetration level on the grid dynamics. For this purpose, the MG inertia, obtained by (2.37) for several

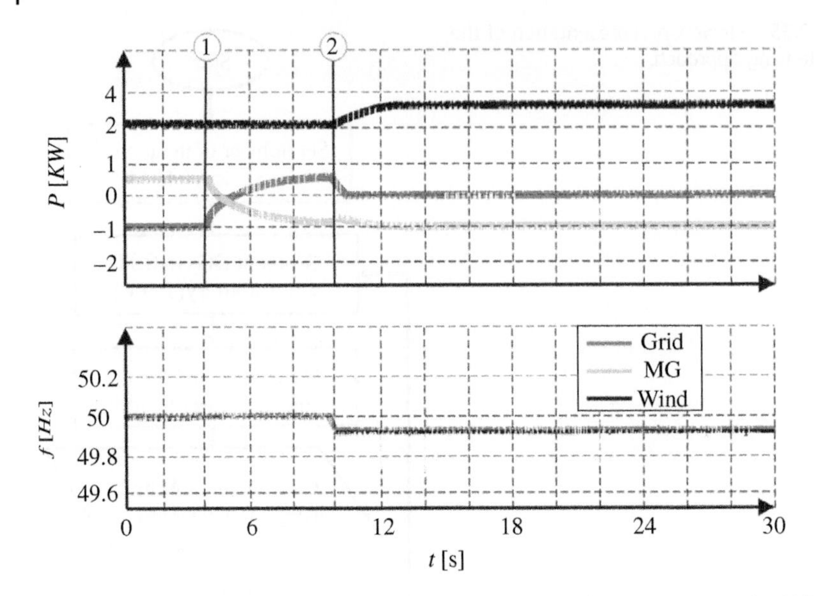

Figure 2.36 Experimental waveforms for real power and frequency output for MG, wind turbine, and host grid.

operating points, maybe plotted against the ratio of the Genset to the MG capacity S_n in Figure 2.37. In Figure 2.37, S_n, is defined as

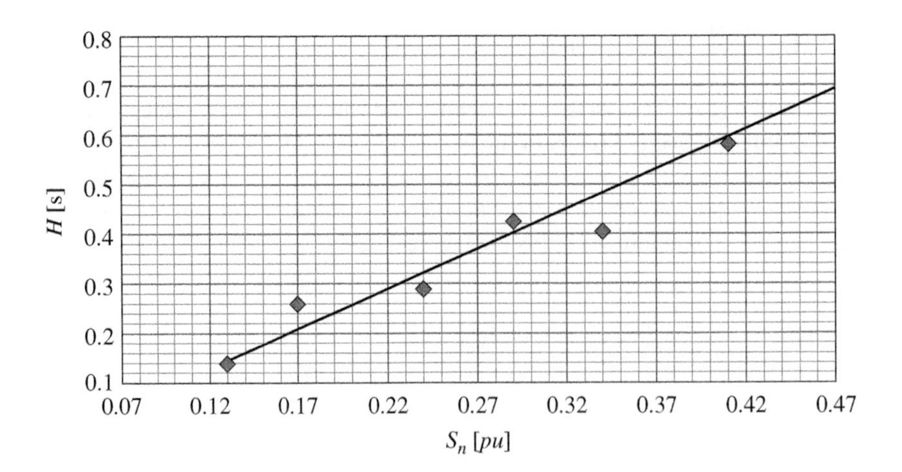

Figure 2.37 Relationship between MG inertia and ratio of Genset to MG capacity.

$$S_n = \frac{P_{Genset}}{P_{MG}} = \frac{P_{Genset}}{P_{Genset} + P_{ESS} + P_{PV} + P_{WT}} \tag{2.41}$$

Actually, S_n in (2.41) changes by the recommitment of the constituent Gensets. For this purpose, two Gensets with the rated capacities of 5 and 10 kW and with, respectively, 0.15-second and 0.21-second inertia constants are employed. To further generate data sets for model derivation, an analog simulator, with inertia constant of 0.1 seconds, is utilized. The figure reveals that there is a direct relationship of form

$$H_{MG} = 1.6109S_n - 0.0644 \tag{2.42}$$

with root square (R^2) of 0.954.

2.5 Summary

This chapter deals with the dynamic equivalencing of MGs-integrated power systems. Several methods are introduced to model the host grid and the distribution network. The host grid modeling relies on the notion of COG dynamics in mechanics to estimate local and global frequency behavior. Using this framework, the power system dynamic behavior is represented by an equivalent model in which the geographical areas interact with the COG through fictitious interconnectors from which the inherent dynamics of power frequency transients are explained using fundamental physics principles. The relationship between the frequency of the COG and the motion of local centers of angle is determined and expressions to compute local frequency deviations following major disturbances are derived. Afterward, experimental-based models of constitute DGs of the UOK-MG in islanded mode are derived. The basic description of the elements, as well as experimental validation of the results, makes the finding of the chapter comprehensive. A realistic model of the MG operating in grid-connected mode is derived to realize a high penetrated grid in Chapters 3, 4, and 6.

References

1. Resende, F. and Peas Lopes, J. (2007). Development of dynamic equivalents for microgrids using system identification theory. In: *Power Tech, 2007 IEEE Lausanne*, 1033–1038. IEEE.
2. Bevrani, H., Ghosh, A., and Ledwich, G. (2010). Renewable energy sources and frequency regulation: survey and new perspectives. *IET Renewable Power Generation* **4** (5): 438–457.

3. Lasseter, R.H. (2011). Smart distribution: coupled microgrids. *Proceedings of the IEEE* **99** (6): 1074–1082.
4. Bevrani, H., François, B., and Ise, T. (2017). *Microgrid Dynamics and Control*. New York: Wiley.
5. Golpîra, H. and Bevrani, H. (2019). Microgrids impact on power system frequency response. *Energy Procedia* **156**: 417–424.
6. Golpîra, H., Atarodi, A., Amini, S. et al. (2020). Optimal energy storage system-based virtual inertia placement: a frequency stability point of view. *IEEE Transactions on Power Systems* **35** (6): 4824–4835. https://doi.org/10.1109/TPWRS.2020.3000324.
7. Annakkage, U., Nair, N.K.C., Liang, Y. et al. (2012). Dynamic system equivalents: a survey of available techniques. *IEEE Transactions on Power Delivery* **27** (1): 411–420. https://doi.org/10.1109/TPWRD.2011.2167351.
8. Milano, F. and Ortega, A. (2017). Frequency divider. *IEEE Transactions on Power Systems* **32** (2): 1493–1501. https://doi.org/10.1109/TPWRS.2016.2569563.
9. Golpîra, H. and Messina, A.R. (2018). A center-of-gravity-based approach to estimate slow power and frequency variations. *IEEE Transactions on Power Systems* **33** (1): 1026–1035.
10. Benda, B.J., Riley, P.O., and Krebs, D.E. (1994). Biomechanical relationship between center of gravity and center of pressure during standing. *IEEE Transactions on Rehabilitation Engineering* **2** (1): 3–10. https://doi.org/10.1109/86.296348.
11. Meriam, J.L. and Kraige, L.G. (2012). *Engineering Mechanics: Dynamics*. New York: Wiley.
12. Ruiz-Vega, D., Messina, A.R., and Pavella, M. (2004). Online assessment and control of transient oscillations damping. *IEEE Transactions on Power Systems* **19** (2): 1038–1047. https://doi.org/10.1109/TPWRS.2004.825909.
13. Kundur, P., Balu, N.J., and Lauby, M.G. (1994). *Power System Stability and Control*. New York: McGraw-Hill.
14. Bevrani, H. (2009). *Robust Power System Frequency Control*. New York: Springer.
15. Morison, G.K., Gao, B., and Kundur, P. (1993). Voltage stability analysis using static and dynamic approaches. *IEEE Transactions on Power Systems* **8** (3): 1159–1171. https://doi.org/10.1109/59.260881.
16. Golpîra, H., Bevrani, H., and Naghshbandy, A.H. (2012). An approach for coordinated automatic voltage regulator and power system stabiliser design in large-scale interconnected power systems considering wind power penetration. *IET Generation, Transmission & Distribution* **6** (1): 39–49. https://doi.org/10.1049/iet-gtd.2011.0411.
17. Golpîra, H., Seifi, H., Messina, A.R., and Haghifam, M.R. (2016). Maximum penetration level of micro-grids in large-scale power systems: frequency stability

viewpoint. *IEEE Transactions on Power Systems* **31** (6): 5163–5171. https://doi.org/10.1109/TPWRS.2016.2538083.

18. Rogers, G. (2000). *Power System Oscillations*. Norwell, MA: Kluwer.

19. Rogers, G. (1999). Power system structure and oscillations. *IEEE Computer Applications in Power* **12** (2): 14–21. https://doi.org/10.1109/67.755641.

20. Wilson, D.H., Hay, K., and Rogers, G.J. (2003). Dynamic model verification using a continuous modal parameter estimator. In: *Power Tech Conference Proceedings, 2003 IEEE Bologna*, vol. **2**, 6. IEEE.

21. Golpîra, H., Messina, A.R., and Bevrani, H. (2019). Emulation of virtual inertia to accommodate higher penetration levels of distributed generation in power grids. *IEEE Transactions on Power Systems* **34** (5): 3384–3394.

22. Wagner, H.M. (1969). Principles of operations research: with applications to managerial decisions. In: *Principles of Operations Research: With Applications to Managerial Decisions*. Englewood Cliffs, NJ: Prentice-Hall.

23. Aien, M., Hajebrahimi, A., and Fotuhi-Firuzabad, M. (2016). A comprehensive review on uncertainty modeling techniques in power system studies. *Renewable and Sustainable Energy Reviews* **57**: 1077–1089.

24. Golpîra, H., Haghifam, M.R., and Seifi, H. (2015). Dynamic power system equivalence considering distributed energy resources using Prony analysis. *International Transactions on Electrical Energy Systems* **25** (8): 1539–1551.

25. Nikkhajoei, H. and Lasseter, R.H. (2009). Distributed generation interface to the CERTS microgrid. *IEEE Transactions on Power Delivery* **24** (3): 1598–1608.

26. Golpîra, H. (2019). Bulk power system frequency stability assessment in presence of microgrids. *Electric Power Systems Research* **174**: 105863.

27. Krishnamurthy, S., Jahns, T., and Lasseter, R. (2008). The operation of diesel gensets in a CERTS microgrid. In: *Power and Energy Society General Meeting-Conversion and Delivery of Electrical Energy in the 21st Century, 2008 IEEE*, 1–8. IEEE.

28. Renjit, A., Illindala, M., Lasseter, R. et al. (2013). Modeling and control of a natural gas generator set in the CERTS microgrid. In: *Energy Conversion Congress and Exposition (ECCE), 2013 IEEE*, 1640–1646. IEEE.

29. Lemmon, M.D. (2009). Advanced Distribution and Control for Hybrid Intelligent Power Systems. Final Project Report, University of Notre Dame: 1–23. *Report Number: UWM-11012009-03.*

30. Krishnamurthy, S. and Lasseter, R. (2009). Control of Wound Field Synchronous Machine Gensets for Operation in a CERTS Microgrid. Consortium for Electric Reliability Technology Solutions (CERTS), Final Report Task 5. Value and Technology Assessment to Enhance the Business Case for the CERTS Microgrid.

31. Lemmon, M. (2010). Comparison of Hardware Tests with SIMULINK Models of UW Microgrid. Technical Report. Univ. Notre Dame, 2010.

32. Gao, Y. and Ai, Q. (2018). A distributed coordinated economic droop control scheme for islanded AC microgrid considering communication system. *Electric Power Systems Research* **160**: 109–118.

33. Karami, M., Seifi, H., and Mohammadian, M. (2016). Seamless control scheme for distributed energy resources in microgrids. *IET Generation, Transmission & Distribution* **10** (11): 2756–2763.

34. Venkataramanan, G., Illindala, M., Houle, C., and Lasseter, R. (2002). Hardware Development of a Laboratory-Scale Microgrid Phase 1: Single Inverter in Island Mode Operation. NREL Report No. SR-560-32527 Golden, CO. National Renewable Energy Laboratory.

35. Chang, C.-C., Gorinevsky, D., and Lall, S. (2014). Dynamical and voltage profile stability of inverter-connected distributed power generation. *IEEE Transactions on Smart Grid* **5** (4): 2093–2105.

3

Stability Assessment of Power Grids with High Microgrid Penetration

The increasing complexity and operation of the electric power grids with high levels of microgrids (MGs) penetration together with higher loading conditions require the development of efficient control and analysis techniques with the ability to extract the relevant system behavior, identify abnormal behavior, and design corrective measures. This need is due to the distributed nature of diverse renewable energy resources in the system and more complex dispatching strategies.

In this chapter, the analysis of the impact of high levels of MG penetration on power system stability is discussed from various points of view such as frequency stability, small-signal stability and voltage performance. Methods for interpreting system dynamics in terms of simplified system representations are developed, and criteria to determine maximum penetration levels are given using deterministic and statistical sensitivity analyzes. Extensions to this framework to capture the impact of large-scale wind and solar photovoltaic (PV) farms are discussed in later sections of this book.

3.1 Introduction

3.1.1 Motivation

Modern power grids face new technical challenges arising from the increasing penetration of distributed energy resources (DERs) and distributed generation (DG) sources, visualized through the MG concept. While low penetration of MGs has a negligible influence on the stability of the host power system, high penetration levels of MGs may significantly affect system stability and raise several reliability concerns at the transmission level [1–4]. The increased demand for electrical power, as well as environmental concerns and pressing need to reduce dependence on fossil fuel, on the other hand, has forced many power system utilities to set an ambitious target for the deployment of DERs/DGs.

Renewable Integrated Power System Stability and Control, First Edition.
Hêmin Golpîra, Arturo Román-Messina, and Hassan Bevrani.

The integration of multiple MGs into the bulk power system requires the development of new analysis tools and methodologies, to assess various aspects of power system dynamic behavior and transmission limitations. These include (i) determining the maximum levels of DGs/MGs generation in the system, (ii) identifying the best system locations to deploy new DERs, and (iii) assessing MGs' control and dispatching strategies to improve system reliability and stability. The first issue is addressed in this chapter. Discussion of issues two and three are deferred until Chapters 4, 6, and 8.

Because modern DGs utilize inverter-based systems to interconnect with the grid, their interactions with the bulk power system are different from those of conventional synchronous generating units. One critical issue is the reduction of system inertia, especially associated with inverter-based DERs [5–7]. Inertia reduction resulting from significant penetration of utility-scale MGs and wind and solar PV farms renders system dynamics faster and thus jeopardizes system stability and reliability [8, 9].

In this chapter, a systematic methodology for the analysis of the impact of distributed MGs on power system dynamic behavior is proposed. The analysis of system dynamics is facilitated by the adoption of a simplified power system model, in which frequency dynamics are represented explicitly through sensitivity relationships.

First, a brief review of the subject is presented. Some important definitions associated with the system frequency response are introduced.

3.1.2 Relations with Previous Literature

A general overview regarding the impact of low inertia on power system stability and operation is provided in [7, 9]. In [10, 11], a trial-and-error based methodology to determine the maximum allowable penetration of wind generation is presented. The transient stability assessment of power systems with high levels of renewable generation is discussed in [12, 13]. On the area of wind penetration, the scenario-based approach of [14] focuses on system inertia and primary reserve values. A simplified frequency model is employed in [15] to study the frequency behavior of an MG-integrated power system. Another scenario-based approach, applied to a part of the Australian grid, is discussed in [16]. In this respect, a framework for assessing renewable integration limits using a scenario-based approach is discussed in [17]. Keyhani and Chatterjee [18], advocated the use of a new automatic generation control (AGC) structure, which tackles intermittency drawbacks associated with high penetration levels of DGs. In yet another approach, a frequency response model is proposed in [19], which can be used to evaluate AGC performance under wind power uncertainty.

A common limitation of approaches that assess the impacts of high MGs/DERs/DGs penetration levels on system stability is that energy penetration levels are based on heuristic considerations. This requires the Transmission System Operators (TSOs) to define a set of expected operating conditions to be assessed during the time-domain simulation procedure. Due to the highly uncertain nature of MGs/DERs/DGs and the high dimensions of modern interconnected power systems, the number of scenarios to be analyzed can become daunting, and this presents a problem to determine a valid set of operating points. As a consequence of this assumption, many recent research works quantify system inertia for the MGs scenarios based on static unit commitment and dispatch modeling studies. Other studies focus on inertia reduction rather than the impact of the MG's representations, and hence the effects of MG dynamics and structure are neglected. This, in turn, causes rendering results inaccurate, thereby offering limited insight into the impact of MGs on system dynamic performance.

3.2 Frequency Stability Assessment

3.2.1 Background on Frequency Indices

Frequency response, as a measure of an interconnection's ability to stabilize frequency following a disturbance, can be assessed using a combination of physically motivated metrics. Typically, these include the rate of change of frequency (*RoCoF*), the frequency response nadir, and the post-disturbance frequency fluctuation over a given time window [20, 21]. These indices are typically employed by TSOs for online monitoring and control systems, to initiate protection or remedial measures such as load shedding, or dispatch reserve power. Accurate and timely estimation of system dynamics helps the associated entities to procure ancillary systems to support the reliable operation of the power grid [22].

3.2.1.1 Rate of Change of Frequency

The *RoCoF* is the time derivative of the frequency (df/dt). The initial *RoCoF* following a sudden torque imbalance is determined by the amount of stored rotational kinetic energy on the system [23]. As discussed in more detail below, the classical swing equation shows that the *RoCoF* is inversely proportional to the rotational inertia. In typical applications, a *RoCoF* standard of 0.5–1 Hz/s is utilized.

3.2.1.2 Frequency Nadir

The *Frequency nadir* is another metric of interest to examine the impact of high levels of inverter-connected renewable on frequency fluctuations following loss of generation events. Formally, the frequency nadir measures the maximum post-contingency frequency deviation after a generator trip and is the point at which frequency is arrested [24]. The *Union for the Coordination of the Transmission of Electricity* (*UCTE*) establishes 49.2 Hz as the minimum allowable post-contingency frequency [20].

3.2.1.3 Delta Frequency Detection

Based on the *North American Electric Reliability Corporation* (*NERC*) resource subcommittee, a frequency event is detected and captured if, during a 15-second rolling time window, the frequency deviation exceeds a threshold. While the threshold is system dependent, a threshold value of 300 mHz is adopted based on [25, 26]. The *delta frequency detection criterion* is a measure of the ability of the generating units to respond to a disturbance over a given time interval.

3.2.2 Frequency Stability Assessment Under High MG Penetration Levels

Conventional power system stability and dynamics studies are mature topics on which a large amount of work has been done during the last two decades. Therefore, it is advisable to represent renewable-integrated power systems stability and control problems based on the well-studied conventional system behavior. The analysis framework can be summarized in four steps:

- *Step 1:* Perform time domain (T-D) simulation of a base conventional system with specific parameters.
- *Step 2:* Extract the frequency dynamics of interest, including *RoCoF*, frequency nadir, and frequency evolution, from the T-D simulation results of *Step 1*.
- *Step 3:* Calculate the sensitivity of the MGs-integrated power grid dynamics with respect to the base conventional system dynamics.
- *Step 4:* Represent the MGs-integrated power grid dynamics based on the base conventional system behavior using sensitivity factors of *Step 3*. The above steps are summarized in the flowchart in Figure 3.1.

3.2.3 Sensitivity Factors

3.2.3.1 Frequency Response

From Chapter 2, any change in the instantaneous system frequency, caused by a system perturbation, can be presented in the COI formulation as [27, 28]:

Figure 3.1 Flowchart representation of the proposed analytical approach to estimate frequency dynamics.

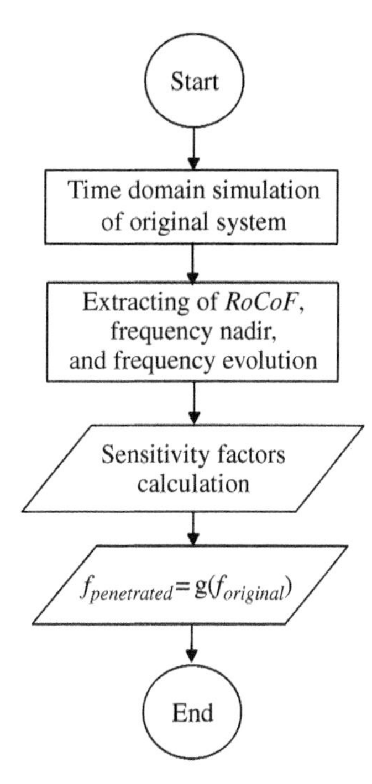

$$\frac{2M_{COI}}{\omega_0}\frac{d^2\theta}{dt^2} = \omega_{COI}(T_{mech} - T_{elec}) \tag{3.1}$$

in which the rotor angle, θ, is defined as

$$\theta = \omega_{COI}t - \omega_0 t + \theta_0 \tag{3.2}$$

and, therefore,

$$2M_{COI}\frac{d(\Delta f_{COI})}{dt} = 2\pi f_{COI}(T_{mech} - T_{elec}) \tag{3.3}$$

Defining now

$$RoCoF = \frac{d(\Delta f_{COI})}{dt} \tag{3.4}$$

and

$$\Delta f_{COI} = \frac{\sum_i M_i \Delta f_i}{\sum M_i} \tag{3.5}$$

the COI frequency response (3.1) can be represented in the form

$$2M_{COI}(RoCoF_{COI}) = 2\pi f_{COI}(T_{mech} - T_{elec}) \tag{3.6}$$

Using the framework shown in Figure 3.1, two power system models are considered next. The first model referred to as the base conventional system, characterized by its inertia constant M_1 and damping coefficient D_1, represents the main (original) system. The second model referred to as the reduced system, characterized by M_2, D_2, is the modified base model after the integration of MGs; in developing this model, it is assumed that M describes the aggregated representation of different generators and load nodes [29].

The frequency response of the base system is then given by

$$2M_1 \frac{d(\Delta f_1)}{dt} = \omega_1(T_{mech} - T_{elec}) \tag{3.7}$$

and for the MG-integrated power system

$$2M_2 \frac{d(\Delta f_2)}{dt} = \omega_2(T_{mech} - T_{elec}) \tag{3.8}$$

Dividing (3.8) by (3.7) gives

$$\frac{M_2}{M_1} \cdot \frac{RoCoF_2}{RoCoF_1} = \frac{2\pi f_2}{2\pi f_1} \tag{3.9}$$

Rearranging the terms in (3.9) yields

$$RoCoF_2 = \frac{f_2}{f_1} \cdot \frac{M_1}{M_2} RoCoF_1 \tag{3.10}$$

Equation (3.10) describes the frequency response of the reduced system as a function of the frequency response and equivalent inertia constants of the base system.

Furthermore, introducing the effect of tie-line power, $P_{tie,ij}$, and the load dynamics, D_i, (3.3) can be rewritten in the following more useful form

$$\dot{f}_i = \frac{1}{2\pi M_i}\left[\Delta P_i - 2\pi D_i f_i - \sum_j P_{tie,ij}\right]; \quad i,j = 1, \ldots, n, \quad i \neq j \tag{3.11}$$

for area i [9]. In (3.11), $P_{tie,ij}$ and n are the tie-line power between areas i and j and the number of areas, respectively.

According to the NERC guideline, the *RoCoF* is defined as the rate of frequency deviation during 500 ms after the inception of a fault. Using this guideline and assuming nominal frequency as 50 Hz, one can rewrite (3.11) as

$$\frac{f_i - 50}{0.5} = \frac{1}{2\pi M_i}\left[\Delta P_i - 2\pi D_i f_i - \sum_j P_{tie,ij}\right] \tag{3.12}$$

and hence,

$$f_i = 50 + \frac{0.5}{2\pi M_i}\left[\Delta P_i - 2\pi D_i f_i - \sum_j P_{tie,ij}\right] \tag{3.13}$$

More precisely, $P_{tie,ij}$ in (3.13) deviates from the pre-fault value and can be explained using Taylor approximations, namely

$$\begin{aligned} P_{tie,0} + \Delta P_{tie} = P_{tie,0} &+ 2\pi P_{\max}(f_i - f_j)\cos(\delta_{i0} - \delta_{j0})\cdot(\delta_i - \delta_j - \delta_{i0} - \delta_{j0}) \\ &+ \left[2\pi P_{\max}(\dot{f}_i - \dot{f}_j)\cos(\delta_{i0} - \delta_{j0})\right. \\ &\left. - 2\pi P_{\max}(f_i - f_j)\sin(\delta_{i0} - \delta_{j0})\right]\cdot(\delta_i - \delta_j - \delta_{i0} - \delta_{j0})^2 \end{aligned} \tag{3.14}$$

Generally, the term $(\delta_i - \delta_j - \delta_{i0} - \delta_{j0})^2$ gets a small value and hence, one can rewrite (3.14) as

$$P_{tie,0} + \Delta P_{tie} = P_{tie,0} + 2\pi P_{\max}\left(f_i - f_j\right)\cos(\delta_{i0} - \delta_{j0})\Delta\delta \tag{3.15}$$

Extension of (3.11)–(3.15) to all the areas in an n-area interconnected power system yields to the $n - 1$ equations, with n unknown frequencies, as

$$F_i\left(f_1, f_2, \ldots, f_j\right) + C_i = 0; \quad j = 1, 2, \ldots, n, \quad i = 1, 2, \ldots, n-1 \tag{3.16}$$

The nth equation to complete the set of (3.16) is formulated based on the COI response (3.1). The frequencies of the areas can then be computed by solving (3.16) together with (3.1) for the MG-integrated power system. Furthermore, the *RoCoF* for the MG-integrated power system can be calculated using (3.10).

3.2.3.2 Delta Frequency Detection

To further proceed with the MG-integrated power system stability assessment, the frequency response of area i in the Laplace domain form of (3.17) is employed [30].

$$\Delta f_i(s) = -\frac{\Delta P_L - \dfrac{2\pi T_{ij}}{s}\Delta f_j(s)}{M_i s + \beta_i + \dfrac{2\pi T_{ij}}{s}} \tag{3.17}$$

In (3.17), β_i is defined as the frequency bias of area i.

By considering ΔP_{Li} in the form of a step function, one could rewrite (3.17) as

$$\Delta f_i(s) = -\frac{\Delta P_L - 2\pi T_{ij}\Delta f_j(s)}{M_i s^2 + \beta_i s + 2\pi T_{ij}} \tag{3.18}$$

in the frequency domain and as

$$\Delta f_i(t) = -\left(\frac{\Delta P_L - 2\pi T_{ij}\Delta f_j}{M_i}\right)\left(\frac{1}{p_1 - p_2}\right)(e^{-p_2 t} - e^{-p_1 t}) \tag{3.19}$$

in the time domain. In (3.19), p_1 and p_2 are the poles of (3.18). Furthermore, at the end of the time interval of interest, the frequency deviation is represented by

$$\Delta f_i(t + \Delta T) = -\left(\frac{\Delta P_L - 2\pi T_{ij}\Delta f_j}{M_i}\right)\left(\frac{1}{p_1 - p_2}\right)\left(e^{-p_2(t+\Delta T)} - e^{-p_1(t+\Delta T)}\right) \tag{3.20}$$

Delta frequency detection criterion imposes that the difference between (3.20) and (3.19) for the main grid as well as the MG-integrated power grid must fulfill

$$\left(\frac{\Delta P_L - 2\pi T_{ij}\Delta f_j}{M_i}\right)\left(\frac{1}{p_1 - p_2}\right)(e^{-p_2 t} - e^{-p_1 t}) - \left(\frac{\Delta P_L - 2\pi T_{ij}\Delta f_j}{M_i}\right)\left(\frac{1}{p_1 - p_2}\right)\left(e^{-p_2(t+\Delta T)} - e^{-p_1(t+\Delta T)}\right) < 0.3. \tag{3.21}$$

Following the same procedure as that of (3.10) specifies the frequency deviation over a given time window for the MG-integrated power grid. For this purpose, all the variables have specific values except for β_i. The steady-state frequency deviations for both the base and reduced systems are employed to calculate β_i. In this way, one could write

$$\beta_{2i} = \beta_{1i} \times \frac{\Delta f_{1i}^{steady\text{-}state}}{\Delta f_{2i}^{steady\text{-}state}} \tag{3.22}$$

Equation (3.22) describes the frequency bias of the MG-integrated power system as a function of the main system characteristics.

3.2.4 Simulation and Results

The effectiveness of (3.1)–(3.22) is investigated on the *New York New England (NYNE)* test system [31]. This system is a reduced-order equivalent of the interconnected New England Test System (NETS) and New York Power System (NYPS), with five geographical regions [32]. Figure 3.2 shows the single line diagram of this system. The study considers two types of loads: constant impedance and induction motors. The test system data are taken from [33].

Eight scenarios, including the rejection of loads and tripping of the generating units, are considered to assess the efficiency of the formulation. As the first example, the accuracy of the formulation is assessed by solving (3.16) for the tripping generator number 13. Observe that the system frequency starts to decline from the nominal value and reaches 49.77 Hz. Figure 3.3 shows the COI frequency responses of the original and the MG-integrated systems, obtained by T-D

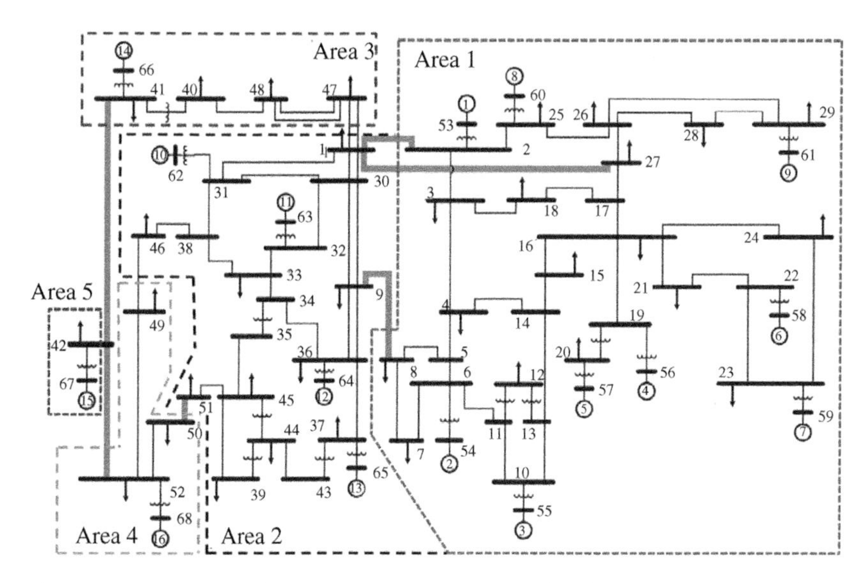

Figure 3.2 Single line diagram of the 68-bus system showing coherent areas and their interconnections.

simulation, for various penetration levels, including 5.6% and 8%. The MG-integrated system is realized by utilizing the MG equivalent model (2.39), developed in Chapter 2, i.e.

$$M^{MG}\frac{df(t)}{dt} = [T_m(t) - T_e(t)][u(\zeta) - u(v)] \tag{3.23}$$

Also of interest, the center of gravity (COG) concept, introduced in Chapter 2, can be employed to estimate the local frequencies shown in Figure 3.3.

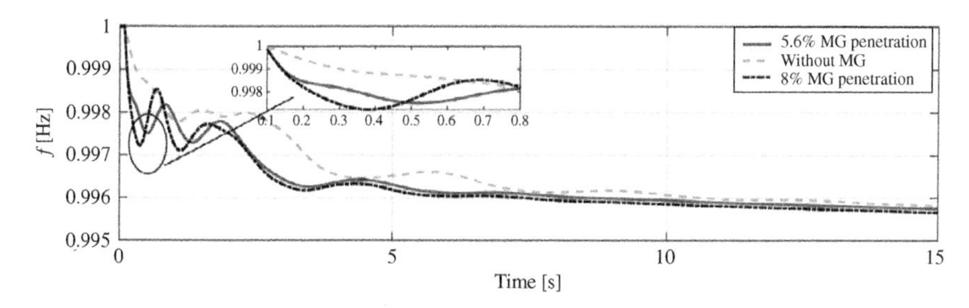

Figure 3.3 Area 2 frequency in response to reduction of COI.

According to (3.16) and for the penetration levels of 5.6%, one could write:

$$0.11f_1 + 0.33f_2 - 0.02f_3 - 0.15f_4 + 0.32f_5 = 0.092 \quad (3.24)$$

$$0.01f_1 + 0.43f_2 - 0.17f_3 - 0.27f_4 + 0.71f_5 = 0.312 \quad (3.25)$$

$$0.67f_1 + 0.03f_2 + 0.07f_3 - 0.83f_4 + 0.21f_5 = 0.762 \quad (3.26)$$

$$0.85f_1 + 0.12f_2 + 0.37f_3 - 0.29f_4 + 0.54f_5 = 0.341 \quad (3.27)$$

To complete the set of equations (3.24)–(3.27), the COI frequency dynamics

$$2M_{COI}\frac{d\left(\frac{\sum M_i \Delta f_i}{\sum M_i}\right)}{dt} = 2\pi f_{COI}(T_{mech} - T_{elec}) \quad (3.28a)$$

is employed. Further manipulation of (3.28a) leads to

$$2M_{COI}\sum_i \frac{M_i \dot{f}_i}{\sum_i M_i} = 2\pi f_{COI}(T_{mech} - T_{elec}) \quad (3.28b)$$

By rearranging the terms in (3.28b) with using the definition of *RoCoF*, i.e. $\dot{f}_i = \frac{\Delta f_i}{0.5}$, one obtains

$$2M_{COI}\sum_i \frac{M_i f_i}{\sum_i M_i} = \pi f_{COI}(T_{mech} - T_{elec}) \quad (3.29)$$

Initializing (3.29) for the given disturbance leads to

$$0.32f_1 + 0.42f_2 + 0.54f_3 + 0.76f_4 + 0.52f_5 = 0.12 \quad (3.30)$$

Solving the set of (3.24)–(3.27) and (3.30) leads to the frequency nadir of 0.9975 pu in Area 2. This indicates a 1.15% error in comparison with those of T-D simulation result. Table 3.1 reports the estimation errors of (3.16) for the penetration levels of 5.6%.

Furthermore, the estimation errors for the penetration levels of 8% are reported in Table 3.2. The results of Tables 3.1 and 3.2 suggest that (3.1)–(3.22) can be effectively used to assess the impact of high penetration levels of MGs on the host grid frequency dynamics.

3.3 Maximum Penetration Level: Frequency Stability

3.3.1 Basic Principle

The frequency metrics in Section 3.2.1 should explicitly follow associated standards to ensure a reliable supply of electricity [20]. In this section, these metrics

Table 3.1 Frequency dynamics error for NYNE test system under 5.6% penetration level.

Scenario	Disturbance	Nadir error (%)	*RoCoF* error (%)	15-seconds error (%)
1	G_{12}	~0	1.99	~0
2	G_{13}	0.3	1.15	0.11
3	L_{14}	0.32	1.22	0.23
4	L_{15}	0.68	0.46	0.18
5	G_{16}	1.00	0.15	0.10
6	L_{37}	~0	~0	0.17
7	L_{42}	1.34	1.04	0.27
8	L_{52}	1.18	0.46	0.12

Table 3.2 Frequency dynamics error for NYNE test system under 8% penetration level.

Scenario	Disturbance	Nadir error (%)	*RoCoF* error (%)	15-seconds error (%)
1	G_{12}	~0	0.81	~0
2	G_{13}	0.13	~0	~0
3	L_{14}	~0	0.38	~0
4	L_{15}	0.68	0.69	0.41
5	G_{16}	~0	~0	0.03
6	L_{37}	~0	0.09	0.07
7	L_{42}	0.74	0.21	0.07
8	L_{52}	1.98	0.12	0.72

are incorporated into (3.1)–(3.22) to establish a systematic methodology for the calculation of the maximum allowable penetration levels of MGs. A schematic depiction of the calculation process is given in Figure 3.4.

3.3.2 Background on MG Modeling

In what follows, the reduced system is realized by employing the developed equivalent model in Chapter 2. More precisely, the model in (3.23) is replaced by some of the conventional synchronous generators.

In Chapter 2 and through (2.42), it was shown experimentally that there is a direct relationship between the inertia constant and the kinetic energy of GENSets divided by the rated VA, as:

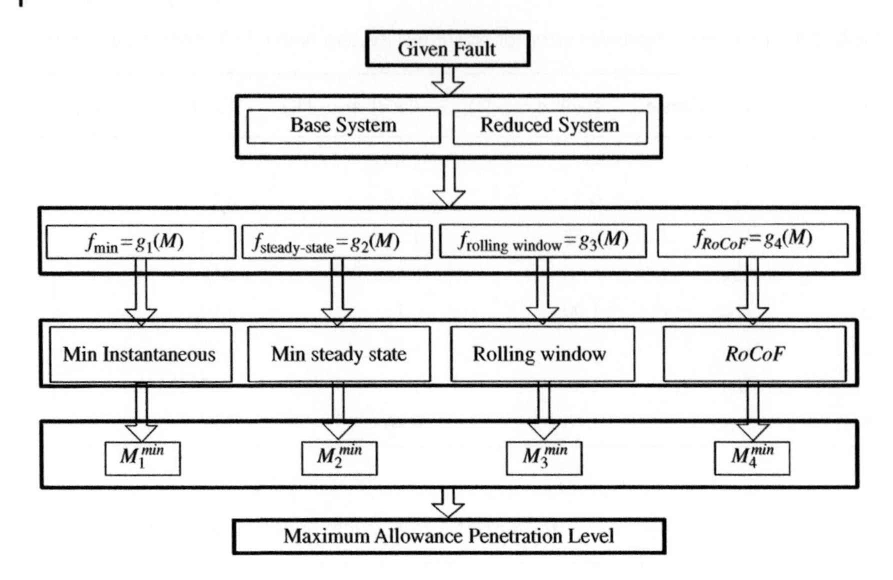

Figure 3.4 Schematic description for the maximum MG penetration level.

$$M^{MG} = 1.6109S_n - 0.0644 \tag{3.31}$$

where $S_n = \frac{P_{Gensets}}{P_{MGs}}$. Here, M^{MG}, P_{Genset}, and P_{MGs} are the MGs inertia, the generation of Genset units, and the generation of the MGs, respectively. To realize high penetration levels of MGs in the power system, multi-MGs (MMGs) structure is considered. It is assumed that each MG in the MMGs configuration has the same structure as that of Figure 2.23, but with the different commitments of the loads and sources. Simulation results suggest that yet there is a direct relationship, with *Root Mean Square Error* (*RMSE*) equal to 0.2165, between the MMGs inertia and the ratio of the GENSets to the MMGs capacities. It should be noted that power systems are robust enough so that small variation of the MMGs parameters, caused by the aforementioned *RMSE*, may not significantly affect the system performance. Therefore, the direct relationship of the form (3.31) can be employed to quantify the penetration levels of MGs.

3.3.3 Minimum Inertia Related to Frequency Nadir

The first impediment to decreasing the rotational inertia arises from the minimum instantaneous frequency which the system can experience during the normal operation. It is noteworthy that a similar criterion, named *point C criterion*, is introduced by NERC in [34]. As stated in Section 3.2.1, the minimum permissible frequency is 49.2 Hz.

To introduce the proposed methodology, assume that the same fault is applied to both, the main and the reduced systems. Accordingly, dividing (3.8) by (3.7) gives:

$$\frac{M_2}{M_1}\frac{d(\Delta f_2)}{d(\Delta f_1)} = \frac{f_2}{f_1} \tag{3.32}$$

Multiplying both sides of (3.32) by Δf_1 results in

$$\frac{M_2}{M_1}\frac{d(\Delta f_2)}{d(\Delta f_1)}\Delta f_1 = \frac{\Delta f_1}{f_1} f_2 \tag{3.33}$$

where the term $\frac{d(\Delta f_2)}{d(\Delta f_1)}$ describes the sensitivity of frequency deviation for the MG-integrated power system to the frequency deviation for the original system.

The minimum permissible inertia can be calculated by utilizing (3.33). Formally, $\frac{d(\Delta f_2)}{d(\Delta f_1)}\Delta f_1$ in (3.33), as the frequency change of the MG-integrated power system, is set to the maximum permissible frequency deviation, i.e. 800 mHz. Further, Δf_1 and f_1 are the post-fault frequency deviation and frequency for the original system, respectively. Moreover, the frequency f_2 is set to 49.2 Hz. Initializing (3.33) by the aforementioned values gives:

$$\frac{M_2}{M_1}(0.8) = \frac{\Delta f_1}{f_1} \times (49.2) \tag{3.34}$$

and then

$$M_2 = M_1 \frac{\Delta f_1}{f_1} \times (61.5) \tag{3.35}$$

Equation (3.35) calculates the minimum rotational inertia which beyond it, the system frequency tends to deviate from the permissible value. It is noteworthy that in the case of frequency increment, (3.35) can be rewritten as the form

$$\frac{M_2}{M_1}(0.8) = \frac{\Delta f_1}{f_1} \times (50.8) \tag{3.36}$$

then

$$M_2 = M_1 \frac{\Delta f_1}{f_1} \times (63.5) \tag{3.37}$$

In deriving (3.35) and (3.37), the amplitude of the disturbance and the kinetic energy of the rotating masses, as two important factors affecting frequency dynamics [35], are taken into account.

Here, it should be noted that (3.35) and (3.37) are derived for a single-area power system or the aggregated inertia for all the areas. However, to calculate the minimum permissible rotational inertia in each area, (3.11) and (3.12) should be used.

Under these considerations, all the parameters in (3.12) are set to the permissible values. More precisely, $(\delta_i - \delta_j - \delta_{io} - \delta_{jo})$ is set to the maximum permissible phase difference between areas, $f_i - f_j$ is set to the maximum permissible frequency deviation and $\left(\dot{f}_i - \dot{f}_j\right)$ will be replaced by an aggregated model of noncoherent generators, namely

$$\dot{f}_i - \dot{f}_j = \frac{1}{2\pi}\frac{M_i + M_j}{2M_iM_j}\left[\frac{M_i}{2M_iM_j}(P_{mi} - P_{ei}) - \frac{M_j}{2M_iM_j}\left(P_{mj} - P_{ej}\right)\right] \tag{3.38}$$

Afterward, the same procedure as those of (3.35) and (3.37) can be followed to calculate the maximum permissible inertia reduction.

3.3.4 Minimum Inertia Related to Delta Frequency Detection

To calculate the minimum rotational inertia in compliance with the rolling window, consider the frequency response of the original system as those of (3.19) and (3.20). By setting ΔT in (3.20) to 15 seconds, one could write

$$\Delta f_i = -\left(\frac{\Delta P_L - 2\pi T_{ij}\Delta f_j}{M_i}\right)\left(\frac{1}{p_1 - p_2}\right)\left(e^{-p_2(t+15)} - e^{-p_1(t+15)}\right) \tag{3.39}$$

According to the delta frequency detection criterion, the difference between (3.39) and (3.19) must fulfill (3.40) for both the original and the MG-integrated power systems, i.e.

$$\begin{aligned}\Bigg|\left(\frac{\Delta P_L - 2\pi T_{ij}\Delta f_j}{M_{1i}}\right)\Bigg[\left(\frac{1}{p_1 - p_2}\right)&\left(e^{-p_2 t} - e^{-p_1 t}\right) \\ &- \left(\frac{1}{p_1 - p_2}\right)\left(e^{-p_1(t+15)} - e^{-p_2(t+15)}\right)\Bigg]\Bigg| \leq 0.3\end{aligned} \tag{3.40}$$

Following the same procedure as done for (3.35) gives

$$\frac{\left(\frac{\Delta P_L - 2\pi T_{ij}\Delta f_j}{M_{1i}}\right)\left[\left(\frac{1}{p_1 - p_2}\right)(e^{-p_2 t} - e^{-p_1 t}) - \left(\frac{1}{p_1 - p_2}\right)\left(e^{-p_2(t+15)} - e^{-p_1(t+15)}\right)\right]}{\left(\frac{\Delta P_L - 2\pi T_{ij}\Delta f_j}{M_{2i}}\right)\left[\left(\frac{1}{p'_1 - p'_2}\right)(e^{-p'_2 t} - e^{-p'_1 t}) - \left(\frac{1}{p'_1 - p'_2}\right)\left(e^{-p'_2(t+15)} - e^{-p'_1(t+15)}\right)\right]} \leq 1 \tag{3.41}$$

To calculate the minimum rotational inertia, (3.41) is tied to the associated standard. In this way, Δf_j is set to the maximum allowable frequency deviation in the viewpoint of the transient or the steady-state value (whichever is the larger). Setting Δf_j to the maximum permissible value is a pessimistic assumption that guarantees the reliable and secure operation of the system for any operating condition. Moreover, (3.41) suggests that except for M and D, the droop characteristic

R significantly affects the maximum penetration levels calculation. The droop characteristic is defined by (2.40) in Chapter 2, i.e.

$$R = \frac{\Delta f}{\Delta p} = \frac{\Delta f}{\Delta t} \times \frac{\Delta t}{\Delta p} \tag{3.42}$$

The DER power-frequency (P/f) grid code determines the permissible value of R in (3.42). Based on the NERC standard [36], a change of 10% per second for the rate of response to a step command to reduce power output is reasonable for a variable generation. This rate is tied to the 15-second-rolling window to calculate the permissible droop value, namely

$$R = \frac{0.3 + 0.5 \text{ Hz}}{15 \text{ second}} \times \frac{1 \text{ second}}{0.1 \text{ pu}} = 0.53 \tag{3.43}$$

By substituting R in (3.22) for the MG-integrated power system, and considering the actual value of R for the original system, one can calculate D in the MG-integrated system. More precisely, while the exact steady-state frequency deviation for the original system is taken from T-D simulation data, it is set to the maximum permissible value for the MG-integrated power system. It is noteworthy that all the parameters in (3.41) have specific values except for M_2 and t. Statistical approaches should be used to calculate M_2. For this purpose, a relationship between the rotational inertias of the original and the MG-integrated power systems is derived.

3.3.5 Minimum Inertia Related to RoCoF

The *RoCoF*, in response to a system perturbation ΔP_L, can be represented by (3.6), as

$$2M(RoCoF) = \Delta P_L \tag{3.44}$$

The TSO's grid code suggests that a requirement for generation units to remain connected to the transmission system during frequency excursions involves a maximum *RoCoF* of 0.5–1 Hz/s. More precisely, one can write

$$\frac{\Delta P_L}{2M} = RoCoF \leq 1 \tag{3.45}$$

Rearranging the terms in (3.45) gives

$$M > \frac{\Delta P_L}{2} \tag{3.46}$$

Physically, (3.46) describes the minimum permissible inertia as a function of the *RoCoF*.

3.3.6 Maximum Penetration Level

The minimum permissible rotational inertia is defined based on the maximum of (3.35), (3.41), and (3.46) as

$$M^{MMGs} = \max \left\{ M_{\min}^{nadir}, M_{\min}^{15-\sec}, M_{\min}^{RoCoF} \right\} \tag{3.47}$$

where $M_{\min}^{nadir}, M_{\min}^{15-\sec}, M_{\min}^{RoCoF}$ are the minimum permissible inertia in compliance with the frequency nadir, 15-second rolling window, and *RoCoF*, respectively. For calculation of the maximum permissible penetration levels of MGs, M^{MMGs} in (3.47) is tied to the equivalent model (3.23). In this way, the same procedure as that of (3.31) is performed to represent the MMGs system inertia, on the base of 200 MVA, based on the Gensets capacity as

$$M_{MMGs} = 1.89 S_n \tag{3.48}$$

where $S_n = \frac{P_{Gensets}}{P_{MMGs}}$. Equation (3.48) suggests that each MMGs with 200 MVA generations, including $200 \times S_n$ MVA Gensets generations, represents $1.89 \times S_n$ second inertia. Accordingly, the maximum penetration levels of MGs are defined by

$$P_{limit} = \frac{P_{MMGs}}{S_n} \tag{3.49}$$

where P_{limit}, P_{MMGs}, and S_n are the maximum allowable penetration level, the generation of the MGs, and the apparent power of the system, respectively.

3.3.7 Simulation and Results

The effectiveness of (3.49) is investigated on two interconnected power systems: NYNE test system of Figure 3.2 and the IEEE 50-machine test system.

3.3.7.1 Analysis Tools

Four tools are utilized to calculate the maximum penetration level: *Matlab 2018b* including power system toolbox (PST) [31] and *PLECS* [37] are used to perform the dynamical simulation; *MAPLE 18* [38] and DSA tool, including the transient stability assessment (TSA) and the small-signal stability assessment (SSSA) [39] softwares, are used to assess stability performance.

3.3.7.2 Dynamical Simulation Results

Figure 3.5 shows the frequency responses of area 2, for different rotational inertia. As shown in this plot, the original system frequency starts to decline following the outage of generator 14 from the nominal value and reaches the minimum value of

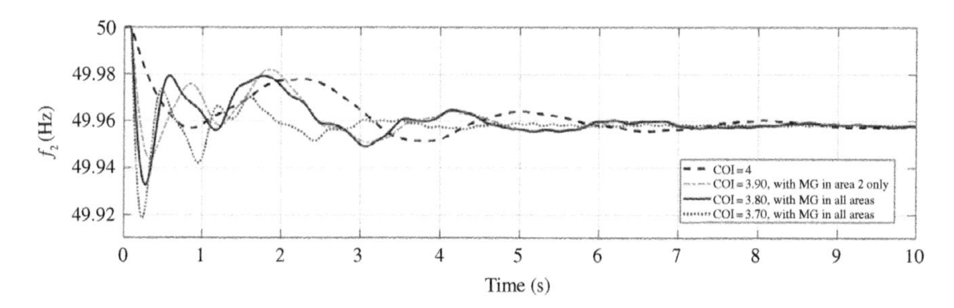

Figure 3.5 Area 2 frequency response in response to reduction of COI.

about 49.96 Hz. Also of interest, the maximum phase differences between areas in (3.12) are calculated using the DSA tool, as shown in Table 3.3.

Moreover, the term $P_{mj} - P_{ej}$ in (3.38) is set to zero which satisfies the acceptable operation in the worst-case condition. On the other hand, $f_i - f_j$ in the MG-integrated power system is set to the maximum permissible value, i.e. 1 Hz. Accordingly, one can write:

$$0.15M_1M_2 + 0.07M_1M_3 - 0.1M_2M_4 - 0.03M_1M_5 = 0 \tag{3.50}$$

$$0.06M_2M_3 + 0.01M_1M_4 - 0.12M_3M_4 - 0.05M_3M_5 = 0.891 \tag{3.51}$$

$$0.08M_3M_1 + 0.06M_3M_2 - 0.07M_1M_4 - 0.09M_4M_5 = 0.711 \tag{3.52}$$

$$0.13M_4M_2 + 0.16M_1M_4 - 0.04M_2M_5 - 0.03M_1M_2 = 0.951 \tag{3.53}$$

Furthermore, the minimum permissible COI constant, $COI_{overall}$, defines the last equation, namely

$$COI_{overall} = M_1 \frac{50 - 49.59}{49.59}(61.5) \tag{3.54}$$

or in the final form of

Table 3.3 Maximum phase angle difference in deg.

	Area 1	Area 2	Area 3	Area 4	Area 5
Area 1	—	13.5	31.5	9.8	11.6
Area 2	13.5	—	16.5	17.2	13
Area 3	31.5	16.5	—	15.6	23.4
Area 4	9.8	17.2	15.6	—	21
Area 5	11.6	13	23.4	21	—

Table 3.4 Minimum permissible rotational inertia in response to tripping G_{14}.

Case study	Area 1	Area 2	Area 3	Area 4	Area 5
Base system	3.66	4	3	3	4.45
Reduced system	1.73	1.89	2.14	1.08	2.06

$$COI_{overall} = 0.51(M_1) = 0.51(3.66) = 1.86 \tag{3.55}$$

Solving the set of (3.50)–(3.55), we obtain the minimum inertia on each area of the system, as reported in Table 3.4.

The minimum permissible inertia in compliance with the 15-second rolling window, on the other hand, is calculated. Representing the MG-integrated power system based on the original system characteristics is done using (3.22), namely

$$\frac{D_{2i} + \frac{1}{R_{2i}}}{D_{1i} + \frac{1}{R_{1i}}} = \frac{0.58}{0.5} = 1.16 \tag{3.56}$$

then

$$D_{2i} + \frac{1}{R_{2i}} = 1.16\left(D_{1i} + \frac{1}{R_{1i}}\right) \tag{3.57}$$

where 0.58 and 0.50 are the steady-state frequency deviations and the maximum allowable deviation, respectively. Substituting (3.57) in (3.41) leads to a time-variant equation with one degree of freedom, M_2. The extracted relationship between the inertias of the original and the MG-integrated power systems in the form of

$$M_2 = 0.94 \ln(M_1) - 0.23 \tag{3.58}$$

is employed to calculate M_2. Initializing (3.58) with the actual values gives

$$M_2 = 0.94 \ln(4) - 0.23 = 1.07 \tag{3.59}$$

The minimum permissible rotational inertia of area 2 is then defined using (3.47), namely

$$M_{2,Critical} = max\{1.89, 1.07, 0.4\} = 1.89 \tag{3.60}$$

Note that 0.4 in (3.60) comes from (3.46). Equation (3.60) suggests that the frequency nadir imposes a lower limit on the rotational inertia. Making use of (3.60) into (3.48) and (3.49) leads to the maximum penetration levels of 11.34% expressed in percentage in terms of the overall system capacity.

The same procedure as those of (3.50)–(3.60) is redone for eight scenarios related to different disturbances. The numerical results are reported in Table 3.5. The

Table 3.5 Maximum penetration level for 16-machine test system.

Scenario	Fault	Minimum instantaneously	15-seconds rolling window	Maximum penetration (%)
1	G_{12}	1.76	0.97	14.23
2	G_{13}	1.20	1.07	12.29
3	G_{14}	1.89	1.07	11.34
4	L_{15}	2.18	0.2	10.94
5	G_{16}	2.49	0.6	10.87
6	L_{37}	3.11	1.53	10.56
7	L_{42}	2.01	1.13	12.01
8	L_{52}	2.12	1.65	12.09

G_i: outage of generator number i; L_i: outage of load number i.

results suggest that the 15-second rolling window could not affect the maximum penetration levels of MGs, as the frequency reaches steady state in less than 10 seconds.

To further confirm the validity of the model, the IEEE 50-machine test system is used. The system is an approximate model of an actual power system. Table 3.6 reports the minimum permissible rotational inertia and the associated maximum penetration levels of MGs for eight different scenarios. In Table 3.6, the main factor that imposes a lower limit on the penetration levels of MGs is underlined.

Table 3.6 Maximum inertia reduction for 50-machine test system.

Scenario	Fault	Minimum instantaneously	15-seconds rolling window	Maximum penetration (%)
1	G_2	65.13	66.16	11.44
2	G_3	63.12	67.41	12.26
3	G_4	71.03	68.86	12.03
4	G_6	60.61	64.71	12.98
5	G_{23}	68.39	70.81	15.71
6	L_{105}	56.61	55.29	19.32
7	L_{144}	58.81	54.13	20.27
8	L_{145}	61.12	63.27	17.3

Of note that $M_{\min}^{RoCoF}$ in (3.47) could not generally affect the penetration level, as the size of disturbance in per unit is negligible in comparison with those of $M_{\min}^{nadir}, M_{\min}^{15-\text{ sec}}$.

3.4 Small-Signal Stability Assessment

3.4.1 Basic Definition

Following the given definition by Kundur et al. [40], small-signal stability refers to the ability of synchronous machines of an interconnected power system to remain in synchronism after being subjected to a small disturbance. Generally, there are two major types of small-signal instability: (i) increase in rotor angle through a non-oscillatory or aperiodic mode due to lack of synchronizing torque and (ii) rotor oscillations of increasing amplitude due to lack of sufficient damping torque.

Generator-turbine rotational inertia plays an important role in providing synchronizing capability whenever a disturbance causes a mismatch between the mechanical input power and the electrical output power [40]. Therefore, decreasing overall system inertia in response to increasing penetration levels of MGs can lead to potential small-signal stability problems, which is addressed in this section. Moreover, it is advantageous to determine whether a particular generator's inertia has a significant impact on a particular inertial oscillation mode. This section addresses this issue using sensitivity analysis for generators.

3.4.2 Key Concept

Derivation of a linear model of a nonlinear system, around a certain operating point, is the key factor in small-signal stability analysis. While the stability of a linearized system is assessed using the eigenvalues, λ, of the state matrix, $\mathbf{A}$, the participation of each state in a given eigenvalue is determined using the right, φ, and the left, ψ, eigenvectors [41, 42]. Following the nomenclature in [41], the ith eigenvalue of the system matrix $\mathbf{A}$, and its corresponding eigenvectors are defined as

$$\begin{aligned} A\varphi_i &= \lambda_i \varphi_i \\ \psi_i A &= \lambda_i \psi_i \end{aligned} \tag{3.61}$$

For a complex eigenvalue that corresponds to an oscillatory mode, the frequency, f, of the oscillation and the damping ratio ς are expressed by [43]:

$$\lambda_i = \sigma_i \pm j\omega_i \tag{3.62}$$

$$f_i = \frac{\omega_i}{2\pi} \quad (3.63)$$

$$\varsigma_i = \frac{\sigma_i}{\sqrt{\sigma_i^2 + \omega_i^2}} \quad (3.64)$$

In (3.62), a positive real part, i.e. $\sigma_i > 0$, corresponds to an oscillation with the increasing amplitude. As the complex pole (3.62) moves towards the right half-plane (RHP), the damping of the system worsens [41]. Here, the damping ratio determines the decline rate of a specific mode.

The problem of interest is to study how the critical modes of a system are affected by penetration levels of MGs. Critical modes of the system are those within the frequency range of 0.01–2 Hz and damping of less than 10% [41]. For analysis purposes, the simple yet efficient method of [41] is applied to the MG-penetrated system.

The analysis framework can be summarized as follows:

- *Step 1:* Calculate the critical modes of the conventional power grid, with no penetration of MGs, using eigenvalue analysis;
- *Step 2:* Repeat the procedure as *Step 1* for the system with different MGs penetration level;
- *Step 3:* Compare the results to assess the impact of MGs penetration level on the small-signal stability;
- *Step 4:* Conduct sensitivity analyses with respect to the displaced generators' inertia.

3.4.3 Simulation and Results

To numerically assess the impact of the penetration levels of MGs on small-signal stability, the NYNE test system of Figure 3.2 is used. The base system with no MGs penetration levels exhibits four inter-area modes of interest, as reported in Table 3.7. Refer to Chapter 2 for more details on how to calculate the modes.

The analysis is reconducted for various MGs penetration levels, and the results are reported in Table 3.8 for the mode which is detrimentally affected by increasing MGs penetration levels, i.e. $0.3590 + 3.7108j$. The results suggest that the oscillatory mode has been adversely affected by increasing MGs penetration levels. This could be justified through the fact that the real part of the eigenvalue moves closer to the RHP as the MGs penetration levels increase. Such justification could be also observed from the reduction in the damping ratio. Of note, the significant reduction of the damping ratio from 20% to 30% MGs penetration levels is mainly caused by displacing some large conventional generators that were in service before the

Table 3.7 Critical modes of the base system.

Mode	Real part	Imaginary part	Frequency	Damping ratio
1	−0.7018	1.9710*j*	0.3137	33.543
2	−0.1184	3.2666*j*	0.5199	3.6236
3	−0.3590	3.7108*j*	0.5906	9.6290
4	−0.1657	4.8915*j*	0.7785	3.3851

Table 3.8 Critical mode detrimentally affected by high MGs.

Penetration level (%)	Real part	Imaginary part	Frequency	Damping ratio
0	−0.3590	3.7108*j*	0.5906	9.6290
10	−0.2816	3.6513*j*	0.5811	7.0690
20	−0.2201	3.5891*j*	0.5712	6.1210
30	−0.1305	3.5705*j*	0.5683	3.6500

30% MGs penetration levels. By displacing those generators, the overall system inertia dramatically reduces, and hence the damping ratio of the adversely impacted mode significantly decreases [41].

To further confirm the validity of the framework, sensitivity analysis is carried out corresponding to the reported mode in Table 3.8. The sensitivity analysis is used to analytically demonstrate the detrimental impacts of the MGs penetration levels on the small-signal stability. The ith eigenvalue sensitivity analysis with respect to the inertia of the jth generator is expressed as

$$\frac{\partial \lambda_i}{\partial H_j} = \frac{\psi_i \frac{\partial A}{\partial H_j} \phi_i^T}{\psi_i^T \phi_i} \tag{3.65}$$

Table 3.9 reports a summary of the sensitivity of the critical mode to the inertia variations of the conventional generators that are being displaced with the MGs in the 20% penetration levels. As the damping of the system modes is determined by the real part of the eigenvalues, the real part is presented in Table 3.9.

The negative real part sensitivity of the eigenvalues in Table 3.9 shows the adverse impact of high MGs penetration levels on system damping. The sensitivity analysis results corroborate the results derived from the full eigenvalue analysis of the system under various MGs penetration levels.

Table 3.9 Eigenvalue sensitivity analysis in the 20% MGs penetration.

Bus number	Inertia	Sensitivity factor
55	4.96	−0.0213
56	4.16	−0.0189
59	4.32	−0.0200
62	2.91	−0.0103
63	2.00	0.0065
64	1.17	0.0094

3.5 Maximum Penetration Level: Small-Signal Stability

3.5.1 Basic Idea

As discussed above, the critical modes of the system are characterized by frequencies in the range of 0.01–2 Hz and damping of less than 10%. Conceptually, the zero damping ratio determines the maximum permissible penetration levels of MGs in the system. The analysis framework to determine the maximum penetration levels of MGs can be summarized in six steps:

- *Step 1:* Extract oscillation modes of the conventional power grid with no penetration of MGs, using eigenvalue analysis.
- *Step 2:* Compute the critical modes in *Step 1*;
- *Step 3:* Repeat the procedure in *Step 1* for systems with different MGs penetration levels;
- *Step 4:* Determine the damping ratios for different MGs penetration levels;
- *Step 5:* Extract a relationship between the MGs penetration levels and the damping ratio of the form

$$\varsigma_i = g(P_{MGs}) \tag{3.66}$$

- *Step 6:* Set the left-hand side of (3.66) equal to zero and solving for P_{MGs}, one can find the maximum penetration levels of MGs.

3.5.2 Simulation and Results

To numerically assess the effectiveness of (3.66) to calculate the maximum penetration levels of MGs, the NYNE test system of Figure 3.2 is employed. Table 3.8

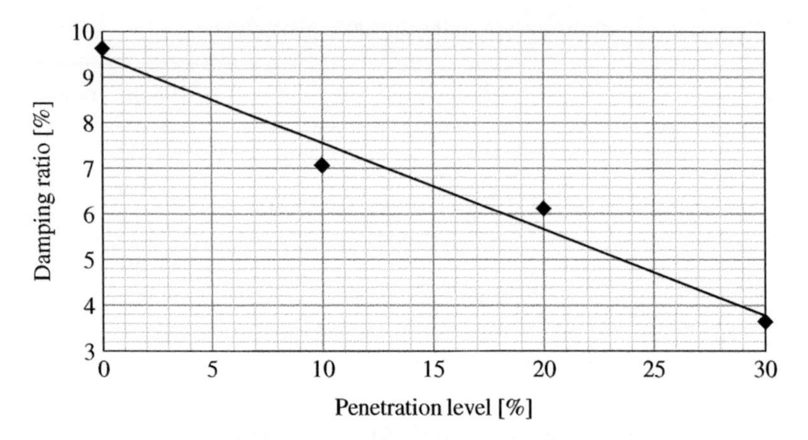

Figure 3.6 Extracted relationship between damping ratio and penetration level.

describes the variations of the damping ratio as a function of penetration levels of MGs. Accordingly, one could derive a relationship in the form of

$$\varsigma_i = -0.1889P_{MGs} + 9.45 \tag{3.67}$$

between the damping ratio and the MGs penetration level, with R^2 of 0.9731. Figure 3.6 visualizes the data reported in Table 3.8 as well as the extracted relationship.

Maximum penetration levels of MGs characterize with zero damping ratio as

$$0 = -0.1889P_{MGs} + 9.45 \tag{3.68}$$

By rearranging the terms in (3.68), one has

$$P_{MGs} = \frac{9.45}{0.1889} = 50.02\% \tag{3.69}$$

Equation (3.69) exhibits that with the 50% penetration levels of MGs, the damping ratio of the mode is zero. To further confirm the validity of (3.69), eigenvalue analysis is conducted for the system with the 50% penetration level of MGs. A comparison of the results reveals that the error of 1.89% is small enough to justify the efficiency of the method.

3.6 Voltage-Based Realization of the MG-Integrated Power Grid

It is well known that MGs are realized in the low and medium voltage levels. Therefore, following the calculation of maximum penetration levels of MGs in

Sections 3.5 and 3.6, this section tries to determine the maximum active power that MGs can inject into each bus of the distribution system without causing steady-state voltage violations.

3.6.1 Key Concepts

The method introduced by Ayres et al. [44] is used to calculate the maximum penetration levels of MGs in the distribution system. The method relies on the Jacobian sensitivities for direct estimation of the maximum amount of active power that MGs can inject into each bus of the system.

Following the nomenclature of [44], assume that a typical bus in the system has the initial voltage of V^0. Installing new MGs in the studied bus affects the voltage which could be represented by

$$V = V^0 + \Delta V_P + \Delta V_Q \tag{3.70}$$

where ΔV_P and ΔV_Q are voltage changes in response to the variations of the injected active and reactive powers, respectively. The first step to calculate ΔV_P and ΔV_Q in (3.70) is to derive the voltage sensitivities related to the active and reactive power injections.

3.6.2 Jacobian Sensitivities

The voltage sensitivity factors can be calculated using the linearized model of the form

$$\begin{bmatrix} \Delta P \\ \Delta Q \end{bmatrix} = \begin{bmatrix} J_{P\theta} & J_{PV} \\ J_{Q\theta} & J_{QV} \end{bmatrix} \begin{bmatrix} \Delta \theta \\ \Delta V \end{bmatrix} \tag{3.71}$$

The elements of the Jacobian matrix (***J***) represent the sensitivities of the power variations, i.e. ΔP, ΔQ, to the voltage variations, i.e. ΔV, $\Delta \theta$.

3.6.2.1 *V-P* Sensitivity

The sensitivity of the voltage to the active power variations in a typical bus, i.e. *V-P* sensitivity, can be calculated by setting $\Delta Q = 0$ in (3.71). In this way, one could write

$$\Delta P = \left(J_{PV} - J_{P\theta} J_{Q\theta}^{-1} J_{QV} \right) \Delta V_P = J_{RPV} \Delta V_P \tag{3.72}$$

By rearranging the terms in (3.72) with the use of the definition of the reduced Jacobian matrix J_{RPV}, one obtains

$$\Delta V_P = J_{RPV}^{-1} \Delta P \tag{3.73}$$

It should be noted that inversion of $J_{Q\theta}$ is feasible only if all the buses are modeled as *PQ* buses. This is the case for distribution systems, where the slack bus is the only voltage control bus [44]. Moreover, MGs in the steady-state studies are formally represented by *PQ* buses since they do not contribute to the voltage control of the system.

3.6.2.2 V-Q Sensitivity

Equation (3.73) is derived assuming the unity power factor. To tackle the power factor into the problem formulation (3.73) should consider *V-Q* sensitivities. Following the same procedure as that of (3.72) gives

$$\Delta Q = \left(J_{QV} - J_{Q\theta} J_{P\theta}^{-1} J_{PV}\right) \Delta V_Q = J_{RQV} \Delta V_Q \tag{3.74}$$

Rearranging the terms in (3.74) gives the final form of

$$\Delta V_Q = J_{RQV}^{-1} \Delta Q \tag{3.75}$$

where J_{RQV} is the reduced Jacobian matrix, which represents the sensitivity of voltage magnitude with respect to the reactive power injection variations. As the penetration levels of MGs express in the term of active power, one could substitute

$$\Delta Q = \Delta P \times \tan\left(\cos^{-1}(pf)\right) \tag{3.76}$$

in (3.75) to derive

$$\Delta V_Q = J_{RQV}^{-1} \Delta P \times \tan\left(\cos^{-1}(pf)\right) \tag{3.77}$$

Considering (3.73) and (3.77), the bus voltage *V* in (3.70) is rewritten as

$$V = V^0 + J_{RPV}^{-1} \Delta P + J_{RQV}^{-1} \Delta P \times \tan\left(\cos^{-1}(pf)\right) \tag{3.78}$$

and then

$$V = V^0 + \left(J_{RPV}^{-1} + J_{RQV}^{-1} \times \tan\left(\cos^{-1}(pf)\right)\right) \Delta P \tag{3.79}$$

By defining

$$S_{PQ} = J_{RPV}^{-1} + J_{RQV}^{-1} \times \tan\left(\cos^{-1}(pf)\right) \tag{3.80}$$

Equation (3.79) gets the final form of

$$V = V^0 + S_{PQ} \Delta P \tag{3.81}$$

Considering the acceptable steady-state voltage, the maximum permissible MGs penetration level in each bus of the distribution system is defined by

$$\Delta P = \frac{\Delta V}{S_{PQ}} \tag{3.82}$$

Indeed (3.82) specifies the capacity of MGs which could be installed in the studied bus without violating the steady-state voltage.

3.6.3 Simulation and Results

The practical applicability of the above framework is assessed using the IEEE 70-Bus test system of Figure 3.7. The system data are taken from [45]. The network active and reactive power demands are 3802.2 kW and 2694.6 kVAr, respectively. In the studies described below, the transformer taps are adjusted to 1.04 pu to

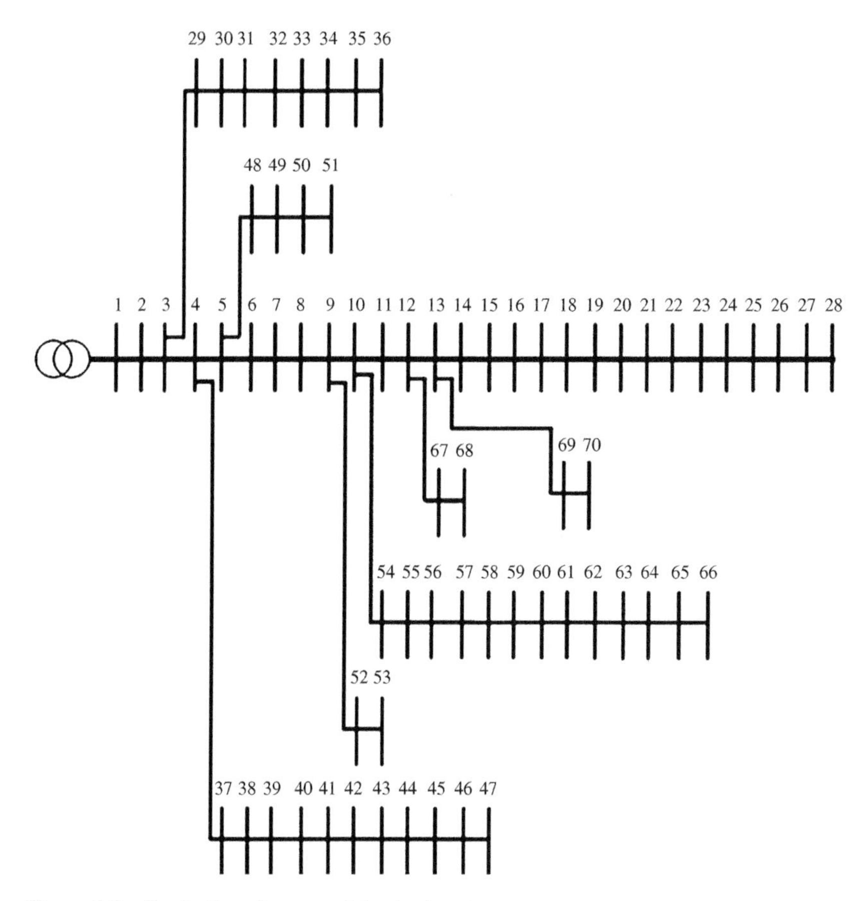

Figure 3.7 Single-line diagram of the test system.

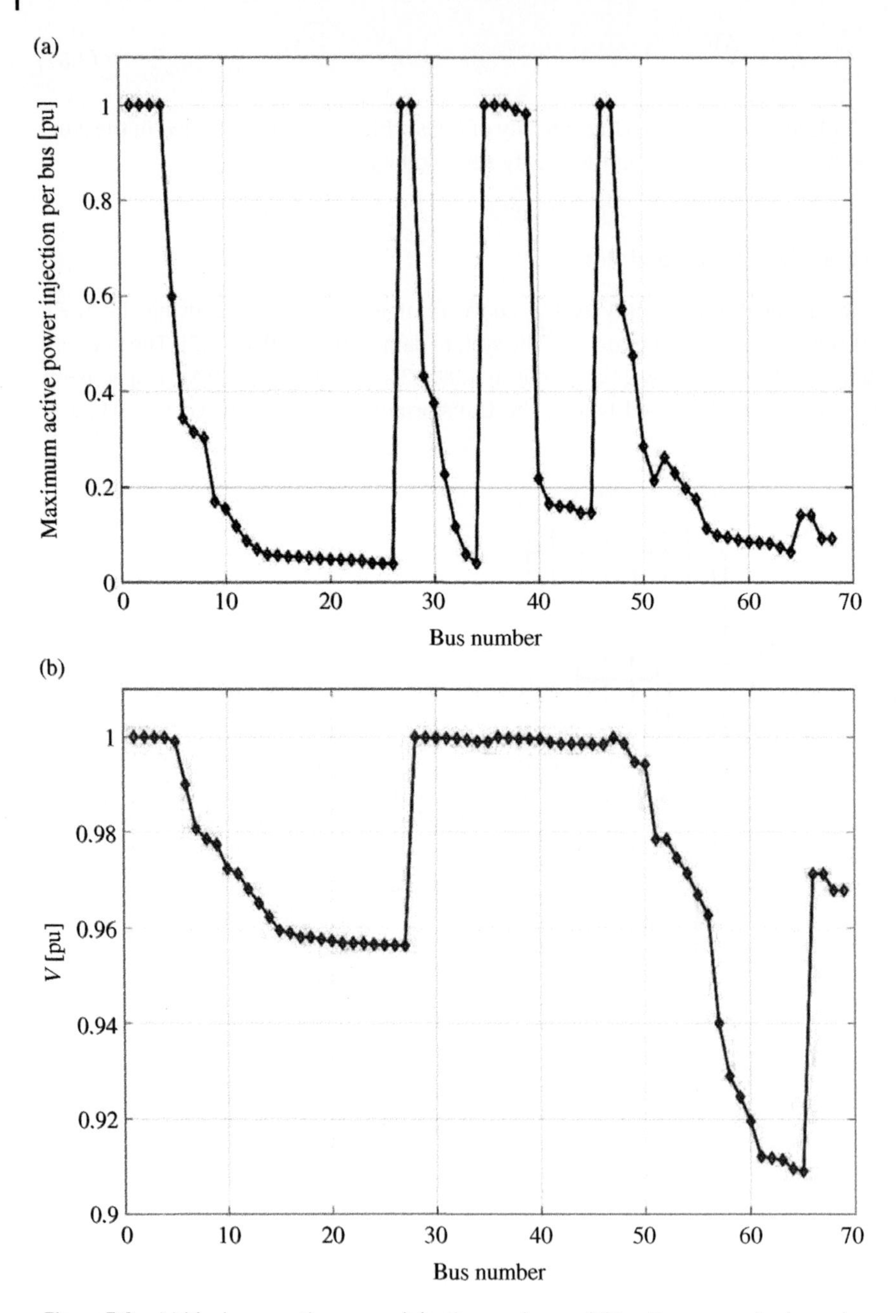

Figure 3.8 (a) Maximum active power injection per bus and (b) voltage magnitude per bus.

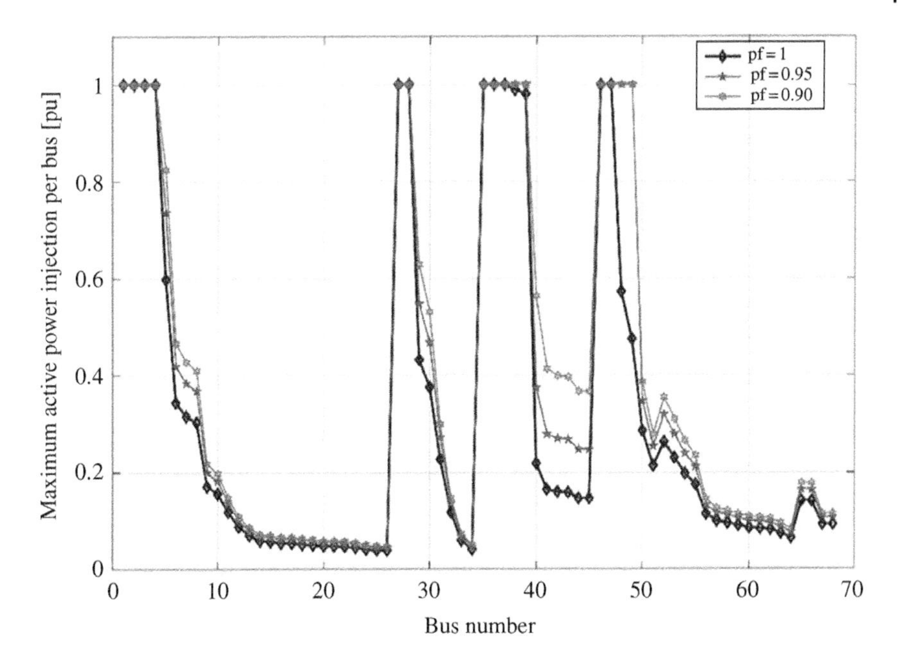

Figure 3.9 Maximum active power injection per bus for various power factor.

maintain the voltage magnitude of every bus within the allowable range of 0.95–1.05 pu, for the case without DGs.

Considering the unity power factor, Figure 3.8a demonstrates the maximum active power injection per bus in the system. For this purpose, the minimum permissible voltage magnitude in (3.80) considers being 0.9 which is an acceptable value for an emergency condition. Figure 3.8b shows the voltage magnitude per bus after installing the maximum capacity on each bus.

Also of interest, Figure 3.9 depicts the effect of power factor on the maximum active power injection per bus. It appropriately demonstrates the effectiveness of (3.80).

3.7 Summary

MG-integrated power system stability analysis from frequency, small-signal, and voltage points of view is discussed in this chapter. Furthermore, some mathematical-based approaches have been discussed to determine the maximum penetration level of MGs in the grid. The methods explicitly rely on the basic power system equations which, in turn, make the proposed frameworks completely independent of the test cases.

References

1. Hatziargyriou, N., Asano, H., Iravani, R., and Marnay, C. (2007). Microgrids: an overview of ongoing research, development, and demonstration projects. *IEEE Power Energy Magazine* **5** (4): 78–94. https://doi.org/10.1109/MPAE.2007.376583.
2. Bevrani, H., François, B., and Ise, T. (2017). *Microgrid Dynamics and Control*. New York: Wiley.
3. Lasseter, R.H. (2002). Microgrids. In: *2002 IEEE Power Engineering Society Winter Meeting. Conference Proceedings (Cat. No. 02CH37309)*, vol. **1**, 305–308. IEEE.
4. Lasseter, R.H. and Paigi, P. (2004). Microgrid: a conceptual solution. In: *Power Electronics Specialists Conference, 2004. PESC 04. 2004 IEEE 35th Annual*, vol. **6**, 4285–4290. IEEE.
5. Golpîra, H., Atarodi, A., Amini, S. et al. (2020). Optimal energy storage system-based virtual inertia placement: a frequency stability point of view. *IEEE Transactions on Power Systems* **35** (6): 4824–4835.
6. Golpîra, H. and Messina, A.R. (2018). A center-of-gravity-based approach to estimate slow power and frequency variations. *IEEE Transactions on Power Systems* **33** (1): 1026–1035.
7. Milano, F., Dörfler, F., Hug, G. et al. (2018). Foundations and challenges of low-inertia systems. *2018 Power Systems Computation Conference (PSCC)*: IEEE, pp. 1–25.
8. Golpîra, H., Seifi, H., Messina, A.R., and Haghifam, M.R. (2016). Maximum penetration level of micro-grids in large-scale power systems: frequency stability viewpoint. *IEEE Transactions on Power Systems* **31** (6): 5163–5171. https://doi.org/10.1109/TPWRS.2016.2538083.
9. Ulbig, A., Borsche, T.S., and Andersson, G. (2014). Impact of low rotational inertia on power system stability and operation. *IFAC Proceedings* **47** (3): 7290–7297.
10. Vittal, E., Keane, A., and O'Malley, M. (2008). Varying penetration ratios of wind turbine technologies for voltage and frequency stability. In: *Power and Energy Society General Meeting-Conversion and Delivery of Electrical Energy in the 21st Century, 2008 IEEE*, 1–6. IEEE.
11. Lalor, G., Mullane, A., and O'Malley, M. (2005). Frequency control and wind turbine technologies. *IEEE Transactions on Power Systems* **20** (4): 1905–1913.
12. Folly, K. and Sheetekela, S. (2009). Impact of fixed and variable speed wind generators on the transient stability of a power system network. In: *Power Systems Conference and Exposition, 2009. PSCE'09. IEEE/PES*, 1–7. IEEE.

13. Mitra, A. and Chatterjee, D. (2013). A sensitivity based approach to assess the impacts of integration of variable speed wind farms on the transient stability of power systems. *Renewable Energy* **60**: 662–671.
14. Fernandez-Bernal, F., Egido, I., and Lobato, E. (2014). Maximum wind power generation in a power system imposed by system inertia and primary reserve requirements. *Wind Energy* **18** (8): 1501–1514.
15. Doherty, R., Mullane, A., Nolan, G. et al. (2010). An assessment of the impact of wind generation on system frequency control. *IEEE Transactions on Power Systems* **25** (1): 452–460.
16. Yan, R., Saha, T.K., Modi, N. et al. (2015). The combined effects of high penetration of wind and PV on power system frequency response. *Applied Energy* **145**: 320–330.
17. Ahmadyar, A.S., Riaz, S., Verbič, G. et al. (2018). A framework for assessing renewable integration limits with respect to frequency performance. *IEEE Transactions on Power Systems* **33** (4): 4444–4453.
18. Keyhani, A. and Chatterjee, A. (2012). Automatic generation control structure for smart power grids. *IEEE Transactions on Smart Grid* **3** (3): 1310–1316.
19. Chen, X., Lin, J., Wan, C. et al. (2019). A unified frequency-domain model for automatic generation control assessment under wind power uncertainty. *IEEE Transactions on Smart Grid* **10** (3): 2936–2947. https://doi.org/10.1109/TSG.2018.2815543.
20. U. O. Handbook (2009). P1–Policy 1: load-frequency control and performance [C], *Version: v3. 0 rev,* vol. 15, no. 01.04.
21. Golpîra, H. (2019). Bulk power system frequency stability assessment in presence of microgrids. *Electric Power Systems Research* **174**: 105863.
22. Golpîra, H. and Bevrani, H. (2019). Microgrids impact on power system frequency response. *Energy Procedia* **156**: 417–424.
23. Karimi, H., Karimi-Ghartemani, M., and Iravani, M.R. (2004). Estimation of frequency and its rate of change for applications in power systems. *IEEE Transactions on Power Delivery* **19** (2): 472–480.
24. Golpîra, H., Messina, A.R., and Bevrani, H. (2019). Emulation of virtual inertia to accommodate higher penetration levels of distributed generation in power grids. *IEEE Transactions on Power Systems* **34** (5): 3384–3394.
25. Illian, H.F. (2011). *Frequency Control Performance Measurement and Requirements.* Berkeley, CA: Lawrence Berkeley National Laboratory.
26. NERC (2013). Standard BAL-003-1 – frequency response and frequency bias setting.
27. Bevrani, H. (2008). *Robust Power System Frequency Control.* New York: Springer.
28. Golpira, H. and Bevrani, H. (2011). Application of GA optimization for automatic generation control design in an interconnected power system. *Energy Conversion and Management* **52** (5): 2247–2255.

29. Golpira, H., Bevrani, H., and Naghshbandy, A.H. (2012). An approach for coordinated automatic voltage regulator–power system stabiliser design in large-scale interconnected power systems considering wind power penetration. *IET Generation, Transmission & Distribution* **6** (1): 39–49.

30. Bevrani, H. and Hiyama, T. (2017). *Intelligent Automatic Generation Control*. USA: CRC Press.

31. Chow, J. and Rogers, G. (2000). Power system toolbox. *Cherry Tree Scientific Software*. https://www.ecse.rpi.edu/~chowj/PST_2020_Aug_10.zip (accessed November 2020).

32. Wilson, D.H., Hay, K., and Rogers, G.J. (2003). Dynamic model verification using a continuous modal parameter estimator. In: *Power Tech Conference Proceedings, 2003 IEEE Bologna*, vol. **2**, 6. IEEE.

33. Rogers, G. (2000). *Power System Oscillations*. Norwell, MA: Kluwer.

34. Miller, N., Shao, M., Venkataraman, S. et al. (2012). Frequency response of California and WECC under high wind and solar conditions. In: *Power and Energy Society General Meeting, 2012 IEEE*, 1–8. IEEE.

35. Baggini, A.B. (2008). *Handbook of Power Quality*. New York: Wiley Online Library.

36. Abraham Ellis, R.W., Zavadil, B., Jacobson, D., and Piwko, R. (2012). 2012 Special Assessment Interconnection Requirements for Variable Generation. *NERC Report*, pp. 6–17.

37. PLECS (n.d.). Web address. http://www.plexim.com (accessed November 2020).

38. MAPLE (2018). Web address. https://maplesoft.com (accessed November 2020).

39. DSA Tools (n.d.). Dynamic security assessment software. Powertech Labs Inc. http://www.dsatools.com/ (accessed November 2020).

40. Kundur, P., Balu, N.J., and Lauby, M.G. (1994). *Power System Stability and Control*. New York: McGraw-Hill.

41. Eftekharnejad, S., Vittal, V., Heydt, G.T. et al. (2013). Small signal stability assessment of power systems with increased penetration of photovoltaic generation: a case study. *IEEE Transactions on Sustainable Energy* **4** (4): 960–967.

42. Messina, A.R. (2009). *Inter-area Oscillations in Power Systems: A Nonlinear and Nonstationary Perspective*. Germany: Springer Science & Business Media.

43. Ugwuanyi, N.S., Kestelyn, X., Thomas, O. et al. (2020). A new fast track to nonlinear modal analysis of power system using normal form. *IEEE Transactions on Power Systems* **35** (4): 3247–3257. https://doi.org/10.1109/TPWRS.2020.2967452.

44. Ayres, H., Freitas, W., De Almeida, M., and Da Silva, L. (2010). Method for determining the maximum allowable penetration level of distributed generation without steady-state voltage violations. *IET Generation, Transmission & Distribution* **4** (4): 495–508.

45. Baran, M.E. and Wu, F.F. (1989). Optimal capacitor placement on radial distribution systems. *IEEE Transactions on Power Delivery* **4** (1): 725–734.

4

Advanced Virtual Inertia Control and Optimal Placement

Power grids worldwide have experienced a significant transformation, which has been predominantly characterized by increased penetration of power electronic-based technologies. Among these new technologies are wind and photovoltaic (PV) generations, various storage technologies, flexible AC transmission systems (FACTS) devices, distributed generations (DGs), high-voltage direct current lines, and responsible loads. With significant integration of power electronic-based technologies, the dynamic behavior of power systems has progressively become more dependent on fast-response power electronic devices, thus, altering the power system dynamic behavior. Accordingly, new stability concerns have arisen which need to be appropriately addressed.

In Chapter 3, a framework for stability assessment of renewable integrated power systems was presented. The impediments to realizing high renewable integrated power systems were also discussed in small-signal and frequency stabilities points of view. This chapter discusses special attributes of the specific power electronic-based technologies to enhance system dynamics. Using this framework, the special attributes of the technologies are modeled using the well-known synchronous generator (SG). Optimal placement of the technologies, as the main step toward the successful implementation of advanced control schemes in Chapters 6 and 8, is also discussed in detail.

4.1 Introduction

Modern power grids face new technical challenges arising from the increasing penetration of power-electronic-interfaced loads, DG, and microgrids (MGs). Paramount among these is the inertia requirement challenge, as inverter-connected renewable sources are increasingly replacing SGs. Unlike the conventional SGs, inverter-based DGs do not provide physical inertia to support the grid. Therefore,

Renewable Integrated Power System Stability and Control, First Edition.
Hêmin Golpîra, Arturo Román-Messina, and Hassan Bevrani.

since SGs are gradually replaced by DGs, transmission system operators (TSOs) are faced with the issue of lack of inertia, which intrinsically leads to a large rate of change of frequency (RoCoF) and power variations in the grid. As a result, the power system is prone to frequency and power fluctuations, and the design of relays, including RoCoF-based relays, should be reconfigured.

As suggested by analytical results in Chapter 3, rotational inertia reduction in the grid may adversely affect frequency dynamics, small-signal stability, and system control and leads to degrade the performance of traditional control schemes. This, in turn, may result in large dynamic deviations and, potentially, load shedding, and instability. Accordingly, the increasing penetration of inverter-interfaced DGs motivates the need to develop additional ancillary control services to improve system dynamics and stability.

To address this issue, the concept of the virtual synchronous generator (VSG), virtual synchronous machine, or synchronverter has been proposed [1, 2]. It is shown that by adding short-term energy storage to emulate the kinetic energy of a rotating mass and mimic the swing equation of an SG in the control scheme, inverters can also provide inertia support for the grid to restrain its frequency fluctuation, in the same way as an SG. Since the principle of these concepts is similar, for convenience sake, all these inverters are referred to as VSG in this chapter.

Manipulation of the inertia constant by the VSGs could enhance system dynamics and stability. This could be realized using inertia-based control schemes. Indeed, inertia provision plays an important role in designing advanced control schemes in modern power grids. However, the successful implementation of advanced control schemes not only depends on the nature of the designed scheme but also the adequacy of the inertia provision sources. In this way, after a discussion on the application of VSGs in power system stability and performance improvement, optimal placement of dispatchable inertia besides the associated realization mechanisms would be emphasized in this chapter and afterward, the inertia-based advanced control schemes are addressed in Chapter 6. Without the loss of generality, while DGs (such as wind farms), high-voltage direct current (HVDC) systems, and energy storage systems (ESSs) could emulate virtual inertia, the ESSs are considered as the source of virtual inertia in this chapter.

4.2 Virtual Synchronous Generator

Power grid control must provide the ability of electric power systems to regain a state of operating equilibrium after being subjected to a physical disturbance, with most system variables, i.e. frequency, voltage, and angle, bounded so that practically the entire system remains intact. Thus, the main control loops are known as frequency control, voltage control, and power oscillation damping control.

Additional flexibility may be required from various control levels so that the system operator can continue to balance supply and demand on the modern power grids. The contribution of DGs and renewable energy sources (RESs) in regulation task refers to the ability of these resources to regulate their power output, by appropriate control action. This can be regarded as adding virtual inertia to the grid and considered a solution. Virtual inertia emulation system requires the inverter to be able to store or release an amount of energy depending on the grid frequency's deviation from its nominal value, analogous to the inertia of a conventional generator. This system is known as VSG and will then operate to emulate desirable dynamics, such as inertia and damping properties, by flexible shaping of its output active and reactive powers.

4.2.1 Concept and Structure

The VSG concept was introduced as a promising solution toward the grid stability and performance issues caused by high penetration of RESs. A VSG can be established by using a DG or short-term ESS coupled with the power electronics inverter/converter with an appropriate control mechanism [2].

The general architecture of a VSG is presented in Figure 4.1. In this scheme, a primary power source such as a DG unit is connected to the grid via a power inverter (VSG system). It is expected to operate as a conventional SG by providing inertia and damping property virtually and by displaying the same reaction as the conventional SG when there is a sudden change of load or disturbance in the system. The VSG control block is expected to regulate the output of the inverter based on the *RoCoF* and the difference between the reference frequency and grid frequency like the way conventional SGs are governed by the swing equation:

$$\Delta P_{VSG} = P_m - P_e = J\Delta\omega' + D\Delta\omega = J\frac{d^2}{dt^2}\delta(t) + D\frac{d}{dt}\delta(t) \tag{4.1}$$

where J is the inertia coefficient, D is the damping coefficient, and δ is the rotor angular position difference from its reference position.

The swing Equation (4.1) describes the relative motion of the rotor in respect to the stator field as a function of time. The generator inertia reacts to the disturbance and plays a significant role in power system stability, which is the most important property of a synchronous machine.

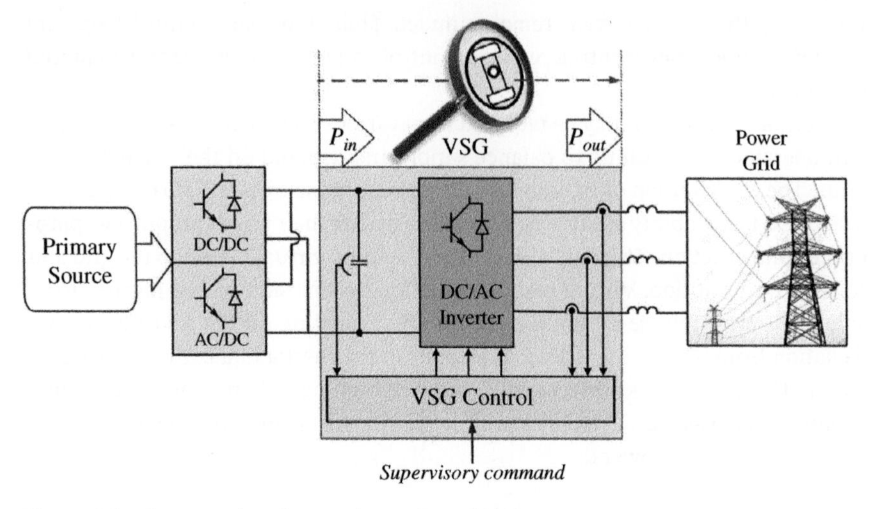

Figure 4.1 Conceptual and general structure of VSG.

4.2.2 Basic Control Scheme and Applications

A typical control scheme of a VSG system is shown in Figure 4.2 [1]. It is an RMS-value-based control scheme without using an additional inner voltage or current control loop. Different from VSG control schemes with an inner voltage loop, this

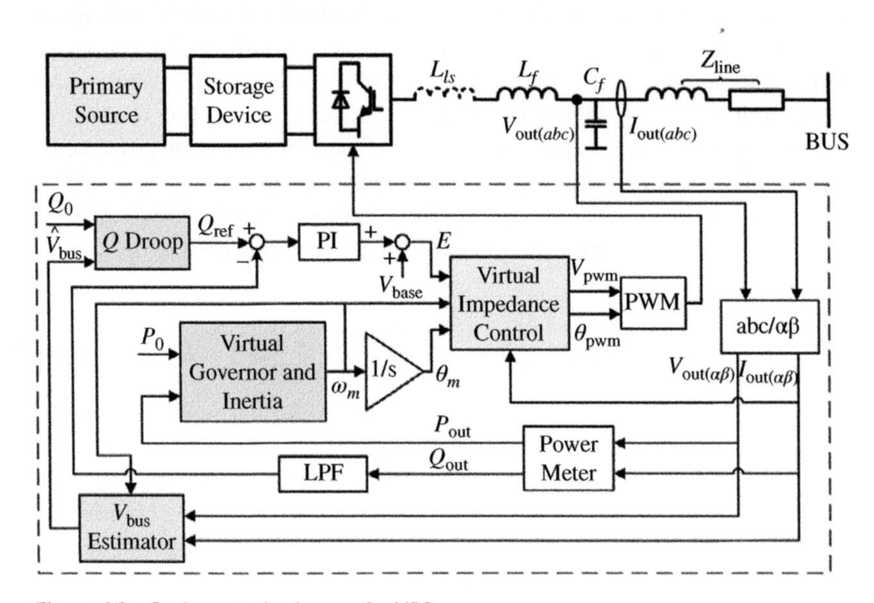

Figure 4.2 Basic control scheme of a VSG.

type of VSG control regulates the output voltage through the reactive power control loop. Owing to the absence of an inner voltage loop, the active power control loop becomes quite simple and robust, and bus voltage deviations become smaller and insensitive to the output impedance [3].

In the power generation part, the "Virtual Governor and Inertia" block is the core of VSG control. In the literature, various approaches using different damping technologies are proposed for this part. A virtual voltage drop over virtual inductance L_{ls} is generated to adjust the equivalent output reactance X of the inverter as shown in (4.2), given by

$$X = \omega_0 \left(L_{LS} + L_f + L_{line} \right) \tag{4.2}$$

The block "V_{bus} estimator" estimates the bus voltage from the measurement of output voltage and current to provide a common reference for the block "Q Droop," which makes a simple linear function between voltage and reactive power [1]. To emulate the steady-state operation of an SG, the governor model is usually simulated in a dispatchable VSG-based power source, as shown in (4.3):

$$P_{in} = P_0 - k_p(\omega_m - \omega_0) \tag{4.3}$$

To mimic the dynamics of an SG, the swing equation should be emulated. If the effect of damper windings is omitted, the swing equation of an SG can be expressed as:

$$P_{in} - P_{out} = J\omega_m \frac{d\omega_m}{dt} \tag{4.4}$$

Combining (4.3) and (4.4) yields

$$P_0 - P_{out} = J\omega_0 \frac{d\omega_m}{dt} + k_p(\omega_m - \omega_0) \tag{4.5}$$

In interpreting the control structure in Figure 4.2, note that the VSG control emulates Equation (4.5) through the "Virtual Governor and Inertia" block.

Apart from the VSG control topology shown in Figure 4.2, several different VSG controls exist in the literature. The VSG control schemes can be categorized into three main groups based on the nature of the output reference from the VSG [4].

1) *Current references-based VSG control*: This group utilizes current references from VSG and allow a quite natural implementation of high-order electrical models of the SG since inverters are controlled to generate the currents that would result from a real SG.
2) *Voltage references-based VSG control*: This group is another possible approach to utilize the voltage references from VSG. The group can be referred to as "voltage references from VSG." Although only a reduced-order model of the SG can

be applied in this approach, but unlike the first group, the utilizing voltage command allows this VSG to function in stand-alone mode.

3) *Power references-based VSG control*: This group emulates the inertia response by tracking the grid frequency without implementing any SG model. The group is called "power references from VSG," as the current reference corresponding to a given power reference is used to control the inverter.

Some applications of VSGs in power systems have been reported in [5–10]. For example, the VSG concept was successfully implemented in [5] for multiterminal HVDC systems to suppress low-frequency oscillation and enhance the power oscillation damping performance of the AC/DC system. A dual VSG-based modular multilevel matrix converter (M3C) control scheme for frequency regulation support of a remote AC grid via low-frequency AC transmission system is proposed in [6]. In [7], a VSG is used to enhance the performance of a stand-alone gas engine generator.

Furthermore, several studies considering the application of VSG in PV systems [8], wind power generation [9], permanent magnet synchronous generators (PMSGs) [10, 11], and doubly fed induction generators (DFIGs) have been reported. Besides, the VSG control can be applied to other grid-tied inverters, such as those in the ESSs [12], bidirectional battery chargers of electrical vehicles (EVs) providing vehicle-to-grid (V2G) services [13], and voltage source converters (VSCs) in high-voltage DC transmission system [5, 14–17]. Generally, the VSG-based DC–AC converter becomes a standard interface for smart grid integration [18].

As emphasized above, the VSGs with emulated inertia capabilities may play a critical role in reducing the *RoCoF* and the frequency nadir of the power system [19]. Owing to this promising feature, it attracts a lot of attention in recent years. These works are mainly targeted at improving the damping of a VSG via fixed parameter or adaptive parameter methods [1], improving its parameter tuning [20], fault ride-through ability, power quality, and applying this concept in various types of grid-interfaced power electronics devices [6, 7, 21, 22].

4.2.3 Application in Power System Dynamic Enhancement

Although the applications of VSGs in an MG have been fully studied, the grid integration performance of VSG is not well understood. In previous VSG studies, the grid is usually modeled as an infinite bus or a single SG. Several works have investigated the interaction between one VSG and one SG in an isolated MG. However, these studies overlook the most interesting case of interaction between the VSGs and multiple SGs.

In this subsection, the impact of high penetration of inverter-based DGs with and without VSG on system dynamic performance is illustrated on the IEEE 9-bus system shown in Figure 4.3. Detailed parameters of generators, governors,

Figure 4.3 IEEE 9-bus system.

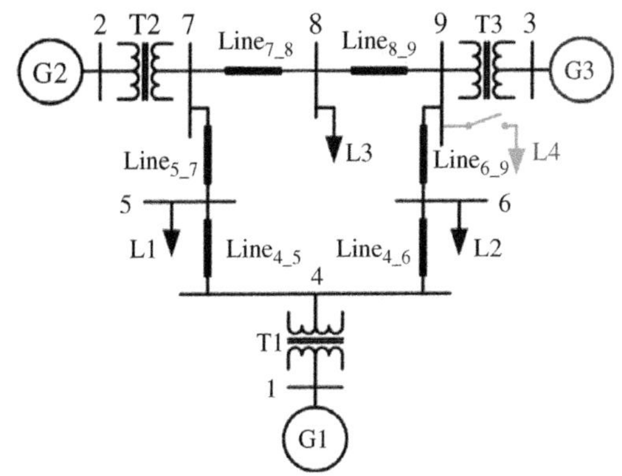

steam turbines, and exciters are taken from [23], except minor changes mentioned in [24]. Power system stabilizers (PSSs) are not included in the system, and the rated power of G1, G2, and G3 are 512, 270, and 125 MVA, respectively.

Three case studies to evaluate the grid integration performance of VSG are considered. In Case 1, it is assumed that all three generators in Figure 4.3 are SGs. Case 2 is the same as Case 1, except that a cluster of coherent grid-following inverter-based DGs with the same overall rated power replaces G3. For the clarity of illustration, the DGs are aggregated into a single DG of 125 MVA. Finally, Case 3 is similar to Case 2, except that the VSG control is used to control the inverter-based DGs.

In these studies, the VSG is designed to have the same inertia and droop coefficient as G3 in Case 1, while the equivalent reactance from the transient emf to the adjacent bus of the G3 node is 0.3 pu, and the damping ratio of VSG is set to 0.9.

4.2.3.1 Scenario 1: 10-MW Load Increase at Bus 9

Figure 4.4 shows the time-domain results, including the frequency and output active power of G1 and G2, when a 10-MW load suddenly connected to Bus 9. The grid integration performance of DGs can be evaluated based on Figure 4.4. Comparison of Cases 1 and 2 shows that increased penetration of inverter-based DG leads to more severe frequency nadir and higher RoCoF after the disturbance.

The results of Case 3 show that these issues can be addressed by the VSG control with a considerable mitigation in RoCoF and frequency nadir. Moreover, Case 3 has a better performance than Case 1, in terms of the improved frequency nadir.

4.2.3.2 Scenario 2: 20-MW Power Command Decrease of G3

The system performance is also investigated following a 20-MW step decrease in power control input of G3. The system responses including the frequency and output active power of G1 and G2 are shown in Figure 4.5. Comparison between the

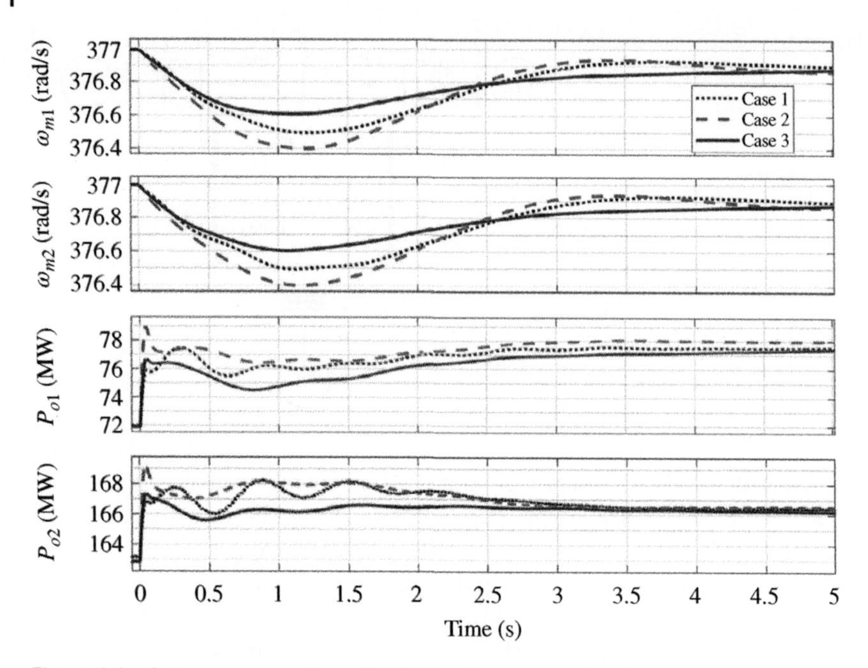

Figure 4.4 System response for a 10-MW load increase at Bus 9.

studied cases further verifies the conclusion obtained with Scenario 1. The VSG improves frequency nadir and reduced transient RoCoF. On the other hand, Case 1 shows completely different dynamics, because changing the power command of an SG is delayed by its slow governor and turbine responses.

4.2.4 Application to Power Grids with HVDC Systems

The large penetration of RESs has added a large power density to the existing power system. This increased power density and efficiency have brought many challenges such as power quality issues, power delivery, and transmission. Numerous topologies and schemes have been suggested to deal with power control and transmission issues of the renewable integrated power grids.

As an important example, consider the test system shown in Figure 4.6, in which the wind farms are located far from the load centers and thus require a long transmission lines. For farther wind farms installations, current trends in research and practice point toward the use of HVDC transmission with VSC-based HVDC transmission being the preferred approach as it displays distinct control and design advantages over traditional line-commutated converter technology [25]. In such installations, the wind farm collector network typically operates at 50 or 60 Hz,

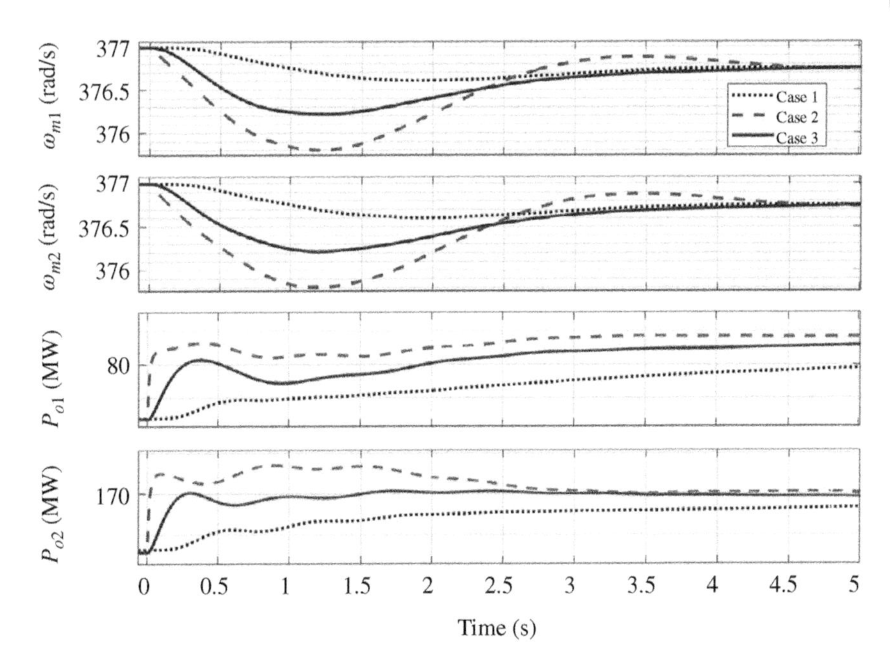

Figure 4.5 System response for a 20-MW step decrease in G3 power command.

which is then converted to HVDC by a wind farm converter station for transmission to a load center converter station (Figure 4.6).

The VSC-HVDC is currently considered as the market leader for wind integration at distances greater than 60–80 km largely due to its established use in point-to-point bulk power transfer [26]. The HVDC system is broadly applied for

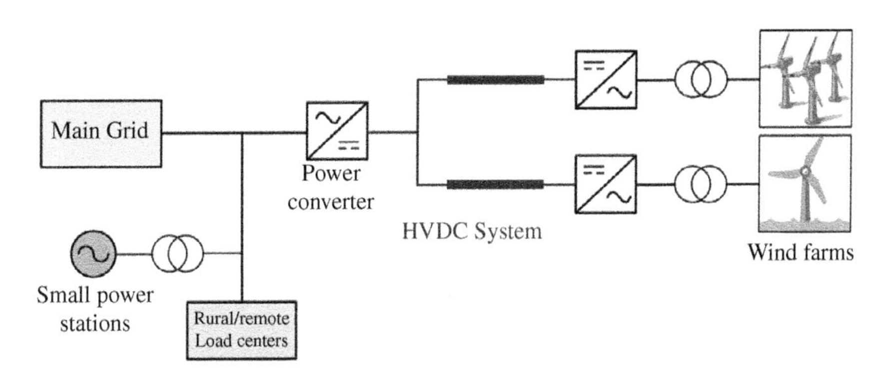

Figure 4.6 Power grid with HVDC system.

point-to-point interconnection of two power systems. It is also well used for interconnecting the offshore wind farm with the onshore stations using a submarine cable.

An HVDC link may improve the system stability of a wide area power system by decoupling of some areas. For instance, consider two interconnected AC grids by an HVDC link. This system exhibits a natural decoupling between the AC grids in terms of both voltage and frequency (the AC grids are interconnected in an asynchronous manner). These kinds of interconnections are primarily aimed at preventing the excursions of oscillations between AC systems, for instance, between a stronger and a weaker electric network [27]. Furthermore, the HVDC transmission technology is currently deemed a desirable solution for long-distance power transmission and offshore RESs.

In contrast, the VSC-HVDC systems have some weak points. The HVDC system is sensitive to faults on the DC lines. When a fault occurs on the DC side of a VSC-HVDC system, the insulated-gate bipolar transistor (IGBT) cannot control and freewheeling diodes work as a bridge rectifier. It is not able to withstand large surge currents and may be damaged before the fault is cleared. Some solutions are proposed; however, additional control and switching devices are needed.

An HVDC system uses the VSC stations that, from a physical inertia point of view, their dynamic behavior is quite different from that of the SGs. Thus, it makes the frequency stability problem in the connected AC grid more serious than the past. For this reason, the HVDC systems are expected to participate in the grid frequency support. This issue is more significant when it is known that the integration of RESs instead of SGs reduces system inertia. Consequently, the system frequency stability and dynamic performance are affected. We can consider this issue from a different perspective. One typical application of HVDC transmission systems is to supply a weak or passive/islanded grid far away from the main grid. With constant power control, the output power from the HVDC-VSC station is constant, and no frequency support is provided to the weak grid. When a significant part of the power in a local gird is supplied by this station, the system inertia becomes lower, and changes in generation or load may cause large-frequency deviations possibly leading to system instability [14]. In response to this challenge, it could be a promising solution to use the VSC station to provide frequency support to the local grid.

There are several works investigating system frequency support through HVDC systems [5, 14, 27–29]. The main idea is to use the VSG concept for grid-tied HVDC converters to emulate the classical SG, thus providing virtual inertia and system frequency support, as shown in Figure 4.7.

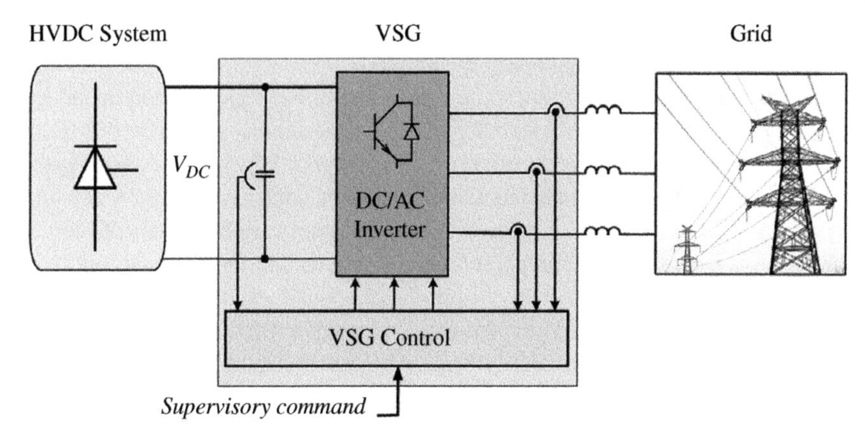

Figure 4.7 HVDC system as a primary source for grid frequency support.

4.3 Dispatchable Inertia Placement

Not only an insufficient level of inertia but also its heterogeneous distribution may render system dynamics faster. This fact, along with the need to economically keeps the system secure, and makes the optimal placement of virtual inertia as a key factor [30]. Accordingly, optimal placement of virtual inertia can be studied in two perspectives: (i) frequency stability and (ii) small-signal stability.

4.3.1 Frequency Dynamics Enhancement

The ever-growing number of frequency incidents, in response to fluctuations of renewable power sources, accompanied by low-level inertia feature jeopardize frequency stability [1, 30]. Following a disturbance, inertial response may positively affect the power imbalance before the activation of traditional frequency control loops. This means that efficient system inertia has a major influence on the frequency stability and performance characteristics such as the *RoCoF* and frequency nadir [31].

4.3.1.1 Background: Literature Review

The effects of ESSs on frequency regulation have been studied in recent research works [32–34]. These works deal with the problem of balancing of generation and load to maintain a constant system frequency and to keep tie-line power flows within some prespecified tolerance. However, these studies consider long-term frequency response as well as steady-state metrics as the main metrics utilized for system resilience analysis, neglecting the impact of inertial response from RESs on system stability. In this direction, some recent studies have investigated the effect of reduced inertia on frequency stability and transient stability of the power system [35–42]. In parallel with these efforts, some recent works, [43–50], investigate ways in which virtual inertia could be emulated, including appropriate control of wind turbines and ESSs. In [51], a framework that addresses various aspects of inertia emulation and control, including how virtual inertia emulation and its location in the system impacts system stability, is proposed. Some questions about the heterogeneous inertial profiles and how the associated negative impacts are reduced by inertia emulation have been raised in [38]. Further, Poolla et al. [30] propose an H_2-based performance metric to determine the optimal placement of virtual inertia. The determination of the optimum size of ESS to provide the primary frequency control is addressed in [52, 53]. However, due to the lifetime concerns, the ESSs cannot effectively participate in the primary regulation. In practice, the ESSs are dispatched using an optimal control strategy, designed to optimize the state-of-charge (SOC) range and the lifetime constraints. Some research works, such as in [54], deal with the optimal placement of virtual inertia in power systems considering network structure. These approaches utilize DC power flow to incorporate network structure into the model.

The main limitation of these approaches is their reliance on static considerations. It is well known, however, that dynamic frequency indices, such as the RoCoF and frequency nadir, are important parameters to assess frequency performance and the development of protection schemes. Based on these considerations, the following subsections formulate the problem of finding virtual inertia locations in terms of dynamic as well as static metrics. The main step toward the optimal placement of virtual inertia in a power grid is to analyze its effects on the frequency behavior; and this could be realized by the appropriate modeling of the virtual inertia.

4.3.1.2 Virtual Inertia Modeling

Main Idea

As discussed in Section 4.2, the VSGs can be used to provide virtual inertia; the VSG-based virtual inertia emulation strategy is based on a similar power-balance-based synchronization mechanism to that defined by the swing equation of a conventional SG [55–57]. This means that the effects of virtual inertia on the frequency indices can be assessed using a dynamical equivalent model obtained by mapping the electromechanical behavior of the ESSs onto the second-order SG model.

To illustrate this idea, let the swing equation of a conventional SG be expressed as:

$$\begin{Bmatrix} \dot{\delta}^s = \omega^s(t) - \omega_0 \\ M\dot{\omega}^s = T_m(t) - T_e(t) \end{Bmatrix} \tag{4.6}$$

where $\dot{\delta}^s$, ω^s, and ω_0 are the mechanical rotor angle, the mechanical rotor angular speed, and the initial angular speed, respectively; M, $T_m(t)$, and $T_e(t)$ are the inertia constant, the mechanical input torque, and the electrical output torque, respectively [58]. Taking the slow electromechanical behavior of the ESS into account, the associated dynamics could be represented by (4.6). The problem of interest, however, is to calculate the equivalent inertia constant and the mechanical input torque.

In what follows, a data-driven approach in which the uncertain behavior of the ESS is accounted for in the swing Eq. (4.6) is proposed.

MUSIC Analysis: Methodology and Application

To introduce the adopted model, assume that the injected power of the ESS to the host grid is a discrete-time signal $P(n)$ of length L. Let the time-varying signal $P(n)$ be decomposed into a summation of K sinusoidal components and noise, as [59, 60]:

$$P(n) = \sum_{k=1}^{K} a_k \cos(n\omega_k + \Phi_k) + w(n) \tag{4.7}$$

where, a_k, Φ_k, ω_k, and $w(n)$ are the magnitude and the initial phase angle, harmonic frequency in radius, and additive white noise, respectively. In the model, a_k and ω_k are assumed to be deterministic and unknown, and Φ_k is unknown and assumed to be random and uniformly distributed in $[-\pi, \pi]$. Alternatively, the model (4.7) can be expressed in the form of noisy complex exponentials as [59]:

$$P(n) = \sum_{k=1}^{K} A_k e^{(jn\omega_k)} + w(n) \tag{4.8}$$

where $A_k = |A_k|e^{j\Phi}{}_k$ is the complex magnitude of the kth-harmonic (noise) signal component. As the MUltiple Signal Classification (MUSIC) algorithm is a noise subspace-based method, it is a good tool to deal with experimental noisy measured signals. Using this framework, the dimensional space is divided into the signal and noise components, which is of high importance to accurately calculate M and T_m (t) in (4.6).

The MUSIC method employs a harmonic model and estimates the frequencies and powers of the signal harmonics. Application of the MUSIC method to the data sequence $P(n)$, gives

$$P_{MUSIC}\left(e^{j\omega}\right) = \frac{1}{\sum_{i=K+1}^{z} |e^{U} s_i|^2} \tag{4.9}$$

where s_i is the eigenvectors associated with the noise subspace that is orthogonal to the signal eigenvector $e = [1\ e^{j\omega}\ ej^{2\omega}\ e^{j(z-1)\omega}]^T$, and e^U denotes the complex-conjugate transpose; z is the dimension of space spanned by $P(n)$. It is worth emphasizing that $P_{MUSIC}(e^{j\omega})$ in (4.9) does not relate to any real power spectrum; rather, the only purpose of this pseudo-spectrum is to generate peaks whose frequencies correspond to those of the dominant frequency components. This feature makes the MUSIC approach interesting to develop an equivalent model based on dominant modes.

For a given signal of interest and using (4.7)–(4.9), the model eigenvalues can be calculated. By knowing the eigenvalues and because the impulse response is the inverse Laplace transform of eigenvalues, one could represent a signal of interest with a predefined model of (4.6). For this purpose, suppose the impulse response of system is $I(t)$; for the input signal $x(t)$, i.e. $x(t) = T_m(t)$ in (4.6), one could write $y(t)$, i.e. $\dot{\omega}$, as:

$$y(t) = x(t)\int_{t=0}^{\infty} I(t)e^{j\omega t}dt \tag{4.10a}$$

where

$$I(t) = y(t) * PS \tag{4.10b}$$

and PS in (4.10b) is the pseudo-spectrum of the signal. Equation (4.10b) reveals that $I(t)$ is obtained from the convolution of $y(t)$ and PS. In the modeling procedure and by measuring the output response of the system $y(t)$ and by knowing $I(t)$, the problem of interest is to calculate $x(t)$ in (4.10a). By calculating T_m (t), the ESS could be replaced by the SG model of (4.6). Using this framework, angular speed ω in (4.10a) is defined based on the dominant frequency components of the pseudo-spectrum in (4.9). Figure 4.8 gives a schematic illustration of this model. In this plot, Figure 4.8a illustrates the process of virtual inertia emulation using the battery ESS, while Figure 4.8b describes a simplified block diagram representation of the equivalent model. The input to the control algorithm is the frequency at the connection point of the inverter f_{cp}, and P_{in} represents the grid injected power.

4.3.1.3 Experimental Verification

The effectiveness of (4.6)–(4.10b) to represent the ESS behavior is illustrated using the existing battery ESS at the University of Kurdistan Micro-Grid (UOK-MG). Figure 4.9 shows a three-phase diagram representation of the UOK-MG.

Figure 4.10 shows the ESS and the main grid power variation behavior, i.e. P_{in} in Figure 4.8, recorded for the UOK-MG. As shown in Figure 4.10, event 1 triggers the charging process of the ESS in response to deviations from the minimum SOC.

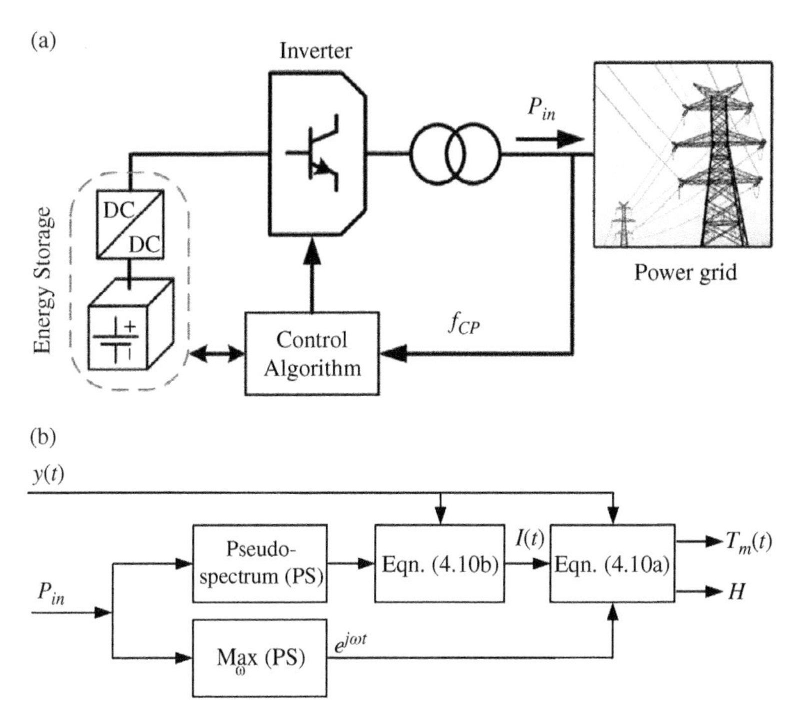

Figure 4.8 Block diagram representation of the proposed modeling process: (a) virtual inertia emulation mechanism and (b) the proposed equivalent model; P_{in}, *y(t), and* f_{cp} represent the grid injected power, the frequency deviation of the ESS, and the frequency at the point of connection of the ESS, respectively.

The main grid power deviation during the charging process in Figure 4.10 is utilized to calculate the pseudo-spectrum of Figure 4.11 which, in turn, is used to estimate the dominant frequency components in (4.9). Setting the frequency deviation of the ESS in Figure 4.12 as $y(t)$ and the pseudo-spectrum of Figure 4.11 as PS for (4.10b) gives

$$\left\{ \begin{array}{c} \dot{\delta}^s = \omega^s - \omega_0 \\ 0.53\dot{\omega}^s = (1 - e^{-0.38t})C - T_e(t) \end{array} \right\} \tag{4.11}$$

where C is a constant value equal to the DC term in (4.9).

In (4.11), note that the constant 0.38 represents the dominant frequency of the pseudo-spectrum of Figure 4.11, which can be introduced into (4.10b) to calculate the 0.53 s inertia constant. Figure 4.12 shows the effectiveness of the equivalent

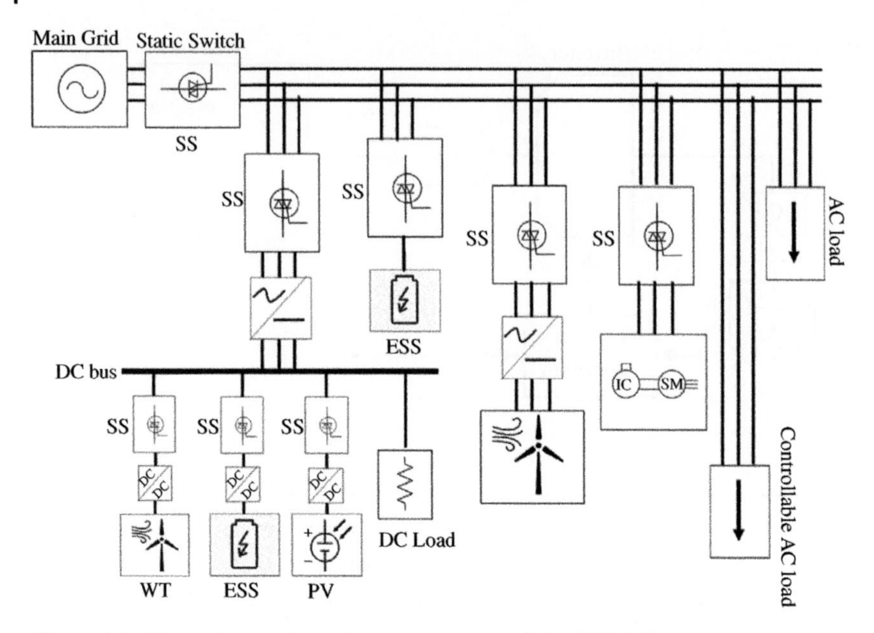

Figure 4.9 Three-phase schematic representation of the UOK-MG.

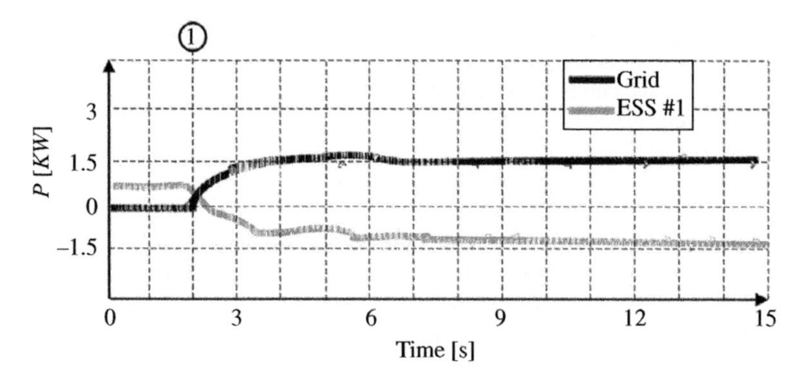

Figure 4.10 The ESS and grid experimental dynamic responses.

model (4.11) to approximate the inertial response behavior of the ESS. To exactly mimic the frequency behavior of ESS using (4.11), the oscillatory behavior of the adopted model can be removed using a 20-sample rolling-averaging window. This approach averages the long-term oscillations, and hence, mitigates the oscillatory behavior beyond the inertial response horizon. Results in Figure 4.12 show that the dynamic behavior of the ESS, especially in the inertial response horizon, can be

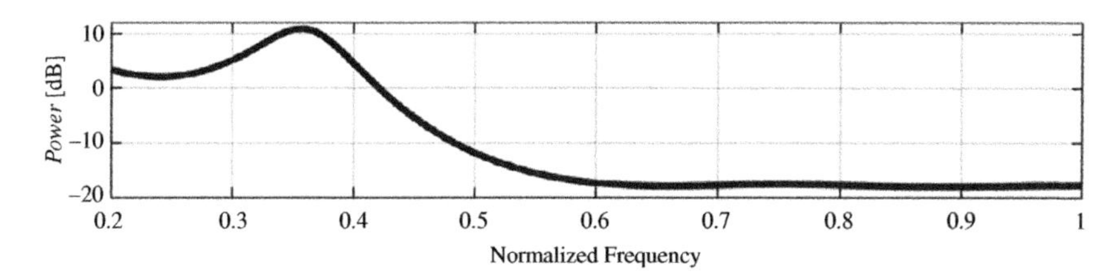

Figure 4.11 The pseudo-spectrum estimation via MUSIC.

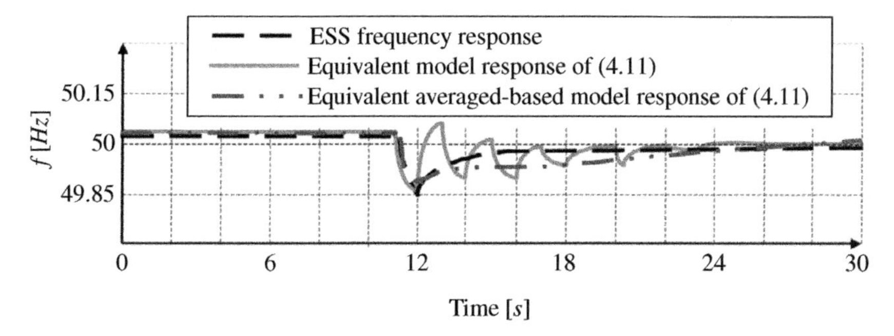

Figure 4.12 Comparison of frequency response computed from the experiment and the equivalent frequency response models.

approximated by the SG model. It should be emphasized that while a conventional SG is slower and less flexible compared to the ESS, the developed dynamical equivalent model of the ESS in the grid-connected mode is not only affected by the fast-inherent features of the ESS but also it is significantly influenced by the external zone, i.e. the host grid, features.

4.3.1.4 Economic Modeling

This section formulates the ESS placement problem as an optimal techno-economic problem.

Costs of Energy Storages and Technologies

There are two main approaches for assessing the cost of ESS technologies: (i) total capital cost (TCC), and (ii) life cycle cost (LCC) [61]. In the TCC approach, all terms associated with the purchase, installation, and delivery of the ESS units, including the power conversion system (PCS) costs (C_{PCS}), costs of ESS (C_{stor}), and balance of plant (BOP) cost (C_{BOP}) can be represented as:

$$C_{cap} = C_{PCS} + C_{BOP} + C_{stor} \times t_{ch} \quad (\$/\text{kW}) \tag{4.12}$$

where t_{ch} is the charging/discharging time. The balance of the ESS, known as the BOP, includes site wiring, interconnecting transformers, and other additional ancillary equipment and is measured on a \$/kW basis [62]. However, the LCC is a more common metric to evaluate and compare different ESS technologies. The annualized LCC is formulated according to (4.13) which considers operation and maintenance costs ($C_{o\&M,a}$), replacement cost ($C_{R,a}$), and annualized TCC:

$$C_{LCC,a} = C_{cap,a} + C_{OM,a} + C_{R,a} \quad (\$/\text{kW yr}) \tag{4.13}$$

in which:

$$CRF = \frac{i(1+i)^T}{(1+i)^T - 1} \tag{4.14}$$

$$C_{cap,a} = TCC \times CRF \quad (\$/(\text{kW yr})) \tag{4.15}$$

$$C_{OM,a} = C_{FOM,a} + C_{VOM,a} \times n_{cycle} \times t_{ch} \quad (\$/\text{kW yr}) \tag{4.16}$$

$$C_{R,a} = CRF \times \sum_{k=1}^{r} (1+i)^{-kt} \times \left(\frac{C_R \times t_{ch}}{\eta_{sys}} \right) \quad (\$/\text{kW yr}) \tag{4.17}$$

where CRF, i, T, r, t, and η_{sys} are the capital recovery factor, interest rate and the lifetime, the number of substitutions in a lifetime, the replacement period, and the overall efficiency, respectively; $C_{FOM,a}$ and $C_{VOM,a}$ define the fixed and variable operation and maintenance costs. The subscript "a" stands for "annualized" costs.

Formulation of the Objective Function and Constraints

Equation (4.13) specifies the annual cost per kilowatt of the installed ESS in compliance with the lifetime. However, for optimal placement of virtual inertia, it is necessary to rewrite the cost function in (4.13) according to the amount of virtual inertia.

The synchronous inertia constant M is defined as the ratio of stored kinetic energy to the rated apparent power of the system as:

$$M = \frac{0.5 J_{VI} \omega^2}{S_{base}} \tag{4.18}$$

where J_{VI}, ω, and S_{base} are the moment of inertia, angular velocity, and rated apparent power, respectively. Since the stored energy in ESSs is usually expressed in volt-ampere hour (VAh_{ESS}), it is needed to express the associated value in terms of Joule. Considering a unity power factor, (4.18) can be rewritten as:

$$KW_{ESS} = KVA_{ESS} = \frac{M^{ESS} S_{base}}{3600\, s} \rightarrow M^{ESS} = \frac{3600\, VAsec_{ESS}}{S_{base}} \tag{4.19}$$

This equation gives the average hourly power that can be injected/absorbed to/from the grid by the ESS. By substituting (4.19) into (4.13), one could write the optimization problem as:

$$\underset{M_i^{ESS}}{minimize} \;\; F\left(M_i^{ESS}\right) = \sum_{i=1}^{n_{ESS}} \left(C_{LCC,a\ i} \times \frac{M_i^{ESS}\, S_{base}}{3600\ sec} \right) \tag{4.20a}$$

$$st: \qquad RoCoF_i \leq RoCoF_{max} \tag{4.20b}$$

$$\Delta f_{nadir\ i} \leq \Delta f_{nadir\ max} \tag{4.20c}$$

$$SOC_{min} \leq SOC_i \leq SOC_{max} \tag{4.20d}$$

where n_{ESS} is the number of ESSs. Moreover, the SOC should remain within an appropriate range which is addressed in (4.20d). The SOC can be calculated as follows [63, 64]:

$$SOC(\Delta t) = SOC(0) - \frac{\int_0^{\Delta t} \zeta p(t)dt}{E_{ESS,rated}} \tag{4.21a}$$

where

$$\zeta = \begin{cases} \zeta_c & p(t) < 0 \\ \dfrac{1}{\zeta_d} & p(t) > 0 \end{cases} \tag{4.21b}$$

and $p(t)$ is battery power which gets negative values for the charging procedure and positive values for the discharging period; $E_{ESS,rated}$, Δt, ζ_c, and ζ_d are the nominal energy capacity, charge/discharge time, and charging and discharging efficiencies of the battery, respectively.

Determining the Bounds of Constraints

Constraints (4.20b) and (4.20c) explain that the optimization problem (4.20a) enforces the RoCoF and frequency nadir in all areas to be less than the standard values. These terms make the optimization problem difficult to deal with as it depends on the dynamical indices. Generally, it is common to specify the lower/upper bounds based on different criteria, including the capacity of equipment or budget. Therefore, the problem of interest here is to rewrite the upper and lower bounds of (4.20b) and (4.20c) in terms of the emulated inertia, i.e. M_i^{ESS}.

Rate of Change of Frequency The *RoCoF* is a meaningful criterion that measures the rate of frequency change following a large loss-of-generation event. Large RoCoF values indicate that less time is available for a system operator to arrest frequency decline. Typically, a time interval from 100 ms to 2 s is defined to measure the RoCoF [65, 66]. The ENTSO standard [65] explains that RoCoF is allowed to get a value between 0.5 and 1 Hz/s.

To obtain dynamic frequency indices based on lower bounds of inequality constraints in (4.20b), the RoCoF can be defined in terms of the classical swing equation of (4.6) as [67]:

$$2M\frac{d\Delta f(t)}{dt} = \Delta P_m(t) - \Delta P_L(t) - \Delta P_{tie}(t) \tag{4.22}$$

where $\Delta P_m(t)$, $\Delta P_L(t)$, and $\Delta P_{tie}(t)$ represent mechanical power, electrical power, and tie-line power changes, respectively. Considering the definition of RoCoF, one could write

$$RoCoF = \frac{\Delta P_m(t) - \Delta P_L(t) - \Delta P_{tie}(t)}{2M} \tag{4.23}$$

The Taylor series expansion of (4.23) about the independent variables of H, ΔP_m, ΔP_L, and ΔP_{tie} gives

$$\begin{aligned}\Delta RoCoF_i &= \frac{\partial RoCoF_i}{\partial \Delta P_{mi}}\Delta\Delta P_{mi} + \frac{\partial RoCoF_i}{\partial \Delta P_{Li}}\Delta\Delta P_{Li} + \frac{\partial RoCoF_i}{\partial \Delta P_{tiei}}\Delta\Delta P_{tiei} + \frac{\partial RoCoF_i}{\partial \Delta M_i}\Delta M_i \\ &= \frac{1}{2M_i}\Delta\Delta P_{mi} + \frac{-1}{2M_i}\Delta\Delta P_{Li} + \frac{-1}{2M_i}\Delta\Delta P_{tiei} + \frac{-(\Delta P_{mi} - \Delta P_{Li} - \Delta P_{tiei})}{2M_i^2}\Delta M_i\end{aligned} \tag{4.24}$$

Since the slow inherent dynamics of interest is only given by the last term in (4.24), the other terms can be neglected. Inserting (4.23) into (4.24) gives

$$RoCoF_i\left(-\frac{\Delta M_i}{M_i}\right) = \Delta RoCoF_i \tag{4.25}$$

It follows that the minimum inertia which guarantees that the RoCoF remains within the permitted range can be calculated as:

$$\begin{aligned} M'^{ESS}_{i,\min} &= \Delta M_{i,\min} = M_i\left(-\frac{\Delta RoCoF_{i,\max}}{RoCoF_i}\right) \\ \rightarrow M'^{ESS}_{i,\min} &= M_i\left(-\frac{RoCoF_{i,\max} - RoCoF_i}{RoCoF_i}\right)\end{aligned} \tag{4.26}$$

where $RoCoF_{i,max}$, is the maximum allowable RoCoF, and $M'^{ESS}_{i,min}$ represents the minimum required inertia that should be emulated by the battery ESS in area i, which complies with the RoCof. It equals to the difference between the desired inertia to enforce the system to follow the standards and the present inertia constant, i.e. $\Delta M_{i,min}$.

Frequency Nadir The minimum instantaneous frequency following a disturbance, known as *frequency nadir*, is mainly dependent on the system total inertia and the capability of the power resources to provide primary frequency response. According to the NERC and the Union for the Coordination of the Transmission of Electricity (UCTE) standards [68, 69], the allowed minimum frequency in a power system during normal operation is 800 mHz. Taking the time dependence of the governor response into account, one can approximate the frequency nadir after a system event as [43]:

$$\Delta f_{nadir} = \frac{(\Delta P_L + \Delta P_{tie})^2 T_d}{4MR} \tag{4.27}$$

where R is the extra power received from the governor and T_d is the response time of the governor.

In deriving (4.27), it is assumed that the mechanical power through the governor increases as a linear function of time with the steady gradient R/T_d [70, 71]. While this is a conservative assumption, Great Britain and Ireland practices show that this is the case for the power increment within 5 and 10 seconds (T_d), respectively, following a contingency [72]. Applying Taylor's expansion to (4.27) gives

$$\begin{aligned}\Delta\Delta f_{nadir,i} &= \frac{\partial \Delta f_{nadir,i}}{\partial \Delta P_{Li}}\Delta\Delta P_{Li} + \frac{\partial \Delta f_{nadir,i}}{\partial \Delta P_{tie,i}}\Delta\Delta P_{tie,i} + \frac{\partial \Delta f_{nadir,i}}{\partial \Delta M_i}\Delta M_i \\ &= \frac{(\Delta P_{Li} + \Delta P_{tiei})T_{di}}{2M_iR_i}\Delta\Delta P_{Li} + \frac{(\Delta P_{Li} + \Delta P_{tiei})T_{di}}{2M_iR_i}\Delta\Delta P_{tiei} \\ &+ \frac{-(\Delta P_{Li} + \Delta P_{tiei})^2 T_{di}}{4M_i^2R_i}\Delta M_i\end{aligned} \tag{4.28}$$

Following the same procedure as that in (4.26), one could rewrite (4.28) in the form:

$$\Delta f_{nadir,i}\left(-\frac{\Delta M_i}{M_i}\right) = \Delta\Delta f_{nadir,i} = \Delta\left(f_{nadir,i} - f_0\right) \tag{4.29}$$

The minimum inertia, i.e. $M_{i,\min}^{''ESS}$, which guarantees frequency nadir to remain within the permitted range is calculated by:

$$M_{i,\min}^{''ESS} = \Delta M_{i,min} = M_i\left(-\frac{\Delta\Delta f_{nadir\ i,max}}{\Delta f_{nadir\ i}}\right) = M_i\left(-\frac{f_{nadir\ \ max} - \Delta f_{nadir\ i}}{\Delta f_{nadir\ i}}\right) \tag{4.30}$$

To simultaneously satisfy both the frequency nadir and the RoCoF standards, the lower bound for virtual inertia in the optimization problem for each area is selected as the maximum value of (4.26) and (4.30), namely:

$$M_{i,min}^{ESS} = max\left\{M_{i,\min}^{'ESS}, M_{i,\min}^{''ESS}\right\} \tag{4.31}$$

Moreover, the overall system inertia has a direct impact on the frequency indices. This means that some considerations should be made regarding overall system inertia and, consequently (4.26) and (4.30) will be completed by adding a new equality constraint. For this purpose, the frequency of the overall center of inertia (COI), which should satisfy strict frequency standards, can be used to determine the overall amount of inertia in the system as:

$$M_{COI} = Q = M\frac{\Delta f_{COI}}{f_{COI}}(61.5) \tag{4.32}$$

where Δf_{COI} and f_{COI} represent the frequency deviation and frequency of the system, without ESS, after the fault, respectively. Details on how to derive (4.32) are given in *Chapter 3*. Formally, Equation (4.32) gives the required amount of inertia constant which guarantees acceptable frequency dynamics of the COI. Of note that Q would be realized by adding the emulated inertia of ESSs to the conventional SGs inertia. Accordingly, the optimization problem (4.20) can be rewritten as:

$$\underset{M_i^{ESS}}{minimize}\ F\left(M_i^{ESS}\right) = \sum_{i=1}^{n_{ESS}}\left(C_{LCC,a\,i}\frac{M_i^{ESS}\,S_{base}}{3600\ sec}\right) \tag{4.33a}$$

$$st: \qquad M_{COI} = Q \tag{4.33b}$$

$$M_{i,\min}^{ESS} \leq M_i^{ESS} \leq M_{i,\max}^{ESS} \tag{4.33c}$$

$$SOC_{min} \leq SOC_i \leq SOC_{max} \tag{4.33d}$$

where the dynamic inequality constraints (4.20b) and (4.20c) are reformulated as the algebraic inequality constraint (4.33c) in terms of the inertia constant. This increases the simplicity and speed of the calculations.

4.3.1.5 Simulation and Results

Linear System

As the first motivating example, a linearized model of a three-area power system is used to assess the efficiency of the proposed formulation. The block diagram of each area is shown in Figure 4.13.

Firstly, a 0.2 per unit load disturbance is applied in areas 1 and 3. As the first scenario, the required virtual inertia is calculated only based on (4.33b) and arbitrary realized through the model of (4.11) in area 1. For simulation purposes, it is assumed that the ESS would be triggered upon the inception of the fault. Comparison of frequency dynamics for the system with and without virtual inertia reveals that while inertia emulation improves the frequency nadirs of areas 1 and 2, the frequency performance of area 3 deteriorates. This, in turn, numerically justifies the need for optimal inertia placement. Within this framework, (4.33a)–(4.33d) may be written as:

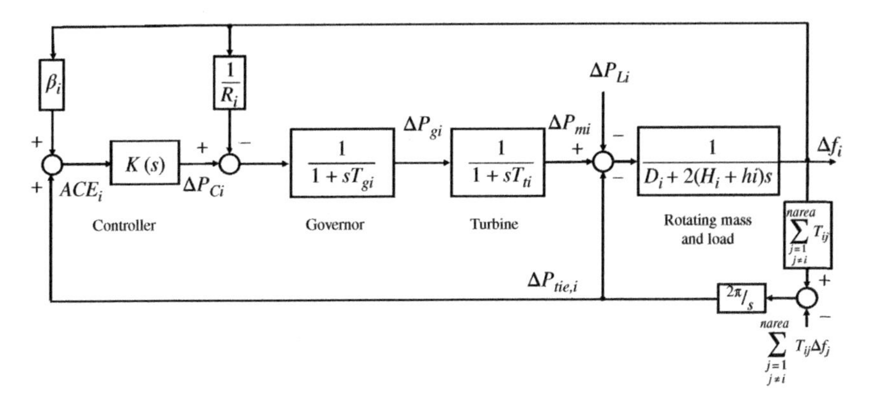

Figure 4.13 Block diagram representation of control area *i*. β_i, R_i, T_{gi}, T_{ti}, and D_i are frequency bias, droop characteristic, governor time constant, turbine time constant, and damping property, respectively.

$$\underset{M_i^{ESS}}{minimize}\ F\left(M_i^{ESS}\right) = \sum_{i=1}^{n_{ESS}} \left(C_{LCC,a\,i} \frac{M_i^{ESS}\, S_{base}}{3600\ sec} \right) \tag{4.34a}$$

$$st: M_{COI} = 0.053 \tag{4.34b}$$

$$0.0129 \le M_1^{ESS} \tag{4.34c}$$

$$0 \le M_2^{ESS} \tag{4.34d}$$

$$0.0225 \le M_3^{ESS} \tag{4.34e}$$

$$30\% \le SOC_i \le 80\% \tag{4.34f}$$

where for instance, the minimum inertia of area 1 in (4.34c) is calculated based on (4.26), (4.30), and (4.31) as:

$$\begin{aligned} M_{1,\min}^{'ESS} &= 0.08335 \times \left(-\frac{1 + 1.1870}{-1.1870} \right) = 0.0129 \\ M_{1,\min}^{''ESS} &= 0 \\ M_{1,\min}^{ESS} &= \max\{0, 0.0129\} = 0.0129 \end{aligned} \tag{4.35}$$

The parameters used in (4.34a) are summarized in Tables 4.1 and 4.2.

The results obtained, using a simple genetic algorithm (GA) with 0.05 and 0.8 mutation and crossover coefficients, respectively, from the optimization of (4.34a) are shown in Table 4.3. To further assess the efficiency of the formulation, Table 4.3 compares the results with those obtained in [30, 73]. Comparison results justify the fact that the dynamic behavior of ESS can significantly affect the optimal placement problem.

Table 4.1 Economical parameters related to the optimization problem [42, 43].

Parameter	Value	Parameter	Value
i (%)	8	$C_{FOM,a}$ (\$/kw-yr)	10
C_{PSC} (\$/kw)	200	$C_{VOM,a}$ (\$/kwh)	5
C_{BOP} (\$/kw)	50	R	2
C_{stor} (\$/kw)	300	t (yr)	6
C_R (\$/kw)	300	η_{sys} (%)	75

Table 4.2 Technical parameters related to the optimization problem [41].

Parameter	Value	Parameter	Value
S_{base} (MVA)	1000	P_{ESS} (MW)	1
$RoCoF_{max}$ (Hz/s)	1	$E_{ESS,\ rated}$ (MVAh)	0.25
$\Delta f_{nadir,max}$ (Hz)	0.8	$\eta_d = \eta_c(\%)$	75
$SOC(0)$	0.5	t (year)	15
SOC_{min}	0.3	t_{ch} (hours)	0.25
SOC_{max}	0.8	n_{cycle}	1000

Table 4.3 Optimization results in three-area system.

Method	M_1^{ESS}	M_2^{ESS}	M_3^{ESS}	$F(M_i^{ESS})$
Formulation (4.33)	0.016	0	0.037	2.5916
Ref. [30]	0.023	0.016	0.022	3.0942
Ref. [73]	0.012	0.009	0.038	2.8735

Further, Figure 4.14 compares the frequency behavior and RoCoF of generators 1, 2, and 3 for three cases of interest: (i) with virtual inertia and according to the formulation of (4.33a)–(4.33d), (ii) with virtual inertia and according to [73], and (iii) with virtual inertia and according to Ref. [30]. It can be seen that while frequency traces of Refs. [30, 73], and the formulation of (4.33a)–(4.33d) meet the RoCoF and frequency nadir standards, (4.33a)–(4.33d) results in a less ESS capacity. Also of interest, modal analysis of the results, as explained in Table 4.4, shows the efficiency of (4.33a)–(4.33d) in comparison with that of Refs. [30, 73].

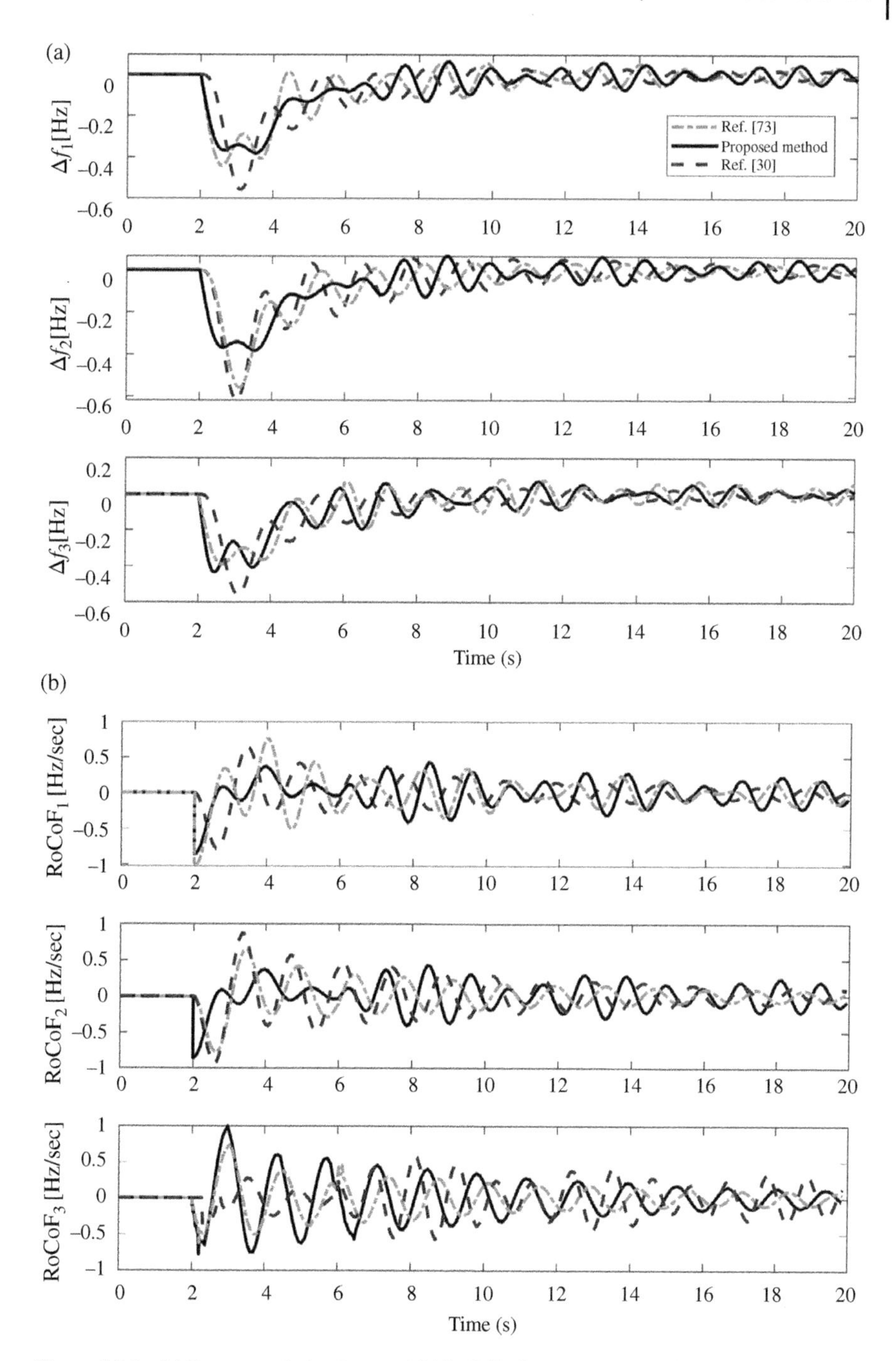

Figure 4.14 (a) Frequency behaviors and (b) RoCoF of generators 1 to 3 of three-area power system for three cases: proposed formulation [30, 73].

Table 4.4 Modal analysis of the systems with virtual inertia.

Method	Proposed method	Ref. [30]	Ref. [73]
Damping	0.2762	0.1265	0.2634

Results show that a lower emulated virtual inertia based on (4.33a)–(4.33d) not only decreases the cost function but also provides better performance in terms of enhanced damping.

Tables 4.3 and 4.4 and Figure 4.14 suggest that with virtual inertia, the results of Refs. [30, 73] seem to fulfill the constraints and are almost the same as the results in (4.33a)–(4.33d). This could be justified through the fact that the set of generator buses for small systems includes a few members to be considered as candidates for ESS installation. Therefore, different algorithms may differ a bit from the capacity point of view rather than the location which in turn causes a negligible difference in the results. To further assess the effects of virtual inertia on the frequency and transient stabilities, two nonlinear systems are used in what follows.

Two-Area Power System In this section, the two-area power system, shown in Figure 4.15, is considered to further demonstrate the efficiency of the optimal placement formulation. Modeling considerations are essentially those described in [74]; all the generating units are modeled with sixth-order synchronous machine models with excitation systems.

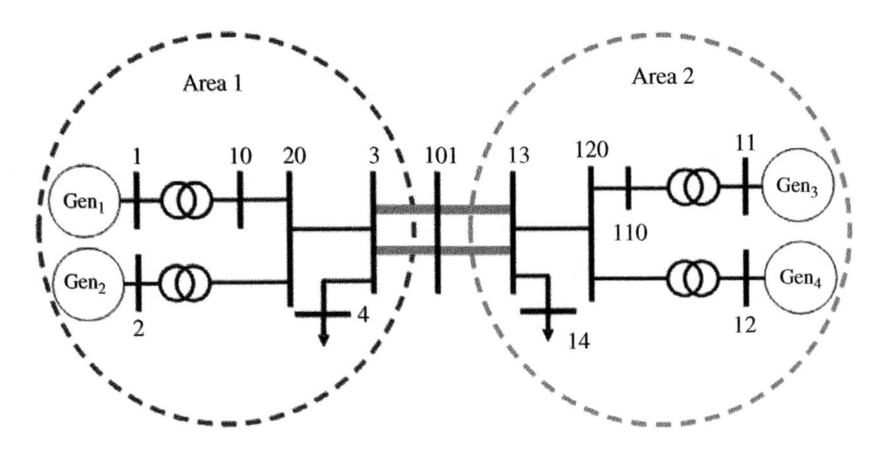

Figure 4.15 Single-line diagram of two-area system.

The test scenario of interest is the outage of generator G_4 in the first second of the simulation. The lower bounds of virtual inertia are calculated according to (4.34a)–(4.34f) and (4.35) as:

$$\underset{M_i^{ESS}}{minimize}\ F\left(M_i^{ESS}\right) = \sum_{i=1}^{n_{ESS}} \left(C_{LCC,a\,i} \frac{M_i^{ESS}\, S_{base}}{3600\ sec} \right) \tag{4.36a}$$

$$st : M_{COI} = 4.973 \tag{4.36b}$$

$$0.3151 \le M_1^{ESS} \tag{4.36c}$$

$$0.3881 \le M_2^{ESS} \tag{4.36d}$$

$$0 \le M_3^{ESS} \tag{4.36e}$$

$$30\% \le SOC_i \le 80\% \tag{4.36f}$$

in which the equality constraint (4.36b) reveals that

$$M_{COI} + M^{ESS} = 4.973 \rightarrow M^{ESS} = 4.973 - 4 = 0.973 \tag{4.37}$$

Solving (4.36) leads to the optimum results of Table 4.5. The results are also compared with those of Refs. [30, 73]. Also of interest, Table 4.6 compares the frequency stability indices for different approaches. The results demonstrate the high efficiency of (4.33a)–(4.33d) to optimally allocate virtual inertia in the system.

The results for the system without virtual inertia suggest that while the frequency nadirs exhibit acceptable performance, the RoCoFs exceed the standard value for some of the generators. The optimal placement of virtual inertia returns the generators with undesired frequency dynamics to the normal region. The efficiency of (4.33a)–(4.33d) is further assessed by time-domain (T-D) simulations of Figure 4.16.

The effect of emulated virtual inertia on transient stability can be assessed using a simple power angle-based stability index η [46]:

$$\eta = \frac{360 - \delta_{max}}{360 + \delta_{max}} \tag{4.38}$$

Table 4.5 Optimization results in two-area system.

Method	M_1^{ESS}	M_2^{ESS}	M_3^{ESS}	$F(M_i^{ESS})$
Proposed method	0.402	0.567	0.004	4.6705
Ref. [30]	0.116	0.332	0.376	6.3104
Ref. [73]	0.212	0.315	0.316	5.7891

Table 4.6 Frequency indicators of two-area system before and after the application of optimal inertia values.

	Without VI		With VI Proposed method		With VI [30]		With VI [73]	
G_i	RoCoF (Hz/s)	Δf_{nadir} (Hz)	RoCoF (Hz/s)	Δf_{nadir} (Hz)	RoCoF (Hz/s)	Δf_{nadir} (Hz)	RoCoF (Hz/s)	Δf_{nadir} (Hz)
1	1.187	0.172	0.989	0.143	0.988	0.126	0.973	0.116
2	1.240	0.160	0.989	0.134	0.981	0.112	0.961	0.162
3	0.706	0.267	0.713	0.254	0.730	0.200	0.786	0.198

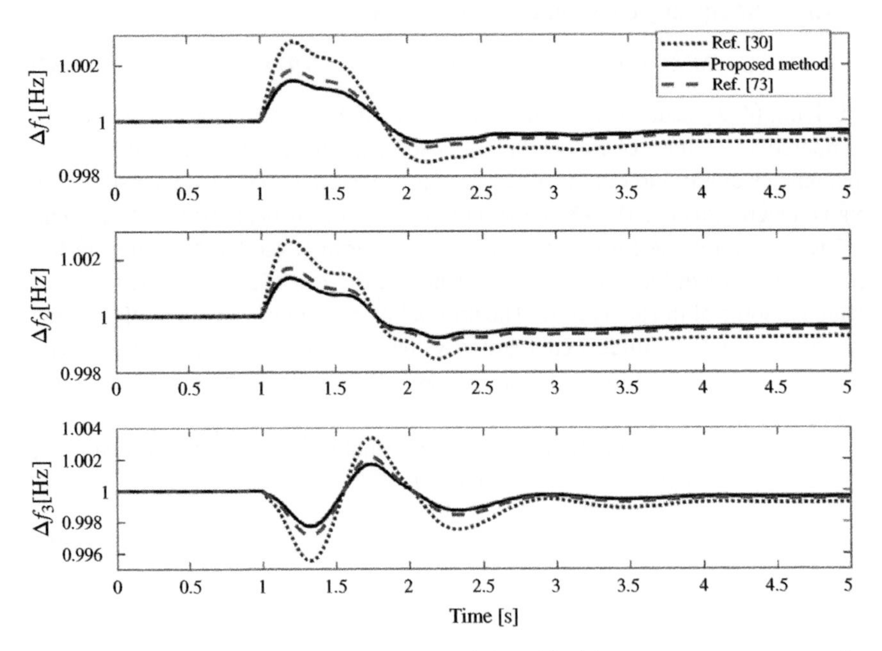

Figure 4.16 Frequency response of generators 1, 2, and 3 of the two-area power system for three cases: proposed formulation [30, 73].

where δ_{max} is the maximum angle separation of any two generators in the system. It is noteworthy that during severe faults, most ESSs, if remain connected and continue to inject active power, get saturated and cannot follow the frequency properly. It is noted here that (4.38) relies on a conservative assumption that the fault is not a severe one, which causes ESSs to be disconnected from the grid. Moreover,

Table 4.7 Transients stability assessment.

Method	δ_{max}	η
Proposed method	11.02	0.94
Ref. [30]	20.8	0.89
Ref. [73]	16.63	0.91

the saturation of ESSs in response to sever faults is neglected. Table 4.7 demonstrates the better performance of the proposed method in comparison with those of Refs. [30, 73].

New York New England System The NYNE test system is used to further illustrate the efficiency of the proposed algorithm for large-scale power systems. A single-line diagram of the system, showing major coherent areas and their interconnections, is shown in Figure 4.17.

Five different contingency scenarios, including tripping of major generating units and load rejection, are considered. Table 4.8 compares the results of the optimization problem, given by (4.33a)–(4.33d), with those of Refs. [30] and

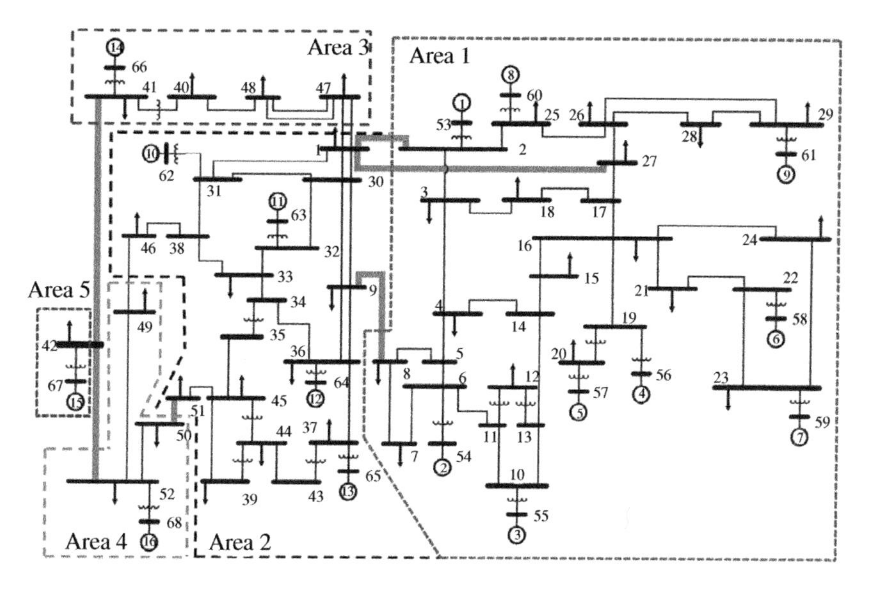

Figure 4.17 Single-line diagram of NYNE test system showing coherent areas and their interconnections.

Table 4.8 Optimization results in NYNE test system.

		M_1^{ESS}	M_2^{ESS}	M_3^{ESS}	M_4^{ESS}	M_5^{ESS}	$F(M_i^{ESS})$
1	Proposed method	0.412	0.432	0.313	0.092	0.111	9.2141
	Ref. [30]	0.506	0.332	0.376	0.201	0.112	12.4031
	Ref. [73]	0.378	0.453	0.306	0.115	0.098	11.0817
2	Proposed method	0.341	0.513	0.209	0.101	0.098	10.8601
	Ref. [30]	0.340	0.712	0.301	0.113	0.160	13.0012
	Ref. [73]	0.300	0.798	0.251	0.098	0.161	12.3140
3	Proposed method	0.474	0.261	0.160	0.261	0.007	9.7516
	Ref. [30]	0.596	0.298	0.267	0.271	0.088	11.2113
	Ref. [73]	0.314	0.351	0.294	0.314	~0	10.0087
4	Proposed method	0.169	0.203	0.617	0.135	0.072	9.8617
	Ref. [30]	0.132	0.512	0.694	0.196	0.209	10.5103
	Ref. [73]	0.100	0.374	0.687	0.096	0.101	10.0102
5	Proposed method	0.613	0.032	0.116	0.076	~0	9.0412
	Ref. [30]	0.743	0.215	0.402	0.031	~0	9.9731
	Ref. [73]	0.412	0.354	0.391	0.116	~0	9.4019

[73]. The results demonstrate the efficiency of (4.33a)–(4.33d) to enhance the frequency dynamics with minimum cost. Furthermore, Figure 4.18 shows the allocation of virtual inertia among the PV buses of the system.

Results show that the proposed optimization scheme works better and more efficiently for larger areas. This can be understood by noting that the set of generator buses for large areas includes many members to be considered as candidates for the ESS placement. As a result, there are many possibilities for placing of the ESSs.

Figure 4.19 compares the frequency dynamics of the system, in response to the outage of generator 1 (Scenario 1), for (4.33a)–(4.33d), and Refs. [30, 73]. It should be noted that while there are negligible deviations between the traces in Figure 4.19, significant differences between the cost functions justify the efficiency of (4.33a)–(4.33d).

Also of interest, the efficiency of (4.33a)–(4.33d) to enhance transient stability is shown in Table 4.9, using (4.38). Table 4.9 shows that the appropriate placement of virtual inertia in the system, considering the dynamical behavior of ESS, could also improve transient stability. This could be justified noting that recent research, for example in Ref. [30], relies on the quasi-steady-state phasors for voltages and currents in transient stability assessment. In other words, they consider constant

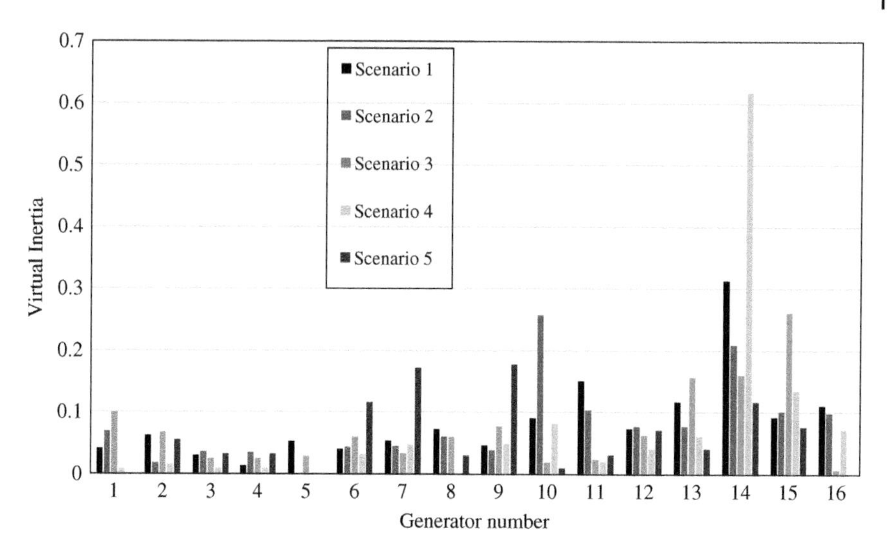

Figure 4.18 Virtual inertia allocation for NYNE test system.

Table 4.9 Optimization results in New England system.

	Method	η_{12}	η_{13}	η_{14}	η_{15}	η_{25}	η_{35}
1	Proposed method	0.92	0.87	0.91	0.89	0.90	0.89
	Ref. [30]	0.85	0.84	0.91	0.90	0.85	0.87
	Ref. [73]	0.87	0.86	0.91	0.89	0.88	0.88
2	Proposed method	0.98	0.83	0.92	0.93	0.95	0.95
	Ref. [73]	0.94	0.82	0.91	0.90	0.91	0.92
	Ref. [30]	0.91	0.84	0.90	0.88	0.89	0.95
3	Proposed method	0.87	0.90	0.90	0.89	0.92	0.92
	Ref. [73]	0.85	0.88	0.90	0.88	0.88	0.88
	Ref. [30]	0.88	0.85	0.87	0.87	0.87	0.88
4	Proposed method	0.96	0.93	0. 93	0.93	0.93	0.93
	Ref. [73]	0.96	0.90	0.91	0.93	0.92	0.89
	Ref. [30]	0.95	0.94	0.93	0.93	0.91	0.85
5	Proposed method	0.94	0.85	0.92	0.92	0.91	0.96
	Ref. [73]	0.89	0.86	0.90	0.90	0.89	0.94
	Ref. [30]	0.93	0.87	0.87	0.87	0.88	0.91

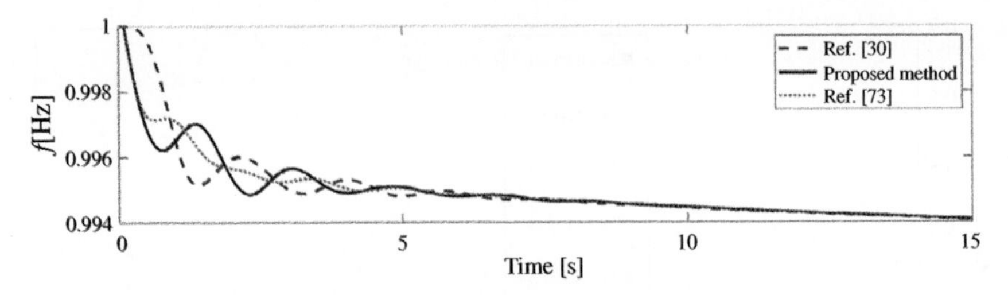

Figure 4.19 Frequency responses of NYNE test system for different approaches.

(nominal) frequency in defining the system impedances which is far from realistic for systems with high penetration of inverter-based ESSs. This point is successfully addressed in (4.33a)–(4.33d) by explicitly representing the dynamic behavior of ESSs in the problem formulation.

4.3.1.6 Sensitivity Analysis

In this section, sensitivity analyses are conducted to understand the effect of operation conditions, including variations of fault magnitudes, operating point, and annualized LCC on the optimization problem. For this purpose, (4.33a) is used to calculate the sensitivity of the cost function to operation condition as:

$$\begin{aligned} F\left(M_i^{ESS}\right) + \Delta F\left(M_i^{ESS}\right) &= \sum_{i=1}^{n_{ESS}} \left(C_{LCC,a\,i} \frac{M_i^{ESS}\, S_{base}}{3600\ sec} \right) + \frac{\partial F\left(M_i^{ESS}\right)}{\partial \Delta P_L} \Delta\Delta P_L \\ &+ \frac{\partial F\left(M_i^{ESS}\right)}{\partial \Delta P_m} \Delta\Delta P_m + \frac{\partial F\left(M_i^{ESS}\right)}{\partial C_{LCC,a\,i}} \Delta C_{LCC,a\,i} \end{aligned} \tag{4.39}$$

According to (4.24) and (4.28), one could write the sensitivity matrix as:

$$\begin{bmatrix} \partial M_{i,\min}^{\prime ESS} \\ \partial M_{i,\min}^{\prime\prime ESS} \end{bmatrix} = \begin{bmatrix} \dfrac{\partial RoCoF_i}{\partial \Delta P_{Li}} & \dfrac{\partial RoCoF_i}{\partial \Delta P_{mi}} \\ \dfrac{\partial \Delta f_{nadir,i}}{\partial \Delta P_{Li}} & 0 \end{bmatrix} \begin{bmatrix} \Delta\Delta P_{Li} \\ \Delta\Delta P_{mi} \end{bmatrix} \tag{4.40}$$

Substituting (4.40) in (4.39) gives

$$\begin{aligned} \Delta F\left(M_i^{ESS}\right) &= \frac{\partial F\left(M_i^{ESS}\right)}{\partial M_i^{ESS}} \frac{\partial M_i^{ESS}}{\partial \Delta P_L} \Delta\Delta P_L + \frac{\partial F\left(M_i^{ESS}\right)}{\partial M_i^{ESS}} \frac{\partial M_i^{ESS}}{\partial \Delta P_m} \Delta\Delta P_m \\ &= \sum_{i=1}^{n_{ESS}} \left(C_{LCC,a\,i} \frac{S_{base}}{3600\ sec} \right) \\ &\times \max \left\{ \frac{\partial RoCoF_i}{\partial \Delta P_{Li}} \Delta\Delta P_L + \frac{\partial RoCoF_i}{\partial \Delta P_{mi}} \Delta\Delta P_m, \frac{\partial \Delta F_{nadir,i}}{\partial \Delta P_{Li}} \Delta\Delta P_{Li} \right\} \\ &+ \sum_{i=1}^{n_{ESS}} \left(\frac{M_i^{ESS} S_{base}}{3600\ sec} \right) \Delta C_{LCC,a\,i} \end{aligned} \tag{4.41}$$

Table 4.10 Sensitivity analysis.

Scenario	Cost function of (4.33a)–(4.33d)	Cost function of (4.41)
2	9.7516	9.5913
3	9.8617	9.9302

The effectiveness of (4.41) is now assessed for the NYNE system. For this purpose, the outage of generator 1 in area 1, i.e. Scenario 1, is considered as the base case for sensitivity analysis. The cost functions for the outage of generator 7 in area 1, i.e. Scenario 2, and generator 11 in area 2, i.e. Scenario 3, are calculated using (4.41). Table 4.10 compares the exact results of (4.33a)–(4.33d) with those of (4.41) which justify the effectiveness of the sensitivity analysis (4.41).

For uncertainty analysis, the equality constraint (4.33b) is represented in the objective function (4.33a) as:

$$\underset{M_i^{ESS}}{minimize}\ F\left(M_i^{ESS}\right) = \sum_{i=1}^{n_{ESS}} \left(C_{LCC,a\,i} \frac{M_i^{ESS} S_{base}}{3600\ sec} \right) + \beta(M_{COI} - Q) \tag{4.42}$$

where β is arbitrary chosen high to enforce the results to follow the equality constraint (4.33b). Considering parametric uncertainty for the inertia constant M_{COI}, one could write (4.42) as:

$$\begin{aligned} \underset{M_i^{ESS}}{minimize}\ F\left(M_i^{ESS}\right) &= \sum_{i=1}^{n_{ESS}} \left(C_{LCC,a\,i} \frac{M_i^{ESS} S_{base}}{3600\ sec} \right) + \beta(M_{COI} + \gamma - Q) \rightarrow \\ \underset{M_i^{ESS}}{minimize}\ F\left(M_i^{ESS}\right) &= \sum_{i=1}^{n_{ESS}} \left(C_{LCC,a\,i} \frac{M_i^{ESS} S_{base}}{3600\ sec} \right) + \beta(M_{COI} - Q) + \beta\gamma \end{aligned} \tag{4.43}$$

where γ is expressed as a percentage of M_{COI}. To deal with uncertainty analysis, a simple interval approach is utilized. This approach assumes that the uncertain parameters take value in a specified interval. It could be reinterpreted as the probabilistic modeling with a uniform probability density function (PDF). In this method, the upper and lower bounds of the uncertain inertia parameter are defined. The aim is to find the lower and upper bounds of the objective function [75]. Using the proposed framework, assume that the maximum variation of the inertia constant is limited to $\pm 5\%$. This means that the interval of interest can be defined as:

$$\gamma = [M_{COI} - 0.05(M_{COI}), M_{COI} + 0.05(M_{COI})] \tag{4.44}$$

which, in turn, requires that $F(M_i^{ESS})$ takes the values:

$$F\left(M_i^{ESS}\right) = [3.8298, 5.5112] \tag{4.45}$$

with a uniform PDF.

4.3.2 Small-Signal Stability

4.3.2.1 Objective Function

The optimal placement of ESS to mitigate undesired frequency dynamics in low-inertia power grids is discussed in Section 4.3.1. In this section, same formulation would be derived from small-signal stability point of view.

The conducted analysis in Chapter 3 indicates that the damping ratio of critical modes of system are detrimentally affected by the increasing of renewable sources penetration, i.e. decreasing inertia. Critical modes of a system are those within the frequency range of 0.01–2 Hz and damping of less than 10%.

To optimally place ESSs in the system according to small-signal stability considerations, the same formulation as in (4.33a)–(4.33d) with some modifications given in (4.46a)–(4.46c) is utilized:

$$\underset{M_i^{ESS}}{minimize}\ F\left(M_i^{ESS}\right) = \sum_{i=1}^{n_{ESS}} \left(C_{LCC,a\,i} \frac{M_i^{ESS}\, S_{base}}{3600\ sec}\right) \tag{4.46a}$$

$$st : M_{COI}^{ss} = K \tag{4.46b}$$

$$SOC_{min} \leq SOC_i \leq SOC_{max} \tag{4.46c}$$

where superscript "ss" stands for small signal. Of note that, M_{COI}^{ss} in (4.46b) includes two terms: (i) overall system inertia (M) and (ii) provided virtual inertia by ESS (M^{ESS}). In (4.46b), K is determined on which the critical mode of interest be characterized by the minimum damping ratio of 10%. For this purpose, a relationship between the inertia and the damping ratio of the desired mode would be derived. In this way, the trend of changing the damping ratio in response to the variation of the inertia is used as the input for the curve fitting tool. Accordingly, one could write

$$\varsigma_i = f(M) \tag{4.47}$$

By setting the left-hand side of (4.47) to 10%, the required inertia, i.e. K in (4.46b), which guarantees the desired damping ratio would be calculated.

To further proceed with the derivation of the objective function, the equality constraint (4.46b) is represented in the objective function (4.46a) as:

$$\underset{M_i^{ESS}}{minimize}\, F\left(M_i^{ESS}\right) = \sum_{i=1}^{n_{ESS}} \left(C_{LCC,a\,i} \frac{M_i^{ESS} S_{base}}{3600\ sec} \right) + \alpha\left(M_{COI}^{ss} - K\right) \tag{4.48}$$

where α is arbitrary chosen high to enforce the results to follow the equality constraint (4.46b). On the other hand, as the damping ratio of a specific mode is influenced by the generating units according to their participation factors, it seems that the participation factors should be reflected in the objective function (4.48). More precisely, the machines with higher participation factors in the studied mode should contribute to the minimization problem (4.48) with lower weights. In this way, one could rewrite (4.48) as:

$$\underset{M_i^{ESS}}{minimize}\, F\left(M_i^{ESS}\right) = \sum_{i=1}^{n_{ESS}} \left(C_{LCC,a\,i} \frac{M_i^{ESS} S_{base}}{3600\ sec} \right) + \alpha \left(M + \sum_i \frac{1}{pf_i} M_i^{ESS} - K \right) \tag{4.49}$$

where pf_i is the participation factor of generator i in the critical mode of interest.

4.3.2.2 Simulation Results

The efficiency of (4.49) is examined on the NYNE benchmark of Figure 4.17. Table 4.11 reports the eigenvalues of the system with no MGs penetration level (details on how to obtain the results are given in Chapter 3). This analysis is performed using the small-signal analysis toolbox (SSAT) which is a part of the DSA Tools software package.

Simulation results reveal that by penetration of 52% MGs in the system, the damping ratio of the third mode, as the worst mode detrimentally affected, reaches to 0.42%. Therefore, the main aim of (4.49) is to allocate ESSs in the generator busses of the system on which the damping ratio increases to 10%. To deal with (4.49), the derivation of participation factors is mandatory. The participation factors of various machines participating in the third studied mode, i.e. $-0.3590 + 3.7108j$, are presented in Figure 4.20. As seen from the figure, the generator number 15 has the highest participation factor in this mode.

Table 4.11 Critical modes of the base system.

Mode	Real part	Imaginary part	Frequency	Damping ratio
1	−0.7018	1.9710j	0.3137	33.543
2	−0.1184	3.2666j	0.5199	3.6236
3	−0.3590	3.7108j	0.5906	9.6290
4	−0.1657	4.8915j	0.7785	3.3851

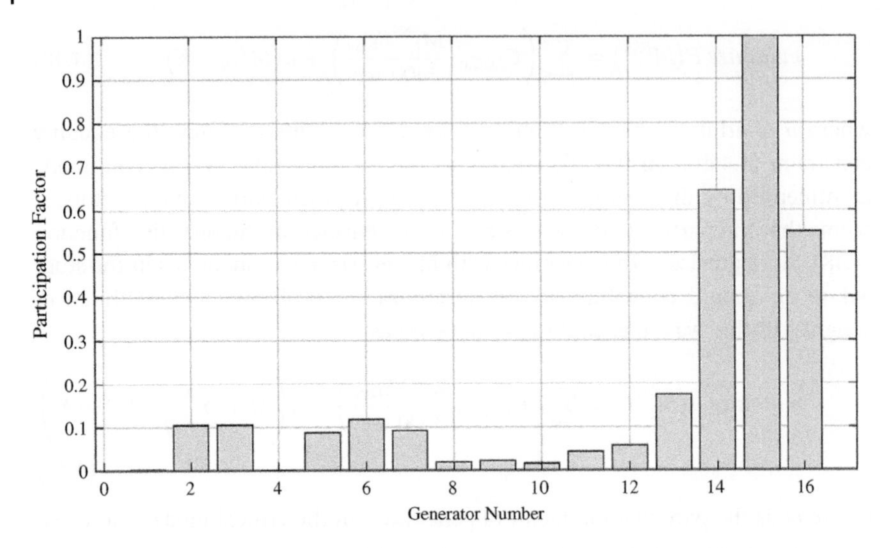

Figure 4.20 Participation factor of the machines participating in the critical mode detrimentally affected by high MGs penetration level.

According to (4.49) and Figure 4.20, it is expected that buses related to generators number 15 and 16 should require a larger ESSs capacity. Figure 4.21 shows the allocation of virtual inertia among the generator buses of the system which in turn justify the effects of the participation factor on the considered objective function.

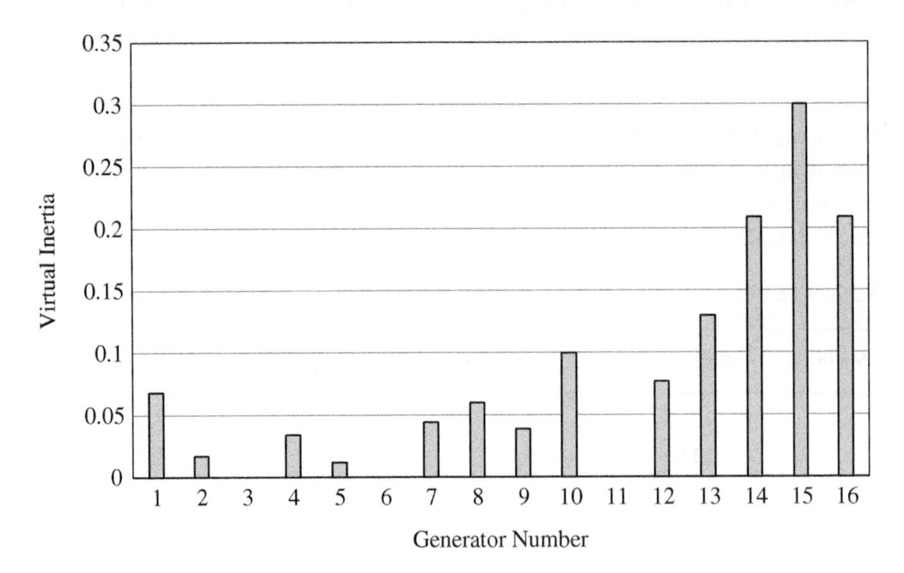

Figure 4.21 Virtual inertia allocation for NYNE test system.

4.4 Summary

The advanced control schemes in modern power grids rely on inertia manipulation in the system to enhance stability. The successful development of advanced control schemes highly depends on the adequacy of the inertia provision sources. In the present chapter, first, the VSG concept and its applications in power grids are explained. Afterward, dispatchable inertia is optimally placed in the system according to dynamical metrics to enhance system stability and dynamic performance.

References

1. Bevrani, H., François, B., and Ise, T. (2017). *Microgrid Dynamics and Control.* New York: Wiley.
2. Bevrani, H., Ise, T., and Miura, Y. (2014). Virtual synchronous generators: a survey and new perspectives. *International Journal of Electrical Power & Energy Systems* **54**: 244–254.
3. Liu, J., Miura, Y., Bevrani, H., and Ise, T. (2016). Enhanced virtual synchronous generator control for parallel inverters in microgrids. *IEEE Transactions on Smart Grid* **8** (5): 2268–2277.
4. D'Arco, S. and Suul, J.A. (2013). Virtual synchronous machines—classification of implementations and analysis of equivalence to droop controllers for microgrids. In: *Proceedings of the 2013 IEEE Grenoble Conference*, 1–7. IEEE.
5. Cao, Y. et al. (2017). A virtual synchronous generator control strategy for VSC-MTDC systems. *IEEE Transactions on Energy Conversion* **33** (2): 750–761.
6. Al-Tameemi, M., Liu, J., Bevrani, H., and Ise, T. (2020). A dual VSG-based M3C control scheme for frequency regulation support of a remote AC grid via low-frequency AC transmission system. *IEEE Access* **8**: 66085–66094.
7. Hlaing, H.S., Liu, J., Miura, Y. et al. (2019). Enhanced performance of a stand-alone gas-engine generator using virtual synchronous generator and energy storage system. *IEEE Access* **7**: 176960–176970.
8. Mishra, S., Pullaguram, D., Buragappu, S.A., and Ramasubramanian, D. (2016). Single-phase synchronverter for a grid-connected roof top photovoltaic system. *IET Renewable Power Generation* **10** (8): 1187–1194.
9. Ma, Y., Cao, W., Yang, L. et al. (2017). Virtual synchronous generator control of full converter wind turbines with short-term energy storage. *IEEE Transactions on Industrial Electronics* **64** (11): 8821–8831.
10. Wang, Y., Meng, J., Zhang, X., and Xu, L. (2015). Control of PMSG-based wind turbines for system inertial response and power oscillation damping. *IEEE Transactions on Sustainable Energy* **6** (2): 565–574.

11. Li, Y., Xu, Z., and Wong, K.P. (2016). Advanced control strategies of PMSG-based wind turbines for system inertia support. *IEEE Transactions on Power Systems* **32** (4): 3027–3037.
12. Lopes, L.A. (2014). Self-tuning virtual synchronous machine: a control strategy for energy storage systems to support dynamic frequency control. *IEEE Transactions on Energy Conversion* **29** (4): 833–840.
13. Suul, J.A., D'Arco, S., and Guidi, G. (2016). Virtual synchronous machine-based control of a single-phase bi-directional battery charger for providing vehicle-to-grid services. *IEEE Transactions on Industry Applications* **52** (4): 3234–3244.
14. Guan, M., Pan, W., Zhang, J. et al. (2015). Synchronous generator emulation control strategy for voltage source converter (VSC) stations. *IEEE Transactions on Power Systems* **30** (6): 3093–3101.
15. Aouini, R., Marinescu, B., Kilani, K.B., and Elleuch, M. (2015). Synchronverter-based emulation and control of HVDC transmission. *IEEE Transactions on Power Systems* **31** (1): 278–286.
16. Dong, S., Chi, Y., and Li, Y. (2015). Active voltage feedback control for hybrid multiterminal HVDC system adopting improved synchronverters. *IEEE Transactions on Power Delivery* **31** (2): 445–455.
17. Li, C., Li, Y., Cao, Y. et al. (2018). Virtual synchronous generator control for damping DC-side resonance of VSC-MTDC system. *IEEE Journal of Emerging and Selected Topics in Power Electronics* **6** (3): 1054–1064.
18. Zhong, Q.-C. (2016). Virtual synchronous machines: a unified interface for grid integration. *IEEE Power Electronics Magazine* **3** (4): 18–27.
19. Bevrani, H. (2009). *Robust Power System Frequency Control*. Springer.
20. Fathi, A., Shafiee, Q., and Bevrani, H. (2018). Robust frequency control of microgrids using an extended virtual synchronous generator. *IEEE Transactions on Power Systems* **33** (6): 6289–6297.
21. Jongudomkarn, J., Liu, J., Yanagisawa, Y. et al. (2020). Model predictive control for indirect boost matrix converter based on virtual synchronous generator. *IEEE Access* **8**: 60364–60381.
22. Terazono, D., Liu, J., Miura, Y. et al. (2020). Grid frequency regulation support from back-to-back motor drive system with virtual-synchronous-generator-based coordinated control. *IEEE Transactions on Power Electronics*: 1. https://doi.org/10.1109/TPEL.2020.3015806.
23. Demetriou, P., Asprou, M., Quiros-Tortos, J., and Kyriakides, E. (2017). IEEE 9-bus modified test system. https://www2.kios.ucy.ac.cy/testsystems/index.php/ieee-9-bus-modified-test-system/ (accessed November 2020).
24. Bevrani, H., Imanaka, M., and Kato, T. (2020). Grid dynamics shaping using flexible controlled power converters. *Energy Reports* **6**: 1490–1495.

25. Madariaga, A., Martín, J., Zamora, I. et al. (2013). Technological trends in electric topologies for offshore wind power plants. *Renewable and Sustainable Energy Reviews* **24**: 32–44.
26. Rodrigues, S., Restrepo, C., Kontos, E. et al. (2015). Trends of offshore wind projects. *Renewable and Sustainable Energy Reviews* **49**: 1114–1135.
27. Castro, L.M. and Acha, E. (2015). On the provision of frequency regulation in low inertia AC grids using HVDC systems. *IEEE Transactions on Smart Grid* **7** (6): 2680–2690.
28. Leon, A.E. (2017). Short-term frequency regulation and inertia emulation using an MMC-based MTDC system. *IEEE Transactions on Power Systems* **33** (3): 2854–2863.
29. Rakhshani, E., Remon, D., Cantarellas, A.M. et al. (2016). Virtual synchronous power strategy for multiple HVDC interconnections of multi-area AGC power systems. *IEEE Transactions on Power Systems* **32** (3): 1665–1677.
30. Poolla, B.K., Bolognani, S., and Dörfler, F. (2017). Optimal placement of virtual inertia in power grids. *IEEE Transactions on Automatic Control* **62** (12): 6209–6220.
31. Golpîra, H. and Bevrani, H. (2019). Microgrids impact on power system frequency response. *Energy Procedia* **156**: 417–424.
32. Lian, B., Sims, A., Yu, D. et al. (2016). Optimizing $LiFePO_4$ battery energy storage systems for frequency response in the UK system. *IEEE Transactions on Sustainable Energy* **8** (1): 385–394.
33. Wu, Z., Gao, D.W., Zhang, H. et al. (2017). Coordinated control strategy of battery energy storage system and PMSG-WTG to enhance system frequency regulation capability. *IEEE Transactions on Sustainable Energy* **8** (3): 1330–1343.
34. Zhang, F., Hu, Z., Xie, X. et al. (2017). Assessment of the effectiveness of energy storage resources in the frequency regulation of a single-area power system. *IEEE Transactions on Power Systems* **32** (5): 3373–3380.
35. Ahmadyar, A.S., Riaz, S., Verbič, G. et al. (2018). A framework for assessing renewable integration limits with respect to frequency performance. *IEEE Transactions on Power Systems* **33** (4): 4444–4453.
36. Golpîra, H., Seifi, H., Messina, A.R., and Haghifam, M.-R. (2016). Maximum penetration level of micro-grids in large-scale power systems: Frequency stability viewpoint. *IEEE Transactions on Power Systems* **31** (6): 5163–5171.
37. Spahic, E., Varma, D., Beck, G. et al. (2016). Impact of reduced system inertia on stable power system operation and an overview of possible solutions. In: *Proceedings of 2016 IEEE Power and Energy Society General Meeting (PESGM)*, 1–5. New York: IEEE.
38. Ulbig, A., Borsche, T.S., and Andersson, G. (2014). Impact of low rotational inertia on power system stability and operation. *IFAC Proceedings Volumes* **47** (3): 7290–7297.

39. Wang, Y., Bayem, H., Giralt-Devant, M. et al. (2014). Methods for assessing available wind primary power reserve. *IEEE Transactions on Sustainable Energy* **6** (1): 272–280.
40. Wang, Y., Delille, G., Bayem, H. et al. (2013). High wind power penetration in isolated power systems—assessment of wind inertial and primary frequency responses. *IEEE Transactions on Power Systems* **28** (3): 2412–2420.
41. Chu, Z., Markovic, U., Hug, G., and Teng, F. (2020). Towards optimal system scheduling with synthetic inertia provision from wind turbines. *IEEE Transactions on Power Systems* **99**: 1-1.
42. Golpîra, H., Haghifam, M.R., and Seifi, H. (2015). Dynamic power system equivalence considering distributed energy resources using Prony analysis. *International Transactions on Electrical Energy Systems* **25** (8): 1539–1551.
43. Golpîra, H., Messina, A.R., and Bevrani, H. (2019). Emulation of virtual inertia to accommodate higher penetration levels of distributed generation in power grids. *IEEE Transactions on Power Systems* **34** (5): 3384–3394.
44. D'Arco, S. and Suul, J.A. (2014). Equivalence of virtual synchronous machines and frequency-droops for converter-based microgrids. *IEEE Transactions on Smart Grid* **5** (1): 394–395.
45. Rakhshani, E., Remon, D., Cantarellas, A.M., and Rodriguez, P. (2016). Analysis of derivative control based virtual inertia in multi-area high-voltage direct current interconnected power systems. *IET Generation, Transmission & Distribution* **10** (6): 1458–1469.
46. Hammad, E., Farraj, A., and Kundur, D. (2019). On effective virtual inertia of storage-based distributed control for transient stability. *IEEE Transactions on Smart Grid* **10** (1): 327–336.
47. Farmer, W.J. and Rix, A. (2019). Optimising power system frequency stability using virtual inertia from inverter-based renewable energy generation. In: *2019 International Conference on Clean Electrical Power (ICCEP)*, 394–404. IEEE.
48. Attya, A., Anaya-Lara, O., and Leithead, W. (2018). Novel concept of renewables association with synchronous generation for enhancing the provision of ancillary services. *Applied Energy* **229**: 1035–1047.
49. Poolla, B.K., Groß, D., and Dörfler, F. (2019). Placement and implementation of grid-forming and grid-following virtual inertia and fast frequency response. *IEEE Transactions on Power Systems* **34** (4): 3035–3046.
50. Esmaili, M., Ghamsari-Yazdel, M., Amjady, N., and Chung, C.Y. (2020). Convex model for controlled islanding in transmission expansion planning to improve frequency stability. *IEEE Transactions on Power Systems* https://doi.org/10.1109/TPWRS.2020.3009435.
51. Borsche, T.S., Liu, T., and Hill, D.J. (2015). Effects of rotational inertia on power system damping and frequency transients. In: *2015 54th IEEE Conference on Decision and Control (CDC)*, 5940–5946. IEEE.

52. Fini, M.H. and Golshan, M.E.H. (2018). Determining optimal virtual inertia and frequency control parameters to preserve the frequency stability in islanded microgrids with high penetration of renewables. *Electric Power Systems Research* **154**: 13–22.

53. Oudalov, A., Chartouni, D., and Ohler, C. (2007). Optimizing a battery energy storage system for primary frequency control. *IEEE Transactions on Power Systems* **22** (3): 1259–1266.

54. Wogrin, S. and Gayme, D.F. (2015). Optimizing storage siting, sizing, and technology portfolios in transmission-constrained networks. *IEEE Transactions on Power Systems* **30** (6): 3304–3313.

55. Mo, O., D'Arco, S., and Suul, J.A. (2016). Evaluation of virtual synchronous machines with dynamic or quasi-stationary machine models. *IEEE Transactions on Industrial Electronics* **64** (7): 5952–5962.

56. Golpîra, H., Atarodi, A., Amini, S. et al. (2020). Optimal energy storage system-based virtual inertia placement: a frequency stability point of view. *IEEE Transactions on Power Systems* **35** (6): 4824–4835.

57. Golpîra, H., Seifi, H., and Haghifam, M.R. (2015). Dynamic equivalencing of an active distribution network for large-scale power system frequency stability studies. *IET Generation, Transmission & Distribution* **9** (15): 2245–2254.

58. Ajala, O., Dominguez-Garcia, A., Sauer, P., and Liberzon, D. (2018). A library of second-order models for synchronous machines. arXiv preprint arXiv:1803.09707.

59. Bollen, M.H. and Gu, I.Y. (2006). *Signal Processing of Power Quality Disturbances*. Wiley.

60. Manolakis, D.G., Ingle, V.K., and Kogon, S.M. (2000). *Statistical and Adaptive Signal Processing: Spectral Estimation, Signal Modeling, Adaptive Filtering, and Array Processing*. Boston: McGraw-Hill.

61. Zakeri, B. and Syri, S. (2015). Electrical energy storage systems: a comparative life cycle cost analysis. *Renewable and Sustainable Energy Reviews* **42**: 569–596.

62. Mongird, K., Viswanathan, V.V., Balducci, P.J. et al. (2019). Energy Storage Technology and Cost Characterization Report. Pacific Northwest National Lab (PNNL), Richland, WA (United States).

63. Teruo, I. (2005). State of charge calculation device and state of charge calculation method. US Patents, US6845332B2.

64. Tang, Z.X. and Lim, Y.S. (2016). Frequency regulation mechanism of energy storage system for the power grid. 4th IET Clean Energy and Technology Conference (CEAT 2016), Kuala Lumpur, 1–8. doi: https://doi.org/10.1049/cp.2016.1272.

65. European Network of Transmission System Operators for Electricity (ENTSOE) (2016). *Frequency Stability Evaluation Criteria for the Synchronous Zone of Continental Europe – Requirements and Impacting Factors*, March.

66. Eto, J.H. (2011). *Use of Frequency Response Metrics to Assess the Planning and Operating Requirements for Reliable Integration of Variable Renewable Generation.* Berkeley, CA: Ernest Orlando Lawrence Berkeley National Laboratory Tech. Rep. LBNL-4142E.

67. Golpîra, H. (2019). Bulk power system frequency stability assessment in presence of microgrids. *Electric Power Systems Research* **174**: 105863.

68. Ekwue, A. and Cory, B. (1984). Transmission system expansion planning by interactive methods. *IEEE Transactions on Power Apparatus and Systems* **7**: 1583–1591.

69. C. E. O. H. ENTSO-E (2009). P1-Policy 1: Load-Frequency Control and Performance, ed: Tech. Rep. 2000-130-003, May 2000.

70. Chávez, H., Baldick, R., and Sharma, S. (2014). Governor rate-constrained OPF for primary frequency control adequacy. *IEEE Transactions on Power Systems* **29** (3): 1473–1480.

71. Teng, F., Trovato, V., and Strbac, G. (2016). Stochastic scheduling with inertia-dependent fast frequency response requirements. *IEEE Transactions on Power Systems* **31** (2): 1557–1566.

72. Grid, N. (2019) Security and quality of supply standards. https://www.nationalgrideso.com/industry-information/codes/security-and-quality-supply-standards (accessed November 2020).

73. Borsche, T. and Dörfler, F. (2017). On placement of synthetic inertia with explicit time-domain constraints. *arXiv preprint arXiv:1705.03244.*

74. Rogers, G. (2012). *Power System Oscillations.* Berlin: Springer Science & Business Media.

75. Aien, M., Hajebrahimi, A., and Fotuhi-Firuzabad, M. (2016). A comprehensive review on uncertainty modeling techniques in power system studies. *Renewable and Sustainable Energy Reviews* **57**: 1077–1089.

5

Wide-Area Voltage Monitoring in High-Renewable Integrated Power Systems

The development of synchronized phasor measurement units (PMUs) technology has given utilities the ability to implement wide-area voltage stability monitoring systems, which provide time-stamped data in near real time. In this sense, selected voltage control areas can be used to implement wide-area voltage–volt–ampere reactive (VAR) control schemes, develop voltage monitoring and prediction schemes, monitor reactive power reserve requirements, and develop (localized) reactive power markets, just to mention a few potential applications.

In recent years, several analysis tools and techniques have emerged which are suitable for application to voltage and VAR monitoring and control at both local and global levels. At a local level, these techniques can be used to develop independent voltage monitoring and control strategies for microgrids (MGs) and distributed generators (DGs) such as those described in Chapter 2; at a global level, they have the potential to be integrated to wide-area voltage control schemes using synchrophasor measurements.

In this chapter, a novel use of measurement-based analysis techniques to identify weakly interacting voltage control areas in power systems with high penetration of distributed energy resources is presented. The proposed technique combines the inherent abilities of graph–theoretical techniques with spectral clustering and visualization methods to identify voltage control areas and reconstruct system behavior using selected measurements. Attention is focused on three main aspects, namely the identification of critical system zones showing a coherent behavior and the associated reactive power sources, the use of spectral analysis to identify the critical buses, and the computation of reduced-order models.

The evaluation of voltage stability problems is done using both static and dynamic techniques. To facilitate comparison between both simulation approaches and allow full comparison between static and dynamic approaches, statistical techniques are being developed to extract modal characteristics directly from time-domain simulations. Connections with other modal identification methods are also investigated and numerical issues are discussed.

Renewable Integrated Power System Stability and Control, First Edition.
Hêmin Golpîra, Arturo Román-Messina, and Hassan Bevrani.

5.1 Introduction

Wide-area area voltage control of large interconnected systems has attracted considerable interest in the last few decades [1–6]. Network voltage control at generators, and dynamic reactive power compensation devices placed at key system locations, among other measures, can support voltage regulation and enhance system transient stability and operating flexibility [7]. The increasing size and complexity of power systems with high penetration levels of MGs/DGs distributed along a large geographical area or located in remote zones, however, make voltage control challenging.

As the number of modern renewable energy sources (RESs) with improved static and dynamic reactive power capability grows, a significant challenge is to integrate them into existing voltage control schemes as well as to develop effective coordination schemes [8–10]. Better detection and forecasting techniques through the concept of wide-area voltage monitoring are also needed to allow full use of control capabilities and realize wide-area monitoring structures.

In the recent past, there has been renewed interest in the use of automatic voltage regulation schemes for wind and solar photovoltaic (PV) farms connected to the bulk power systems [9–11]. Studies show that static VAR compensators (SVCs) located adjacent to large wind and PV farms can be used to maximize reactive power reserves and improve voltage profiles. Voltage–VAR control can contribute to the overall power system angle and voltage stability and result in improved system operation and security. This is a subject that is receiving increasing attention.

The integration of advanced voltage–VAR controls in the wind and solar generators, on the other hand, raises several complex issues [8, 11]. First, farm-level voltage control introduces a hierarchical control system that needs to be optimized to improve the overall system response to system perturbations. Further, coordination with other nearby generators or network reactive power compensation devices may be needed as in many cases, wind farms require additional reactive power support, especially during transient conditions.

One of the critical issues in modeling the inverter-based generators is reactive power capability. Some interesting phenomena of voltage and reactive power responses from solar PV generators have been observed and investigated, such as high voltages under normal conditions, high transient voltages, and sustained oscillations following a fault or change in the control characteristics of wind and solar PV farms [1]. These problems may cause further reliability concerns such as overload of subtransmission and distribution facilities, unexpected generation tripping for overvoltage or under-excitation, and even system-wide transient instability.

Experience with the application of primary and secondary voltage control in power systems in European countries shows that coordination of reactive sources may result in enhanced system wide-area control and reliability [2, 12–14]. To avoid undesirable interactions, voltage control at the various levels should temporally and spatially independent. This requires splitting the system into non-interacting zones in which voltage is controlled individually.

Large-scale coordination of reactive power sources is challenging due to a large number of control characteristics, the location and type of controllers, and the characteristics of each device. Issues such as reserve capacity control and the efficient utilization and coordination of reactive power sources must be addressed to achieve fast automatic voltage control and keep the capacitive output margin against system contingencies.

This chapter discusses the experience in the development of data-driven analysis techniques to identify and update voltage control zones and the associated reactive power resources. A systematic methodology for the identification of voltage control zones is first introduced. The proposed procedure consists of three main steps: (i) the identification of strongly connected buses showing coherent behavior, (ii) the identification of generators and SVCs participating in the critical zones, and (iii) the determination of distance measures indicating relationships between bus voltage magnitudes and reactive power sources. These methods are suitable for large-scale applications and can be used to coordinate multiple available reactive compensation devices, including SVCs, synchronous condensers, generator excitation systems, and modern RESs equipped with closed-loop voltage control schemes.

The design methodology is demonstrated on a complex test system with significant wind penetration in which several SVCs are used to control system voltage. Results show that properly coordinated reactive power sources may have an important impact on system dynamic behavior.

5.2 Voltage Control Areas: A Background

Recent years have witnessed the development and application of wide-area voltage control systems with the ability to monitor voltage deviations at key transmission buses, update set points of major closed-loop controllers, and coordinate reactive power sources to regulate network voltages. Due to the local nature of voltage behavior, a major issue in these hierarchical control schemes pertains to the identification of nearly independent voltage control zones.

A general review of the voltage control structures is presented in Chapter 1. In this chapter, a brief overview of these methods, in the context of voltage monitoring techniques is given. The discussion begins with a review of fundamental concepts in the development of practical wide-area voltage control of power systems. Then, the need for data-driven approaches to identify voltage control zones is established.

5.2.1 Voltage Sensitivities

A widely used criterion to determine voltage control areas is the notion of voltage and reactive power sensitivities. Following Kundur [15] and adopting the nomenclature in Chapter 3, the linearized steady-state system power voltage equations can be written as:

$$\begin{bmatrix} \Delta \boldsymbol{P} \\ \Delta \mathbf{Q} \end{bmatrix} = \begin{bmatrix} \boldsymbol{J}_{P\theta} & \boldsymbol{J}_{PV} \\ \boldsymbol{J}_{Q\theta} & \boldsymbol{J}_{QV} \end{bmatrix} \begin{bmatrix} \Delta \boldsymbol{\theta} \\ \Delta \mathbf{V} \end{bmatrix} \tag{5.1}$$

where $\boldsymbol{J}_{P\theta}$, $\boldsymbol{J}_{PV}$, $\boldsymbol{J}_{Q\theta}$, and $\boldsymbol{J}_{QV}$ are blocks of the Jacobian matrix and represent sensitivities between power flow and bus voltage changes; $\Delta \boldsymbol{P}$ and $\Delta \mathbf{Q}$ are the incremental changes in bus real power and bus reactive power injection; $\Delta \boldsymbol{\theta}$ and $\Delta \mathbf{V}$ are the incremental changes in bus voltage angle and bus voltage magnitude.

Making $\Delta \boldsymbol{P} = \mathbf{0}$ in (5.1), one has that

$$\Delta \mathbf{Q} = \left[\boldsymbol{J}_{QV} - \boldsymbol{J}_{Q\theta}(\boldsymbol{J}_{P\theta})^{-1}\boldsymbol{J}_{PV}\right]\Delta \mathbf{V} = \left(\left.\frac{\partial \mathbf{Q}}{\partial \mathbf{V}}\right|_{\mathbf{V}^o}\right)\Delta \mathbf{V} = \boldsymbol{J}_{QV_R}\Delta \mathbf{V} \tag{5.2}$$

and, therefore,

$$\Delta \mathbf{V} = \left(\boldsymbol{J}_{QV_R}\right)^{-1}\Delta \mathbf{Q} = \left(\left.\frac{\partial \mathbf{V}}{\partial \mathbf{Q}}\right|_{\mathbf{V}^o,\mathbf{Q}^o}\right)\Delta \mathbf{Q} = \boldsymbol{J}_{VQ_R}\Delta \mathbf{Q}$$

where matrix $\boldsymbol{J}_{QV_R}$ is called the reduced steady-state Jacobian matrix of the system and $\boldsymbol{J}_{VQ_R}$ is a sensitivity matrix with coefficients:

$$\boldsymbol{J}_{VQ_R} = \left[J_{VQ_{R_{ij}}}\right] = \begin{bmatrix} \frac{\partial V_1}{\partial Q_1} & \frac{\partial V_1}{\partial Q_2} & \cdots & \frac{\partial V_1}{\partial Q_n} \\ \frac{\partial V_2}{\partial Q_1} & \frac{\partial V_2}{\partial Q_2} & \cdots & \frac{\partial V_2}{\partial Q_n} \\ \vdots & \vdots & \vdots & \vdots \\ \frac{\partial V_n}{\partial Q_1} & \frac{\partial V_n}{\partial Q_2} & \cdots & \frac{\partial V_n}{\partial Q_n} \end{bmatrix}$$

where n is the total number of nodes, and in the interest of simplicity, all buses are assumed to be PQ buses.

Simplified approaches for calculating sensitivity relations are described in Ref. [15]. Further, Ref. [16] describes alternative approaches to calculate sensitivity relationships based on the active–reactive decoupling characteristics in the power flow computations.

In the same vein as Lagonotte et al. [4] assume now that a dominant node (a pilot point or *pilot node*), V_{p_k} is determined that represents the average or dominant system behavior for area k. Valuable insight into the influence of changes in the magnitude of V_{p_k} on reactive power deviations of nearby generators, SVCs, and RES in area k can be obtained by writing the sensitivity of bus reactive power injection at generator i to changes in the bus voltage deviations of the pilot bus at bus k as:

$$\frac{\Delta Q_{g_i}}{\Delta V_{p_k}} = \left(\left.\frac{\partial Q_{g_i}}{\partial V_{HV_i}}\right|_{V^o,Q^o}\right)\left(\left.\frac{\partial V_{HV_i}}{\partial V_{p_k}}\right|_{Q^o}\right), i \in k \tag{5.3}$$

where V_{HV_i} is the high side bus of the machine at bus i and Q_{g_i} represents reactive power injections at bus i; the terms $\frac{\partial Q_{g_i}}{\partial V_{HV_i}}, \frac{\partial V_{HV_i}}{\partial V_{p_k}}$ represent sensitivity coefficients.

Numerically, the coefficients $\frac{\partial Q_{g_i}}{\partial V_{HV_i}}$ can be obtained directly from the load flow Jacobian matrix; the second term, $\frac{\partial V_{HV_i}}{\partial V_{p_k}}$, represents an electrical distance and is a subproduct of the calculation of electrical distances in Section 5.2.2. Sensitivity matrices are real and nonsymmetrical and reflect the propagation of voltage variation following reactive power injection at a bus [17]. Variations of this model using other system formulations are described in [4, 15–17], and references therein.

5.2.2 Electrical Distances

In Ref. [4], an approach based upon the application of electrical distance sensitivities was introduced to estimate the magnitude of coupling between two nodes as well as to identify non-overlapping voltage control areas.

Using the same notation as in Eq. (5.3), the electrical attenuation coefficient, α_{ij}, between nodes i and j can be defined as:

$$\alpha_{ij} = \frac{\Delta V_i}{\Delta V_j} = \left(\frac{\partial V_i}{\partial Q_j}\right)\left(\frac{\partial Q_j}{\partial V_j}\right) = \left(\frac{\partial V_i}{\partial Q_j}\right) / \left(\frac{\partial V_j}{\partial Q_j}\right) \tag{5.4}$$

or $\Delta V_i = \alpha_{ij} \Delta V_j$, where, in general, $\alpha_{ij} \neq \alpha_{ji}$. Physically, $\alpha_{ij} \approx 1$, when buses i, j are electrically close and has a small value when they are electrically distant. Other approaches to determining approximate sensitivity relations based on the load flow equations are described in [4, 17].

An attenuation matrix can then be obtained from the notion of entropy or information theory. The electrical pairwise distance, d_{ij}, between buses i and j can be defined as $d_{ij} = d_{ji} = -\log(\alpha_{ij}.\,\alpha_{ji})$ that has two important properties, positivity and symmetry.

Extending this approach to the multidimensional case, one can define the matrix of attenuation between all the buses, $\boldsymbol{A}$, as:

$$\boldsymbol{A} = \left[\alpha_{ij}\right] = \begin{bmatrix} \alpha_{11} & \alpha_{12} & \cdots & \alpha_{1n} \\ \alpha_{21} & \alpha_{22} & \cdots & \\ & & & \alpha_{2n} \\ \vdots & & \ddots & \\ \alpha_{n1} & \alpha_{n2} & \cdots & \alpha_{nn} \end{bmatrix} \tag{5.5}$$

As discussed in the practical application of the method in Section 5.5, the attenuation matrix is numerically sparse and a threshold value can be defined to construct a truncated attenuation matrix $\hat{\boldsymbol{A}}$ (i.e. not all sites are connected).

In the context indicated later, the attenuation matrix $\boldsymbol{A}$ can be reinterpreted in terms of a connected, weighted graph, whose nodes or vertices represent physical nodes or sensor locations and the edges represent distance relationships. Edges with a value above a predefined threshold can be retained to reduce the computational burden and improve understanding of the underlying connectivity or the required communication structure.

This interpretation provides the opportunity to obtain distance or attenuation measures, directly from observational or measured data as explained in Section 5.4, and makes the algorithms well suited for efficiently applying graph-based clustering procedures. A second desired application is the evaluation of hierarchical or decentralized communication structures.

In the case of power networks, several graph-based system representations have been proposed and used to divide data into clusters based on concepts such as electrical distance [18] or spectral characteristics [19]. A limitation of these formulations is their inability to represent the time-ordering of the data, especially from PMU data recordings, which makes them more appropriate to represent and characterize static structures rather than dynamic information. Further many graph-based representations are based on several simplifying assumptions, many of which make them unsuitable for representing dynamic data.

These observations motivate the view that a dynamic framework is needed to capture both structural and temporal information in measured data.

5.2.3 Reactive Control Zones and Pilot Nodes

Most existing wide-area voltage control schemes rely on the decomposition or division of a large power system into loosely interacting voltage control areas, in which the voltages within each area are independently controlled using large or dominant reactive power resources in an automated fashion [2].

The approach involves three main steps:

1) Subdivide the attenuation data $\boldsymbol{A}$ into subsets or clusters which are pairwise disjoint and connected;
2) Select a robust node or bus that represents the average or dominant system behavior;
3) Determine sensitivity relations between dominant bus voltage behavior and reactive power sources.

These approaches are usually hierarchical; first, a voltage control zone associated with the largest sensitivities is determined. Then, the remaining zones are determined using an iterative approach [1, 2, 4].

5.2.3.1 Selection of Optimal Pilot Buses

The determination of reactive zones and pilot nodes in voltage control schemes can be formulated as a problem of clustering distance measures. In this sense, the determination of pilot bus voltages for area k, V_{p_k} can be posed as the solution of an optimization problem (refer to Figure 5.3):

$$pilot_{node_{area_k}} = min\left[\sum_{i,j\epsilon k}\alpha_{ij}\right] \tag{5.6}$$

where several constrained optimization techniques can be utilized.

Thus, for instance, with reference to Figure 5.1, node 1 is selected as a candidate pilot bus for area i, if $\alpha_{tot_1} = \sum_{k=1}^{5}\alpha_{1k} < \alpha_{tot_2} < \ldots < \alpha_{tot_5}$ subject to the constraint that the electrical distance with other pilot buses should be small to avoid interactions.

From a practical standpoint, two main criteria are often used to select bus pilot buses [2, 4, 12]:

1) Pilot buses should be chosen among the strongest nodes in the system using criteria such as the maximum short-circuit current.
2) Coupling between pilot nodes associated with different control areas should be low to avoid interactions among control systems.

It is worthwhile noting that the determination of voltage pilot nodes involves finding a solution to a changing problem, in which system topology and operating conditions are always changing. This is an issue that must be addressed from a dynamic perspective.

5.2.3.2 Selection of Control Plants

Generating units and flexible AC transmission system (FACTS) devices participating in wide-area control schemes are often selected on heuristic criteria requiring

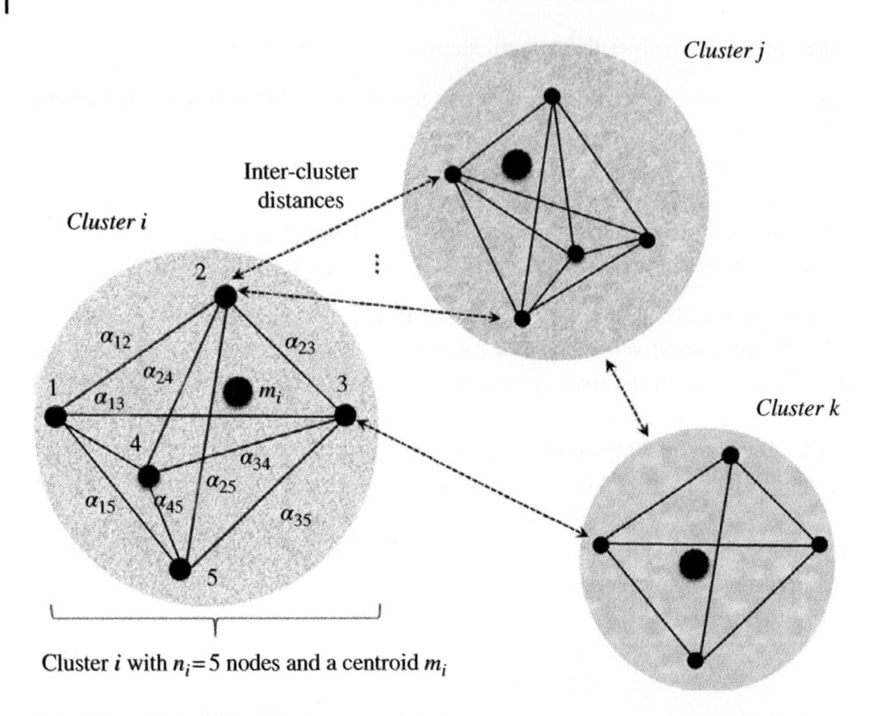

Figure 5.1 Illustration of clusters associated with weighted, connected graphs. Each node within cluster *i* has several connections (dense connection) with other members of the cluster and fewer links (sparse connections) with nodes outside the cluster.

that they belong to the control area under analysis, the size of their reactive power capability, and the highest electrical coupling with the pilot nodes (see Eq. (5.3)).

These methods, however, are commonly characterized by several limitations:

- The methods do not fully recognize the dynamic nature of system behavior;
- In many applications, not only frequently pilot nodes need to be reselected following topology changes but also the control areas and the associated reactive power sources need to be updated;
- Communication and bandwidth structure and requirements cannot be determined.

This has motivated the development of alternative, measurement-based techniques.

5.2.4 Other Approaches

Recently, approaches to address the challenge of determining reactive zones have been proposed. In Ref. [19], spectral clustering was proposed to discover clusters in power networks from the orthogonal structure of spectral embedding. Other approaches to identify clusters include Refs. [20, 21]. In the former approach, the authors discussed the problem of determining voltage–VAR control areas using spectral graph clustering applied to the system graph representation obtained from the power system flow equations. By examining the spectral properties of a graph-based representation, it was possible to identify the location and number of VAR–voltage control areas. In most of these applications, however, use is made of spatial information which precludes their application in a real-time setting.

In Sun et al. [22], an adaptive zone-division-based automatic voltage control system based on the concept of VAR, control space was introduced and applied to the Chinese power grid. A key feature of this approach is that the control zones are no longer fixed but are reconfigured online and updated following variations in the grid structure. Yet another approach is the use of online voltage stability assessment techniques to simultaneously identify voltage control areas and reactive power reserves [23]. Such identification techniques are invaluable when dealing with frequent changes in system behavior or system structure.

5.3 Data-driven Approaches

Real-time voltage and reactive power measurements from time-synchronized wide-area monitoring systems (WAMSs) provide the opportunity to analyze and cluster motion trajectories, thus offering a useful complementary approach to conventional static analysis. One potential drawback of these data is that they may involve various timescales and differ in magnitude and nature. In practical applications, determining reactive control areas directly from time-domain simulations may be difficult since no sensitivity information is readily available, and therefore measures of similarity (coherency) are needed.

Further, with a wide array of multimodal, multichannel PMU data being routinely acquired for system monitoring, there is a pressing need for quantitative tools to combine these varied channels of information [21]. The goal of these measuring and analysis techniques is to combine complementary measurements such as voltage and reactive power measurements while improving on the predictive ability of any individual modality.

In the following subsections, wide-area measurement-based structures and control strategies to decompose the system into an arbitrary number of voltage control zones are introduced that may complement static approaches. Then, data-driven techniques are proposed and tested to identify and characterize voltage control zones.

5.3.1 Wide-Area Voltage and Reactive Power Regulation

Figure 5.2 shows a component diagram of a wide-area voltage monitoring system adopted in the analysis that incorporates ideas from previous work [24]. The voltage control structure is hierarchical and consists of two major levels or measurement hierarchies: primary (local) and secondary (regional).

Inputs to the WAMS include time series PMU data such as bus voltage magnitudes and angles, and reactive power of wind and PV generators, major generating plants, and SVCs. The outputs of the system may be used to select and update PMU or voltage control areas, select sensor locations, or be used as inputs to a voltage monitoring and stability assessment or prediction modules with the ability to determine reduced-order models, and trajectory classification and visualization.

The ability to select loosely or non-interacting voltage control areas for system monitoring from sensor measurements such as supervisory control and data acquisition (SCADA) system and phasor voltages is important for at least three reasons. First, detecting voltage and reactive power deviations in near real-time can help track reactive reserves and make decisions about reactive power management. Second, the detection of voltage and reactive power deviations can produce

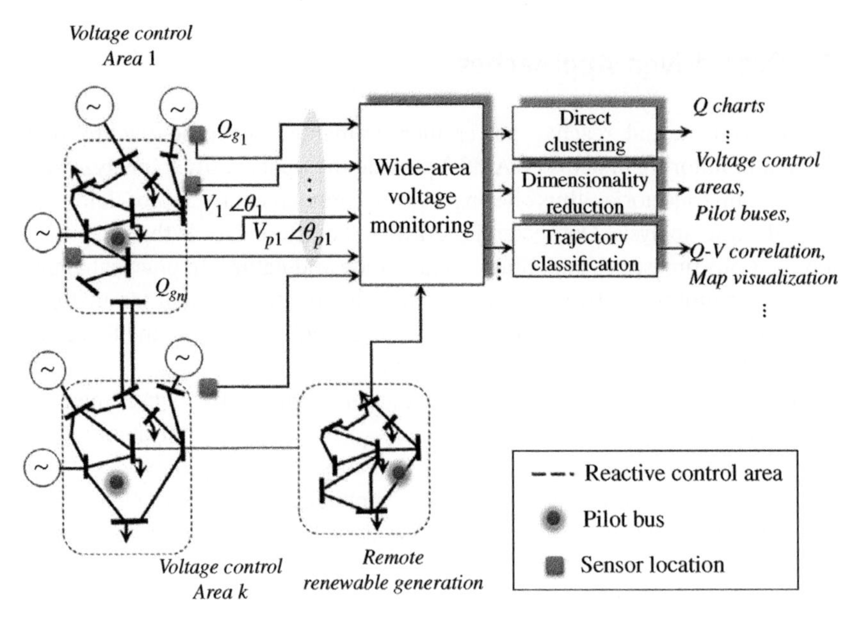

Figure 5.2 Overview of the steps in computing voltage control areas. Here, V_{p_j} represents the set point of pilot buses, V_j and θ_j are the bus voltage magnitude and phases, and Q_{g_j} represent reactive power output from major generators and network reactive power compensation devices.

knowledge about the strength of interactions between multimodal data, i.e. voltage and reactive power. Further, measurement techniques give some insight into the nature of control structure and communications required for efficient implementation of practical WAMSs, especially in power systems with RESs [25].

Three main activities are of interest here:

1) Direct trajectory clustering;
2) Dimensionality reduction and spectral clustering; and
3) Trajectory classification and visualization.

The issues of dimensionality reduction and spectral clustering are addressed in Sections 5.4.1 and 5.4.2. The issue of direct clustering is deferred to Section 5.4.3.

5.3.2 PMU-Based Voltage Monitoring

Recent attempts to develop WAMS using synchrophasor technology allow for efficient network partitioning and the identification of critical voltage areas. In the bulk of these applications, each data type, i.e. voltage or reactive power measurements is analyzed independently and then combined to determine its relationship with other data modalities. Cross-information between data types, however, is lost or the sequential combination of the independent model may result in inaccurate system characterization.

Processing of large volumes of voltage and reactive power observations or simulations, on the other hand, raises significant challenges as:

- Data represent different physical units. Simple data normalization techniques may affect results;
- Bus voltage magnitudes from different physical regions may show similar behavior thus obscuring physical interpretation. Phase information may be needed to improve the extraction of coherent structures;
- Practical voltage control systems should only consider selected voltage system behavior (model reduction) from which global system behavior can be reconstructed;
- Correlation techniques are also needed to identify common features between different datasets.

5.4 Theoretical Framework

In this section, an approach based on spectral analysis techniques is introduced to examine dynamic trends and phase relationships between key system signals from measured data. Drawing on graph-based techniques and spectral analysis methods, a technique based on the notion of diffusion maps (DMs) is used to express an ensemble of measured data as a nonlinear combination of modal coordinates from which voltage control zones can be determined.

First, the notion of dynamic trajectories in the context of WAMS is introduced.

5.4.1 Dynamic Trajectories

Measured dynamic data have a useful interpretation in terms of spatiotemporal motion trajectories and associated graphs that evolve with time, as shown schematically in Figure 5.3. Referring to Figure 5.3, assume that $x_k(t_j)$ denotes a sequence of observations of a simulated or measured transient process at locations x_k, $k = 1,...,m$, and time t_j, $j = 1,...,N$, where the m locations represent sensors [26].

At each time instance t_j, the instantaneous (pointwise) distance between two motion trajectories (time sequences) $x_i(t_j)$ and $x_j(t_j)$, is given by:

$$d_{ij}\left(x_i, x_j, t_j\right) = \underbrace{\left\| x_i\left(t_j\right) - x_j\left(t_j\right) \right\|}_{\substack{\textit{Distance} \\ \textit{function}}} \tag{5.7}$$

where $x_i(t_j)$ and $x_j(t_j)$ represent the instantaneous values of the time sequences at time instance t_j (a time slice in the spatiotemporal representation in Figure 5.1). In practice, distances can be obtained for a time window of interest and a full distance

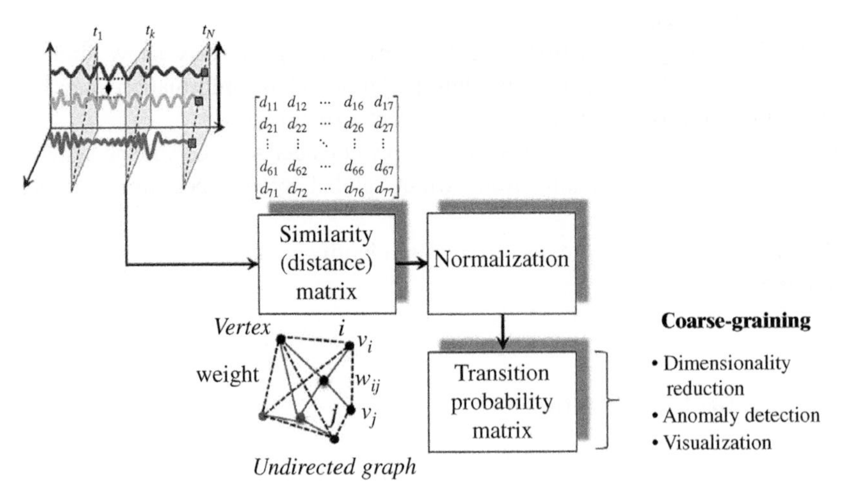

Figure 5.3 Illustration of distance (similarity) matrices and the associated undirected graphs.

matrix can then be defined as $\boldsymbol{D} = [d_{ij}]$. In this case, however, the distance definition does not account for the time-ordering of the signals.

This allows the application of graph-clustering algorithms or coarse-grained methods to distance (similarity) matrices.

5.4.2 Spectral Graph Theory

Similarity or distance matrices based on (5.7) have an interesting interpretation in terms of weighted, undirected graphs, where $d_{ij} = d_{ji}$ [27]. Referring back to Figure 5.3, a weighted graph $G(V, w)$ with m vertexes $V = \{v_1 \quad v_2 \dots \quad v_m\}$ and nonnegative weights $w_{ij} = w(v_i, v_j)$ can be constructed by computing a pairwise similarity distance between trajectories. Each measurement (or sensor) location corresponds to a node or vertex, while the edges (weights) represent the interaction strength or affinity between trajectories [24, 27, 28].

Using the distance between motions or trajectories, the weight on the graph edges is given by:

$$w_{ij} = \left\| \boldsymbol{x}_i(t) - \boldsymbol{x}_j(t) \right\|$$

with

$$w_{ii} = \sum_{j=1}^{m} w_{ij}$$

An $m \times m$ distance matrix, $\boldsymbol{W} = [w_{ij}]$ can now be defined whose nodes correspond to the dynamic trajectories [29–32]. Figure 5.3, which is excerpted from Ref. [33], illustrates the definition of vertices and weights in the concepts of the intended application. Similar interpretations can be found in the context of spatial filtering or the modeling of spatial processes [28].

As implied in Figure 5.3, the calculation of weights, w_{ij}, requires consideration of all pathways on the graph. In practical applications, such as those associated with dynamic trajectories, however, transition probabilities between different nodes may be small and a threshold can be used to obtain a sparse representation of the graph.

Coupled with efficient visualization techniques, graphs can also be used to display efficiently coherent groups.

5.4.3 Kernel Methods

Graph-based kernel methods have been recently explored to extract dominant behavior from measured data [29]. Given a measurement matrix $\boldsymbol{X}$ in (5.1), a positive-definite matrix kernel (a Gaussian kernel) $\boldsymbol{K}$ can be obtained whose (i, j)th element is given by:

$$[\boldsymbol{K}] = [k_{ij}] = \exp\left(-\frac{\|\boldsymbol{x}_i(t) - \boldsymbol{x}_j(t)\|}{\varepsilon_i\varepsilon_j}\right), 1 \leq i, j \leq m \tag{5.8}$$

where the kernel bandwidth ε_i, ε_j controls the hop size of the random walk and can be automatically tuned based on the distribution of matrix $\boldsymbol{A}$. The use of other kernels is discussed elsewhere.

Physically, matrix $\boldsymbol{K}$ can be interpreted as the adjacency matrix of a graph, each of whose nodes represents one sensor or trajectory. The kernel bandwidth scale, ε_i, ε_j modulates the notion of distance in (5.7) and can be tuned to cluster system behavior. If $\varepsilon_i\varepsilon_j \gg \max_{ij}\|x_i(t) - x_j(t)\|$, then for all edges $\{i, j\}$, $k_{ij} \approx 1$; for low kernel bandwidth values, $\varepsilon_i\varepsilon_j \to 0$, the pairwise distances k_{ij} become increasingly similar and large fluctuations in the density of points in the high-dimensional space are smoothed out.

From the matrix kernel in (5.8), one now can define a diagonal matrix $\boldsymbol{D} = [d_{ij}]$ whose entries are the row sums of $\boldsymbol{K}$, namely:

$$\boldsymbol{D} = [d_{ij}] = \begin{bmatrix} \sum_{j=1}^{m} K_{1j} & 0 & \dots & 0 \\ 0 & \sum_{j=1}^{m} K_{2j} & \dots & 0 \\ 0 & 0 & \dots & \sum_{j=1}^{m} K_{mj} \end{bmatrix}$$

where $d_{ii} = \sum_{j=1}^{m} K_{ij}$ is the degree of node $\boldsymbol{x}_i$.

The distance matrix has an interesting interpretation in terms of an undirected probabilistic graph. In this concept, the transition probability p_{ij} from i to j can now be obtained as $p_{ij} = k_{ij}/\sum_{k=1}^{m} k_{ik}$. The right-stochastic Markov matrix or probability matrix, $\boldsymbol{M}$, can then be defined as:

$$\boldsymbol{M} = [m_{ij}] = \boldsymbol{D}^{-1}\boldsymbol{K} = \begin{bmatrix} \frac{k_{11}}{\sum_{j=1}^{m} K_{1j}} & \frac{k_{12}}{\sum_{j=1}^{m} K_{1j}} & \cdots & \frac{k_{1m}}{\sum_{j=1}^{m} K_{1j}} \\ \frac{k_{21}}{\sum_{j=1}^{m} K_{2j}} & \frac{k_{22}}{\sum_{j=1}^{m} K_{2j}} & \cdots & \frac{k_{2m}}{\sum_{j=1}^{m} K_{2j}} \\ \vdots & \vdots & \ddots & \vdots \\ \frac{k_{m1}}{\sum_{j=1}^{m} K_{mj}} & \frac{k_{m2}}{\sum_{j=1}^{m} K_{mj}} & \cdots & \frac{k_{mm}}{\sum_{j=1}^{m} K_{mj}} \end{bmatrix} \tag{5.9}$$

in which the matrix element $m_{ij} = p_{ij}$ can be interpreted as the probability p_{ij} of hopping from point i to point j in t steps of a discrete random walk [32].

Physically, the transition probability matrix, $\boldsymbol{M}$, defines the random walk of a particle on the graph. Formally, suppose that the initial probability of the particle being at a vertex v_j is $p_j^0 (j = 1, \dots, n)$. It follows that the probability of the trajectory

v_j taking the edge w_{ij} is $m_{ij}p_j^o$. The extension to the multivariate case follows along the same lines.

Paramount to the automated extraction of low-dimensional representations is the adaptive computation of the Gaussian kernel widths, ε. In general, the scale parameter is related to the statistics and the geometry of the data points. Following Ref. [29], let $\boldsymbol{X}_j$ denotes a cloud of points around $\boldsymbol{x}_i$. The variance of the distance between the point $\boldsymbol{x}_i$ to all the points $\boldsymbol{x}_j \in \boldsymbol{X}_j$ is given by Ref. [30]:

$$\varepsilon_i = \sum_{x_i \in \boldsymbol{X}_j} \frac{\left\| \left\| \boldsymbol{x}_i - \boldsymbol{x}_j \right\| - \hat{\boldsymbol{X}}_i \right\|^2}{|\boldsymbol{X}_i|}$$

where

$$\hat{\boldsymbol{X}}_i = \sum\nolimits_{x_i \in X} \left\| \boldsymbol{x}_i - \boldsymbol{x}_j \right\| / |\boldsymbol{X}_i|.$$

It is easy to prove that matrix $\boldsymbol{M}$ is nonnegative, unsymmetrical, and invariant to the observation modality and is resilient to measurement noise. It is a stochastic transition (probability) matrix with the following important properties:

$$\begin{cases} m_{ij} \geq 0 \\ \sum\nolimits_j m_{ij} = 1 \, for \, j = 1, \ldots, n \text{ (row stochastic)} \end{cases}$$

Diagonalization of $\boldsymbol{M}$ produces an ordered set of eigenvectors and eigenvalues $\{(\boldsymbol{\phi}_i, \lambda_i), i = 1, \ldots, d\}$ with $\lambda_1 = 1$, and $\lambda_1 \geq \lambda_2 \geq \ldots \geq \lambda_d$. Once a distance matrix is obtained, spectral techniques can be used to obtain reduced-order representations as well as to identify and eliminate nonrelevant clusters. The main goal of cluster analysis is data reduction, by subdividing a set of objects into a hierarchical arrangement of homogeneous subgroups. A significant outcome is reduced complexity with a minimal loss of information, which fosters a better understanding of the analyzed data.

Examples of this class of algorithms are nonlinear reduction techniques such as DMs [29], the Mei–Sheila algorithm [33], and the Markov clustering (MCL) algorithm described in Ref. [34]. Two approaches discussed here, spectral clustering and the MCL algorithm, show promise for automated identification of coherent patterns and model reduction.

5.4.3.1 Markov Matrices

A key step of spectral analysis algorithms is the projection into a reduced-order subspace. Let matrix $\boldsymbol{M}$ have a set of eigenvalues λ_j with associated eigenvectors, ψ_j. From linear system theory, the eigenvectors of matrix $\boldsymbol{M}$ in (5.9) satisfy

$$\left(\boldsymbol{M} - \lambda_j \boldsymbol{I}\right)\psi_j = \boldsymbol{0}, j = 1, \ldots, n \tag{5.10}$$

For purposes of analysis, it is convenient to perform spectral analysis on a similar matrix, $\boldsymbol{M}_s = [m_{s_{ij}}] = \boldsymbol{D}^{-1/2}\boldsymbol{K}\boldsymbol{D}^{1/2}$ (a symmetric matrix) that results in a symmetric eigenvalue problem. From (5.10), it follows readily that:

$$\boldsymbol{M} = \boldsymbol{D}^{-1}\boldsymbol{K} = \boldsymbol{D}^{-1/2}\left(\boldsymbol{D}^{-1/2}\boldsymbol{K}\boldsymbol{D}^{1/2}\right)\boldsymbol{D}^{-1/2} = \boldsymbol{D}^{-1/2}\boldsymbol{M}_s\boldsymbol{D}^{-1/2} \tag{5.11}$$

and thus

$$\boldsymbol{M}_s = \boldsymbol{D}^{-1/2}\boldsymbol{K}\boldsymbol{D}^{1/2} = \boldsymbol{D}^{-1/2}\boldsymbol{D}\boldsymbol{M}\boldsymbol{D}^{-1/2} = \boldsymbol{D}^{1/2}\boldsymbol{M}\boldsymbol{D}^{-1/2}$$

is a normalized affinity matrix (a normalized kernel), where use has been made of the identity $\boldsymbol{K} = \boldsymbol{D}\boldsymbol{M}$.

Since matrices $\boldsymbol{M}$ and $\boldsymbol{M}_s$ are related by a similarity transformation, they share the same eigenvalues. To prove this, let $\boldsymbol{\psi}$ denote the eigenvalues of $\boldsymbol{M}$ and $\boldsymbol{\Phi}$ denote those of $\boldsymbol{M}_s$. Observing that $\boldsymbol{M}_s = \boldsymbol{D}^{-1/2}\boldsymbol{K}\boldsymbol{D}^{-1/2}$, it can be inferred that $\boldsymbol{M}_s$ is a symmetric matrix, which allows the use of special techniques for calculating the associated singular values. Griffith [31] gives an interesting interpretation of these matrices in the context of spatial analysis and filtering.

Collecting all eigenvalues yields

$$\boldsymbol{M}\boldsymbol{\Psi} = \lambda_j\boldsymbol{\Psi}$$
$$\boldsymbol{M}_s\boldsymbol{\Phi} = \lambda_j\boldsymbol{\Phi}$$

with $\boldsymbol{\Psi} = [\boldsymbol{\psi}_o \quad \boldsymbol{\psi}_1 \quad \cdots \quad \boldsymbol{\psi}_{m-1}]$ and $\boldsymbol{\Phi} = [\boldsymbol{\phi}_o \quad \boldsymbol{\phi}_1 \quad \cdots \quad \boldsymbol{\phi}_{m-1}]$.

From these relationships, the eigenvalues λ_j of $\boldsymbol{M}_s$ satisfy

$$(\boldsymbol{M}_s - \lambda_j\boldsymbol{I})\boldsymbol{\phi}_j = \boldsymbol{0} = \left(\boldsymbol{D}^{1/2}\boldsymbol{M}\boldsymbol{D}^{-1/2} - \lambda_j\boldsymbol{I}\right)\boldsymbol{\phi}_j = (\boldsymbol{M} - \lambda_j\boldsymbol{I})\underbrace{\boldsymbol{D}^{-1/2}\boldsymbol{\phi}_j}_{\boldsymbol{\psi}_j} = \boldsymbol{0}$$

from which it follows that $\boldsymbol{\phi}_j = \boldsymbol{D}^{1/2}\boldsymbol{\psi}_j$ is an eigenvector of $\boldsymbol{M}_s$ with eigenvalue λ_j, and matrices $\boldsymbol{M}$ and $\boldsymbol{M}_s$ share the same eigenvalues.

Moreover, because of the symmetry conditions, the eigenvalues are real and satisfy $\lambda_o = 1$, and $\lambda_o > \lambda_j > \lambda_{j+1}$, $j = 1, 2, 3, \ldots, n$. Noting further that $\boldsymbol{M}_s$ is a row-stochastic matrix $\left(\sum_j m_{s_{ij}} = 1\right)$, the solution of the system $\boldsymbol{M}_s\boldsymbol{\phi}_o = \lambda_o\boldsymbol{\phi}_o$, for $\lambda_o = 1$ gives the constant vector, $\boldsymbol{\phi}_o = [1 \quad 1\ldots \quad 1]^T$, which is often associated with the overall trend or average behavior.

The mapping from the original space into the new DM space is now defined at time t, as the map:

$$\boldsymbol{\Psi}_\varepsilon(\boldsymbol{x}_i) : \boldsymbol{x}_i \rightarrow \left[\lambda_1^t\psi_1(i), \quad \lambda_2^t\psi_2(i), \ldots, \quad \lambda_{N-1}^t\psi_{N-1}(i)\right]^T \in \Re^{N\times 1} \tag{5.12}$$

where $\psi_m(i)$ denotes the ith element of $\boldsymbol{\psi}_m$.

In practice, a d-dimensional DM can be defined as:

$$\boldsymbol{\Psi}_\varepsilon^d(\boldsymbol{x}_i) : \boldsymbol{X} \rightarrow [\lambda_1\psi_1(i), \quad \lambda_2\psi_2(i), \ldots, \quad \lambda_d\psi_d(i)]^T \in \Re^d \tag{5.13}$$

where $d << N$ is the number of relevant coordinates or subset of dominant eigenvectors (the intrinsic dimensionality). Typically, two to three coordinates suffice to capture relevant system behavior.

By truncating the spatial patterns, dimensionality reduction can be performed. Figure 5.4 gives a schematic illustration of this model, showing three main activities or features: (i) nonlinear mapping of the system dynamics, (ii) the calculation of a low-dimensional embedding, and (iii) inverse mapping and model reconstruction.

Associated with the spectral model, the diffusion distance can now be defined in terms of the forward probabilities $\boldsymbol{M}$ as:

$$\boldsymbol{D}_{ij} = \sum_{j=1}^{m} \frac{\left(m_{ir} - m_{jr}\right)}{\Psi(\mathrm{x}_r)} \quad ;$$
$$\Psi(\mathbf{x}_r) = \frac{\sum_{j=1}^{m} M_{jm}}{\sum_{k=1}^{m} \sum_{j=1}^{m} M_{jk}} \tag{5.14}$$

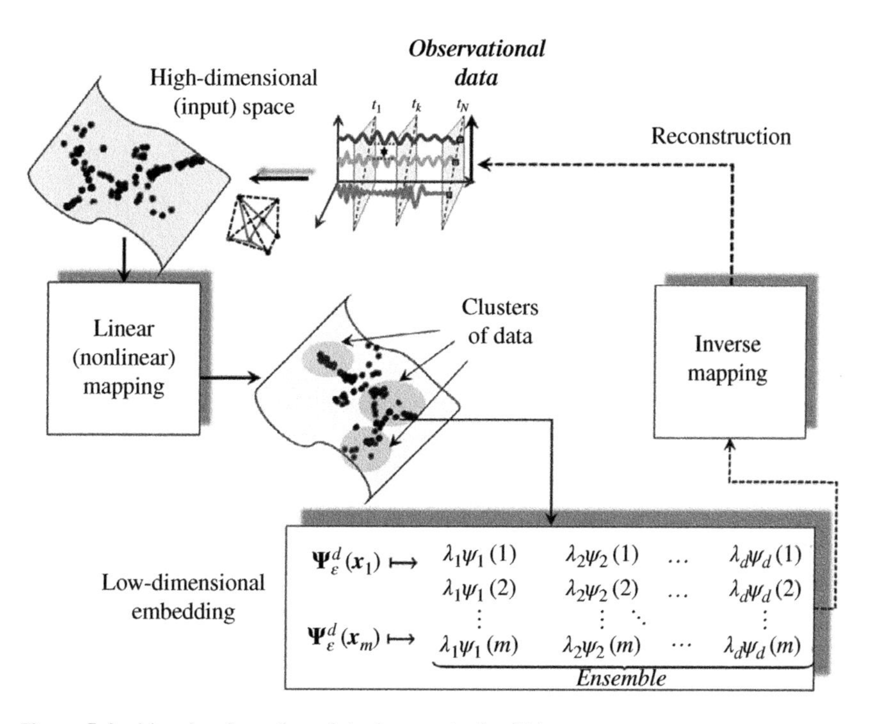

Figure 5.4 Mapping from the original space to the DM space.

Physically, the diffusion distance is small if there are many high-probability paths of length t between two points.

Several remarks regarding time-coupled DMs are in order [33]:

- The low-dimensional approximation to the original model in physical space (reconstruction) can be expressed as:

$$\hat{X} = [\hat{x}_1 \quad \hat{x}_2 \quad \dots \quad \hat{x}_d] = \sum_{j=o}^{m} a_j \psi_j^T = \underbrace{\sum_{j=o}^{d} a_j(t) \psi_j^T}_{Relevant\ system\ behavior} + \underbrace{\sum_{k=d+1}^{m} a_k(t) \psi_k^T}_{non-essential\ coordinates}$$

- The eigenvectors associated with the largest singular values correspond to slow modes governing the long-time system evolution, i.e. ψ_j, $j = 1, \dots, d$;
- Numerical experience with large spatiotemporal models shows that $\psi_o \psi_o^T > \psi_1 \psi_1^T > \dots > \psi_d \psi_d^T$, and $\psi_i \psi_j^T \approx 0, i, j = 1, \dots, d, i \neq j$; In practice, a spectral gap can be observed at λ_d, such that:

$$\underbrace{\lambda_1 \geq \lambda_2 \geq \dots \geq \lambda_d}_{Slow\ motion} \gg \lambda_{d+1} \geq \dots \lambda_m$$

Now, attention is turned to other graph-based clustering algorithms based on the Markov matrix.

5.4.3.2 The Markov Clustering Algorithm

The MCL algorithm is an unsupervised graph-based algorithm for clustering graphs based on simulation of stochastic flow in the graph [34]. Given a stochastic transition matrix, $\boldsymbol{M}$, the MCL method iteratively applies two operators called expansion and inflation until convergence is obtained. Both expansion and inflation are operators that map the space of column stochastic matrices onto itself. Additionally, a pruning step is performed at the end of each inflation step to save memory. By combining this approach with spectral analysis, a reduced-order model can be obtained.

The inflation value parameter of the MCL algorithm is used to control the granularity of the tightness of these clusters. As noted by Refs. [35, 36], the inflation is equivalent to taking the Hadamard power of a matrix followed by a scaling step to ensure the matrix is stochastic again.

The method is based on the analysis on a transition network which is obtained by (i) mapping the dynamic trajectories onto a discrete set of microstates and (ii) building a transition network in which the nodes are the microstates and a link is placed between them if two microstates are visited one after the other along the trajectory.

To formalize the adopted model, let $\boldsymbol{X} \in \Re^{m \times N}$ be the matrix of measurements corresponding to a given operating scenario and $\boldsymbol{M} \in \Re^{m \times N}$ be the transition or Markov matrix. The MCL algorithm is summarized in Table 5.1.

In practical applications, the resulting models, however, may be difficult to interpret and visualize, especially in the case of several dynamic trajectories. To facilitate understanding of the resulting graphs in the developed algorithms, the eigenvalues and eigenvectors of the steady-state matrix are calculated as $(\boldsymbol{M}_{\mathrm{MCL}}(r) - \lambda_{\mathrm{MCL}}\boldsymbol{I})\boldsymbol{\phi}_{\mathrm{MCL}} = \boldsymbol{0}$. This allows direct comparison with spectral analysis techniques in Subsection 5.4.2.

As discussed in Ref. [37], the parameter r determines the granularity or tightness of the clustering. This approach can be used to identify states sharing common, dynamically meaningful characteristics such as coherency or similar timescales. Experience shows that a low value of r may be enough to capture the dynamics of interest ($r = 4$ has been found to give good results in this research). Several variations of this method have been discussed in the literature.

Several properties are inherent to this model such as the following two cases:

Table 5.1 The Markov clustering algorithm.

Given a trajectory matrix $\boldsymbol{X} \in \Re^{m \times N}$,

1) Obtain the undirected graph using the procedures in Sections 5.4.1 and 5.4.2;
2) Build a squared transition matrix, $\boldsymbol{M} = \boldsymbol{P} = \left[p_{ji}\right] \in \Re^{m \times m}$, using the procedure in Section 5.4.3, in which each element p_{ji} represents the transition probability from node j to node i;
3) Normalize the matrix to ensure that the matrix is stochastic again;
4) *Expand* by taking the ith power of the matrix (a Hadamard product):

$$(\boldsymbol{M})^i = \underbrace{\boldsymbol{M} \times \boldsymbol{M} \times \ldots}_{i \text{ times}}$$

   and normalize each column to one;
5) *Inflate* each of the columns of matrix $\boldsymbol{M}$ with power coefficient r as:

$$(\boldsymbol{\Gamma}_r\boldsymbol{M})_{pq} = \frac{\left(M_{pq}\right)^r}{\sum_{q=1}^{m}\left(M_{pq}\right)^r}$$

   where $\boldsymbol{\Gamma}_r$ is the inflation operator;
6) Repeat steps 3 through 5 until MCL converges to a steady-state matrix $\boldsymbol{M}_{\mathrm{MCL}}(r)$;
7) Compute the first few right eigenvalues and eigenvectors of the resulting matrix $\boldsymbol{M}_{\mathrm{MCL}}(r)$, using spectral analysis of matrix $\boldsymbol{M}_{\mathrm{MCL}}(r)$; this results in a set of real eigenvalues and eigenvectors $\mathbf{u}_i$.

1) Like the case of the spectral representation in DMs, the first eigenvector has a unit value and can be associated with average behavior;
2) The second and third eigenvector capture dominant system behavior.

5.4.4 Spatiotemporal Clustering

Spatiotemporal clustering methods with the ability to extract and monitor clusters are well suited for automated voltage monitoring. They can be considered as direct clustering techniques, since they can be applied directly to measured data. Among several clustering techniques, *C*-means clustering offers a powerful means to cluster dynamic trajectories directly from measurements or simulations [33].

Fuzzy clustering can be posed as the solution of the energy function [38, 39]:

$$f(U,\mathbf{v}) = \sum_{k=1}^{K}\sum_{i=1}^{n} u_{ki}^{m}\|x_i - C_k\|^2 \tag{5.15}$$

where U is the fuzzy c-partition of the data, $\mathbf{v}$ is the vector of centers, $\mathbf{v} = [C_1 \quad C_2 \cdots \quad C_c]$, the C_k are the cluster centers, the u_{ki}^{m} represent associated membership likelihoods (the degree of membership), n is the number of clusters, m is a weighting exponent, $0 \le u_{ki} \le 1$, of the trajectory x_i being associated with the cluster center C_k [40].

Using this view, the similarity that a point (node) shares with a given cluster can be represented by a function whose values vary between 0 and 1, subject to the constraints:

$$\sum_{k=1}^{K} u_{ki} = 1, i = 1, \dots, n$$
$$u_{ki} \ge 0, k = 1, \dots, K, i = 1, \dots, n$$

More formally, given n dynamic trajectories, x_i, $i = 1, \dots, n$, the problem can be stated as follows:

$$\begin{aligned} & f(U,\mathbf{v}) = \sum_{k=1}^{K}\sum_{i=1}^{n} u_{ki}^{m}\|x_i - C_k\|^2, \\ & \text{subject to} \sum_{k=1}^{K} u_{ki} = 1, i = 1, \dots, n \\ & u_{ki} \ge 0, \quad k = 1, \dots, K, i = 1, \dots, n \end{aligned} \tag{5.16}$$

Outputs of this algorithm are the clusters and the time evolution of the centroids. Combined with statistical quality control charts, the method can be used to detect anomalous operation and islanding conditions.

5.5 Case Study

To illustrate the utility of data-driven techniques in aiding voltage monitoring and system partitioning, a fully represented model of a large-scale 5449 bus network was developed for the study. Figure 5.5 shows a simplified schematic of the study region showing interconnections with the boundary systems. The study region of interest consists of 174 buses, 104 generators, three major SVCs, and several wind farms. The test system has transmission levels 115, 230, and 400 kV; the total area load is 3263 MW.

The study simulates the entire test system, but results are limited to the study region.

5.5.1 Sensitivity Studies

Sensitivity studies are first conducted to assess the influence of system structure and control action on the nature and distribution of critical modes. In this analysis, the output of a large-scale power flow analysis program is used to determine bus voltages and reactive power sources having a strong influence on system behavior.

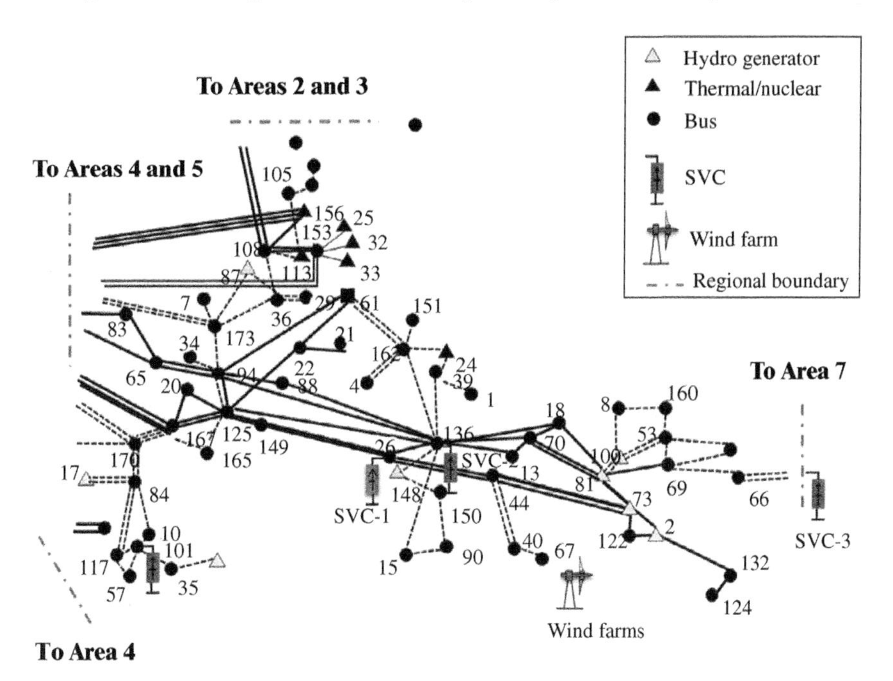

Figure 5.5 Single-line diagram of the study region showing the 230-/400-kV transmission system and its interconnection with other regional systems.

To permit comparison with well-established procedures, two main analysis techniques have been considered: (i) conventional short-circuit studies and (ii) sensitivity studies based on Section 5.2.

In the first stage, short-circuit studies were conducted to determine the main buses and generators involved in the identification of voltage control zones for both the pre-contingency and post-contingency cases [2]. Then, sensitivity coefficients are computed from the Jacobian equations.

For the sake of comparison, Table 5.2 lists the maximum short-circuit levels for major 400-kV transmission buses within the study region extracted from the base case power flow solution ranked in decreasing order of magnitude. As discussed in Ref. [2], pilot nodes are usually chosen based on robustness and coupling characteristics. Candidate pilot nodes are expected to be strong and must exhibit low coupling with other candidate nodes leading to dynamic interactions between reactive power control loops.

The strongest buses identified include buses in the 400-/230-kV network in the northeastern region of the study area (153, 33, 25, 32, 61, 94, and 108) along with a pocket of buses at the center and southeastern portions of the study system (136, 44, 73).

Table 5.2 Short-circuit levels (MVA) for the 400-kV transmission network.

Bus number	Short-circuit level (MVA)
153	19 495
33	19 063
25	18 859
32	18 859
108	18 843
73	16 809
94	14 372
65	13 982
126	13 549
167	13 409
136	13 074
61	10 405
44	9157

Table 5.3 The largest attenuation values α_{ij} computed using (5.4). Base operating case.

Node	Node (attenuation value α_{ij})
153	25 (0.069), 33(0.073), 154 (0.029), 32(0.069), 108(1.449), 156 (4.297)
73	81 (1.755), 122 (1.372), 2 (2.484), 44 (2.17), 18 (2.375), 70 (2.357), 136 (2.82)
136	13 (1.067), 26 (0.355), 18 (1.133), 44 (1.834), 81 (1.805), 126 (1.894), 22 (3.10)
94	65 (0.3601), 88 (1.006), 126 (1.649), 136 (2.135)
108	113 (2.254), 153 (1.449), 25 (1.518), 33 (1.523), 32 (1.518), 156 (2.848)
61	108 (4.51), 162 (4.346), 151 (4.623), 152 (4.623), 24 (4.96), 22 (5.24), 153 (1.449), 32 (1.518)
167	183 (1.253), 186 (2.717), 167 (1.426), 170 (1.974), 168 (1.949), 126 (3.026), 94 (4.335)
126	94 (1.649), 65 (1.98), 22 (1.206), 136 (1.894), 26 (1.889)

Table 5.3, in turn, shows the electrical distances α_{ij} computed using the sensitivity relation:

$$\alpha_{ij} = \frac{\Delta V_i}{\Delta V_j} = \left(\frac{\partial V_i}{\partial Q_j}\right) / \left(\frac{\partial V_j}{\partial Q_j}\right)$$

Results are found to be in good agreement with short-circuit simulations in Table 5.2.

Sensitivity studies were conducted to calculate the attenuation values for the network under consideration. A sample of these results is shown in Figure 5.6 indicating that the attenuation matrix is numerically sparse. Carefully analyzed and

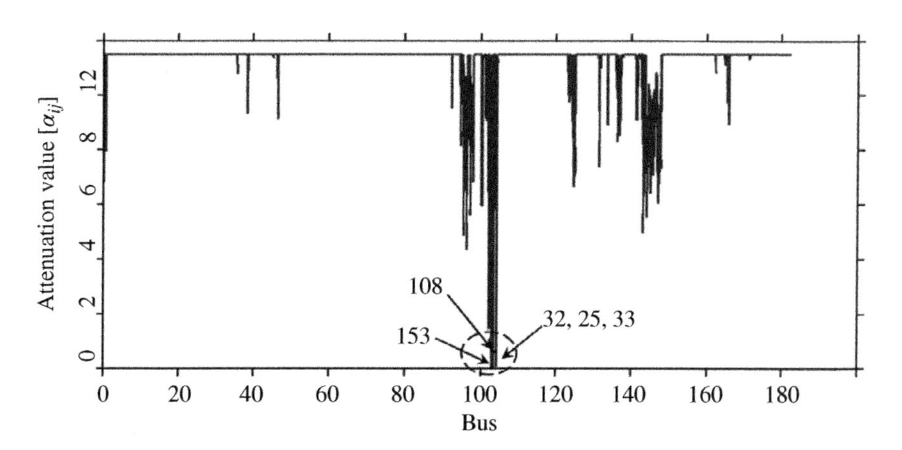

Figure 5.6 Sensitivity coefficients associated with bus 153 (unnormalized values).

Table 5.4 Candidate pilot buses selected using the information from electrical distances.

Zone	Bus (A_{tot_i})
1	153 (51 413), 32 (51 438), 25 (51 438), 33 (51 440)
2	73 (48 905), 81 (48 168), 122 (49 746), 2 (50 421)
3	136 (47 464), 26 (47 724), 18 (47 837), 13 (47 865), 94 (47 956), 126 (48 041), 65 (48 193), 88 (48 294), 44 (48 917), 22 (49 287)
4	167 (49 666)
5	61 (51 574)
6	108 (50 733)
7	156 (51 472)

sorted, electrical distances can be used to divide the system into several voltage control zones. Candidate pilot buses selected from this analysis include buses 153, 73, 136, 167, 61, 108, and 156 (refer to Table 5.4).

Based on the results of this exploratory analysis, seven major voltage control areas or electrical zones were initially identified for the regional voltage control scheme:

Zone 1 consists of major transmission and generation buses in the northeastern part of the system. These include buses 32, 25, 153, and 33. From the analytical results in Table 5.3, bus 153 is selected as the pilot bus for monitoring voltage within this zone.

Zone 2 groups buses in the southernmost part of the study regions and includes buses 2, 73, and 122. Bus 73 is selected as the pilot bus.

Zone 3 consists of buses 44, 26, 18, 70, 13, 36, 94, 65, 22, and 126. Within this zone, it is possible to identify several groups or subclusters. The analysis of electrical distances in Table 5.3 identifies bus 136 as a candidate pilot bus.

Zone 4 located in the northern portion of the study area comprises bus 167 and a pocket of buses in areas 4 and 5.

Zone 5 includes bus 61 and other lower-voltage buses.

Zone 6 comprises bus 108 and neighboring buses.

Zone 7 comprises bus 156 and neighboring buses.

Figure 5.7 shows the spatial distributions and principal transmission resources associated with these voltage control zones. In practice, clusters can be combined if they share certain properties or lack of generation sources.

Further, Table 5.5 lists the bus pilot nodes selected for analysis taking into account generation resources, suggesting that bigger areas could be obtained by relaxing the initial design criteria.

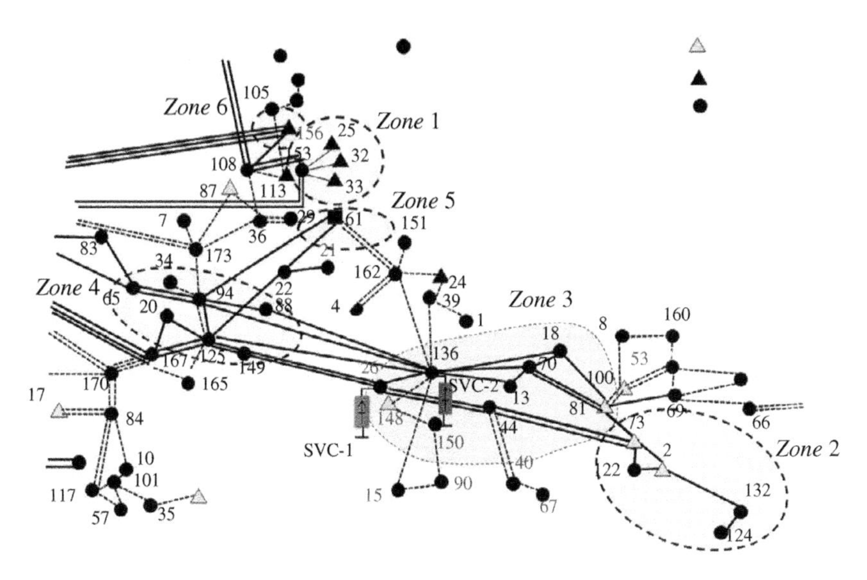

Figure 5.7 Estimated spatial distribution of voltage control areas based on sensitivity analysis.

Table 5.5 Bus pilot nodes selected from the identification of reactive power sources for each voltage control area.

Zone	Pilot node
1	153
2	73
3	136
4	167
5	61

In the studies that follow, complementary studies using data-driven techniques are conducted. Primary goals are to examine the inclusion of renewable generation to existing reactive control schemes.

5.5.2 Data-Driven Analysis

Selected system measurements were used to develop measurement-based approaches to voltage–VAR partition. The data used in this analysis consist of

Table 5.6 Selected system measurements.

Data modality	**Features**
Bus voltage magnitudes ($\boldsymbol{X}_V$)	173 signals
Bus voltage magnitude phases ($\boldsymbol{X}_\theta$)	173 signals
Generator reactive power output signals ($\boldsymbol{X}_{Qg}$)	105 generator reactive power output signals
SVC reactive power output signals ($\boldsymbol{X}_{SVC}$)	Three reactive power output signals (SVC-1, SVC-2, and SVC-3)

bus voltage magnitudes and phases, and generator and SVC reactive output power. Table 5.6 summarizes the main characteristics of signals selected for analysis.

For each of these scenarios, a set of six contingencies of interest was examined for system characterization (see Table 5.7).

Based on these results, the snapshot (measurement) data is defined as:

$$\begin{aligned}
\boldsymbol{X}_V &= [\boldsymbol{V}_1 \quad \boldsymbol{V}_{2}\ldots \quad \boldsymbol{V}_{173}]^T = [\boldsymbol{X}_{\text{bus}_{\text{volt}}} \quad \boldsymbol{X}_{\text{bu}_{\text{volt}_{\text{wfs}}}}]^T \\
\boldsymbol{X}_\theta &= [\boldsymbol{\theta}_1 \quad \boldsymbol{\theta}_{2}\ldots \quad \boldsymbol{\theta}_{173}]^T \\
\boldsymbol{X}_Q &= [\boldsymbol{Q}_1 \quad \boldsymbol{Q}_{2}\ldots \quad \boldsymbol{Q}_{105}]^T = [\boldsymbol{X}_{Q_{\text{gen}}} \quad \boldsymbol{X}_{Q_{\text{svc}}}]^T
\end{aligned} \tag{5.17}$$

Table 5.7 Contingency scenarios selected for analysis.

Contingency scenario	**Description**	**Remarks**
CE01	Generation outage	No-fault, generation outage, in Area 7
CE02	Generation outage	No-fault, generation outage in Area 3
CE03	Single-line outage	No-fault, line tripping. Circuit 1 of tie-line between Area 6 and Area 7 (study area)
CE04	Single-line outage	No-fault, line tripping. Tie-line between Area 6 and Area 3
CE05	Load shedding	2% load shedding at bus 100
CE06	Double-line outage	No-fault, double-line tripping

in which wfs denotes wind farms, and $\boldsymbol{V}_j = [V_j(t_0) \quad V_j(t_1) \quad \dots \quad V_j(t_N)]$, $\boldsymbol{V}_j = [V_j(t_0) \quad V_j(t_1) \quad \dots \quad V_j(t_N)]$, where $\boldsymbol{Q}_j = [V_j(t_0) \quad V_j(t_1) \quad \dots \quad V_j(t_N)]$, and $N = 3603$. In this case, the data need to be concatenated.

Figures 5.8–5.11 show selected simulations for CE06 in Table 5.8. These simulations are representative of other system responses and are selected to stimulate linear system response. $N = 3600$ snapshots corresponding to 30 sec are used for all contingency scenarios in Table 5.7.

5.5.3 Measurement-Based Reactive Control Areas

5.5.3.1 Diffusion Maps

Earlier studies using DMs showed that projection of measured data onto a low-dimensional manifold can capture system motion with only a limited number of interacting degrees of freedom (eigenvectors) [29]. The goal of this section is to examine the application of nonlinear reduction methods to determine reactive power control areas.

Following the general approach outlined in Section 5.4.2, a two-step hierarchical procedure for identifying voltage control areas and the associated reactive power sources was explored. The first step identifies a reduced-order model using the relevant DM coordinates a k-means clustering technique. In the second step,

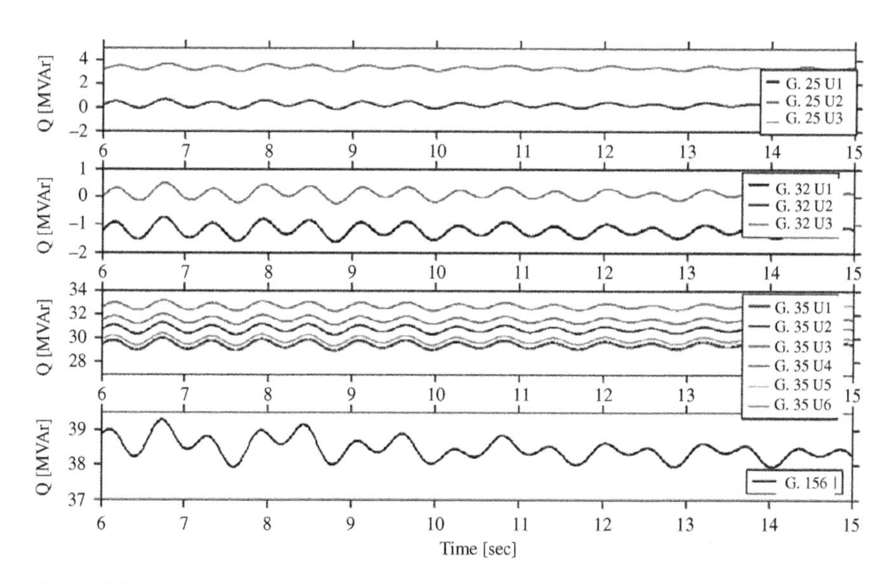

Figure 5.8 Reactive power deviations following a double-line outage for contingency scenario CE06.

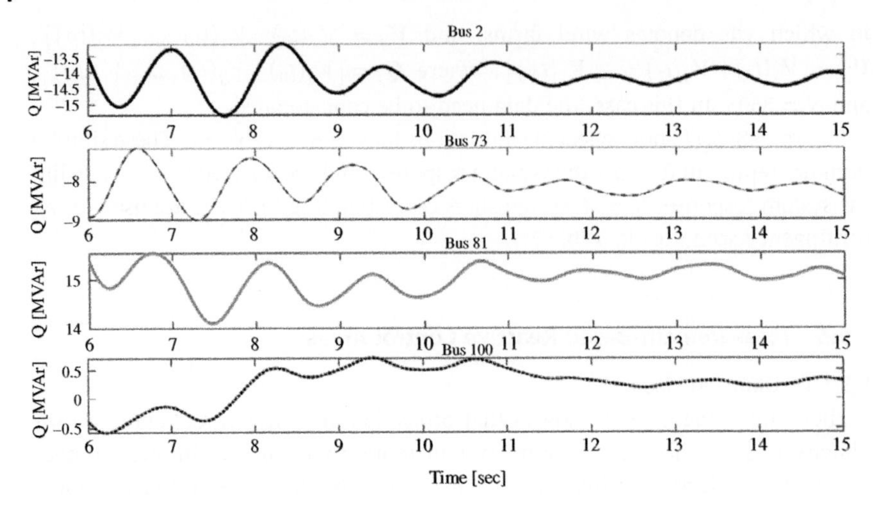

Figure 5.9 Reactive power output for generators in zone 2 for the contingency scenario CE06.

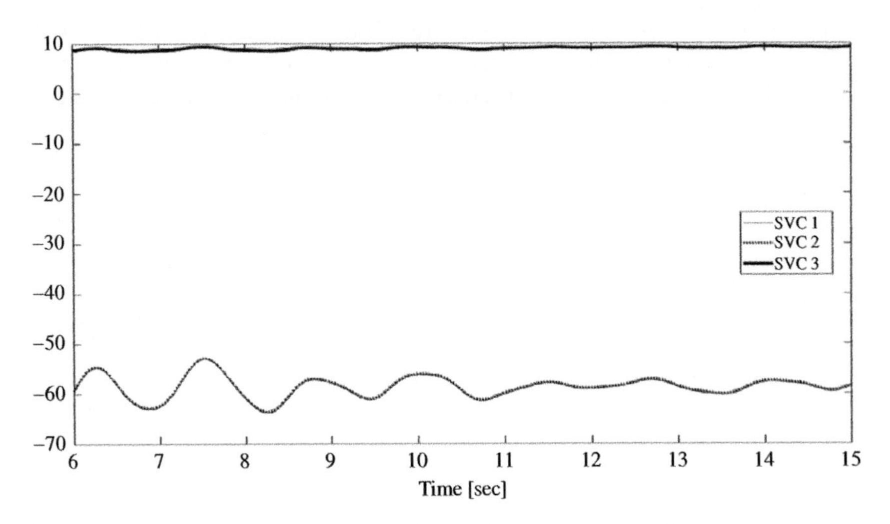

Figure 5.10 SVC reactive output power for contingency scenario CE06.

generators, renewable generation, and SVCs having a large strength of interaction with the selected voltage control areas are determined.

The whole process can be summarized as follows:

1) Construct the Markov matrix $\boldsymbol{M}$ from the observational data, $\boldsymbol{X}$;
2) Perform spectral analysis of the Markov probability transition matrix. Determine the intrinsic dimensionality and modal properties;

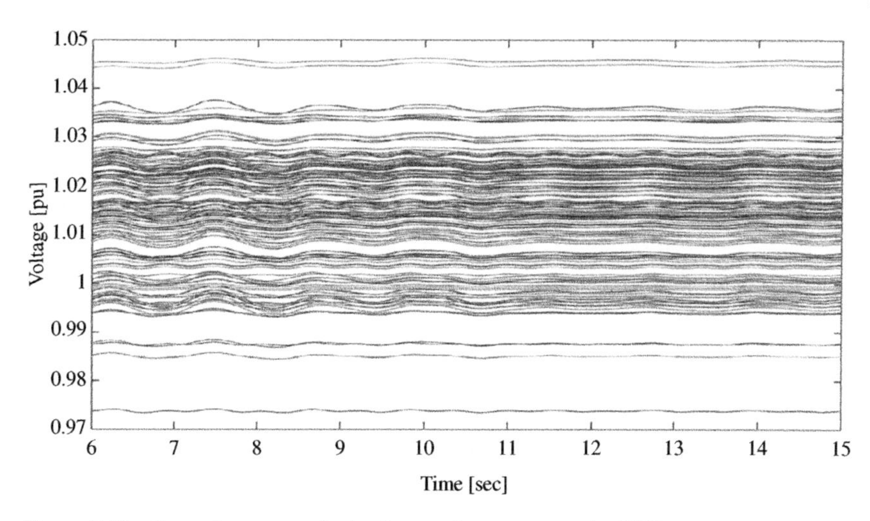

Figure 5.11 Bus voltage magnitudes for contingency scenario CE06.

Table 5.8 Clustering of bus voltage signals using *c*-means.

Cluster	Buses
1	2, 122, 132, 124
2	40, 67
3	8, 53, 69, 160
4	13, 18, 26, 73, 44, 81, 100, 136
5	15, 90, 150
6	1, 4, 24, 39, 151, 162,
7	17, 84, 101, 117, 110,170
8	13, 18, 26, 70, 73, 81, 100, 136, 148,
9	25, 32, 33, 153
10	34, 88, 94, 125, 149, 167

$m = 10$ and contingency scenario CE06.

3) Construct the mapping from the original space to the DM space (refer to Figure 5.4):

$$\Psi_{\varepsilon}^{d} = \{\psi_1 \quad \psi_2 \cdots \quad \psi_d\},$$

4) Cluster the DM embedding using k-means.

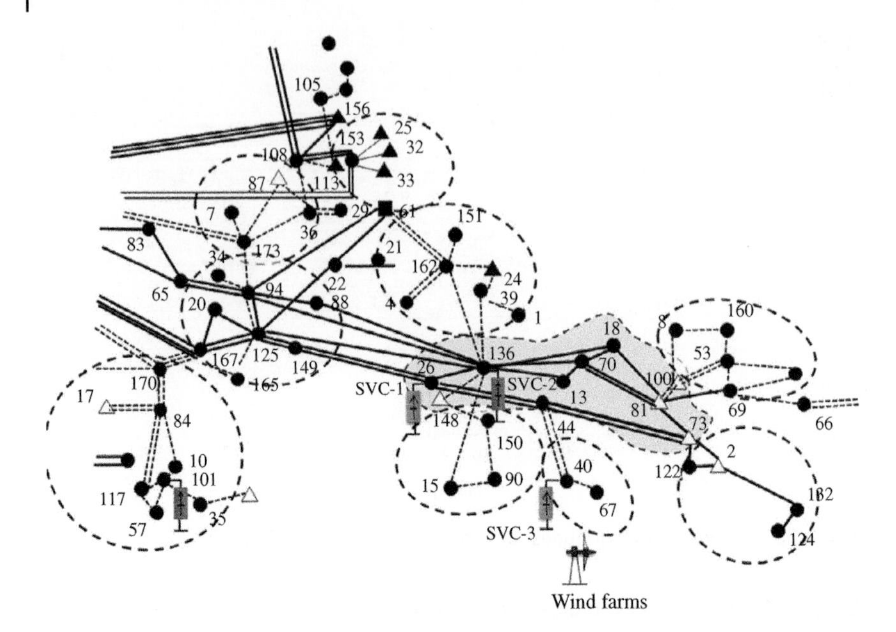

Figure 5.12 Approximate boundaries and geographical locations for clusters extracted using DMs. Modified case with an SVC at bus 40 and a large cluster of WFs connected at bus 67.

Figure 5.12 shows a schematic representation of the study system showing the approximate location of 10 voltage control zones for contingency scenario CE06 above determined using this procedure. As discussed below, similar results are obtained for other contingency scenarios which are not discussed here.

For the ease of visualization, 60 out of the original 173 buses in the database are selected for display. Also, the base case scenario was modified to include an additional SVC (SVC 3) as well as the representation of a cluster of wind farms in the vicinity of bus 67. While not discussed in detail, the identified voltage control zones largely coincide with the critical areas determined for other contingency scenarios.

By further adjusting the diffusion kernels, ϵ_i, ϵ_j voltage control areas can be further subdivided into two or more control areas or merged into a bigger area. Observe that two control areas in the neighborhood of major SVCs are identified associated with the location of large wind farms. Control strategies associated with this voltage control zone are discussed in Chapter 7.

5.5.4 Direct Clustering

The problem of direct clustering of measured bus voltage deviations is addressed by comparing the performance of c-means fuzzy analysis with the results of diffusion-based clustering in Figure 5.12. Table 5.8 gives the clusters determined using c-means fuzzy clustering. For this comparison, the number of clusters, m, was chosen based on the previous application of diffusion methods in Figure 5.12. Results are found to be in good agreement.

To verify the accuracy and suitability of the models and the appropriateness of the selected operating scenarios, comparisons are provided with other analysis methods. Dynamic mode decomposition is first used to determine the buses having the largest participation in the oscillations. These methods use the voltage-based mode shapes of the dominant oscillatory modes extracted from selected bus voltage magnitude recordings.

Table 5.9 shows the top 10 entries determined using the dynamic mode decomposition (DMD) method in Ref. [41]. It is noted that DMD results are highly consistent regardless of the location or severity of the contingencies analyzed. Also, Figure 5.13 gives a comparison between the candidate pilot bus locations determined conventional sensitivity analyses in Table 5.4 with those identified using DMD in Table 5.9.

For comparison, Figure 5.14 shows the leading eigenvector of the DMs for contingency scenarios CE04 and CE05; it is observed that while the modes excited depend on the location of the disturbance and the structure of the network, reduced-order models can still accurately capture relevant system behavior.

Table 5.9 Top 10 (unnormalized) entries of the voltage-based DMD eigenvector (dominant mode).

Contingency scenario	Bus (magnitude of mode shape)
CE01	153 (0.259), 32 (0.220), 31 (0.175), 24 (0.156), 152 (0.156), 108 (0.146), 156 (0.141), 113 (0.130), 33 (0.128), 25 (0.111), 36 (0.102), 106 (0.126), 95 (0.083), 87 (0.086)
CE02	32 (0.229), 153 (0.229), 156 (0.195), 111 (0.172), 31 (0.166), 108 (0.158), 24 (0.137), 110 (0.153), 157 (0.152), 152 (0.139)
CE03	32 (0.227), 153 (0.226), 111 (0.168), 31 (0.162), 107 (0.154), 24 (0.139), 110 (0.152), 156 (0.149), 25 (0.132), 109 (0.135)
CE04	32 (0.212), 153 (0.212), 111 (0.158), 155 (0.173), 31 (0.161), 107 (0.147), 157 (0.141), 110 (0.144), 152 (0.135), 24 (0.135)
CE05	159 (0.132), 65 (0.123), 54 (0.108), 7 (0.106), 68 (0.103), 25 (0.093), 33 (0.094), 8 (0.088), 99 (0.094), 160 (0.096)

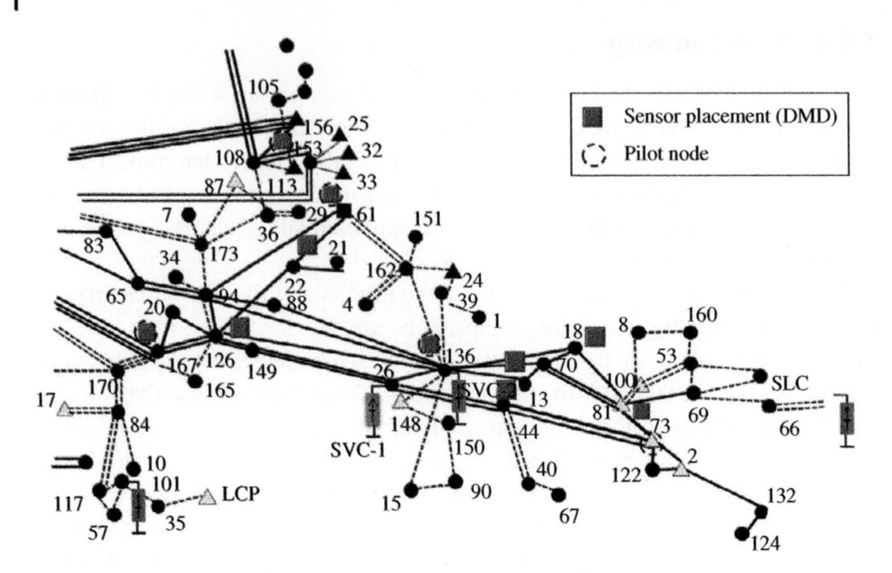

Figure 5.13 The candidate pilot node locations. Empty dashed circles show pilot node locations from Table 5.5. Filled gray squares show the candidate node locations obtained from DM information in Table 5.9.

Similar results are obtained using the MCL algorithm in Figure 5.15. In this case, the leading eigenvector is obtained solving the standard eigenvector equation:

$$(\boldsymbol{M}_{\text{MCL}}(r) - \lambda_{\text{MCL}_2}\boldsymbol{I})\boldsymbol{\phi}_{\text{MCL}_2} = \boldsymbol{0}$$

where the power coefficient r is set to $r = 4$, and $\boldsymbol{\phi}_{\text{MCL}_2}$ denotes the second eigenvector in the spectral decomposition.

5.5.5 Correlation Analysis

The performed studies have shown that spectral techniques can be used to identify voltage coherent areas and the associated dominant buses. This section focuses on how to extend data-driven techniques to account for interactions between bus voltage deviations and reactive power sources.

The simplest approach to determine correlation measures between two data matrices $\boldsymbol{X}$ and $\boldsymbol{Y}$ can be obtained from the correlation matrix $\hat{\boldsymbol{C}} = \boldsymbol{X}^T\boldsymbol{Y}/N$. Reactive power and voltage data, however, are intrinsically dissimilar in nature and some sort of scaling or normalization may be required. Also, data fusion may be needed to allow for the analysis of true interactions between the data modalities.

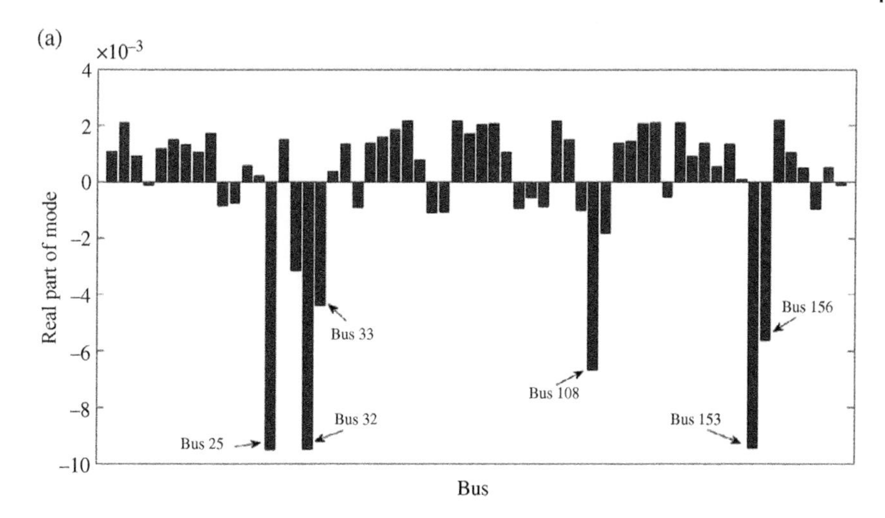

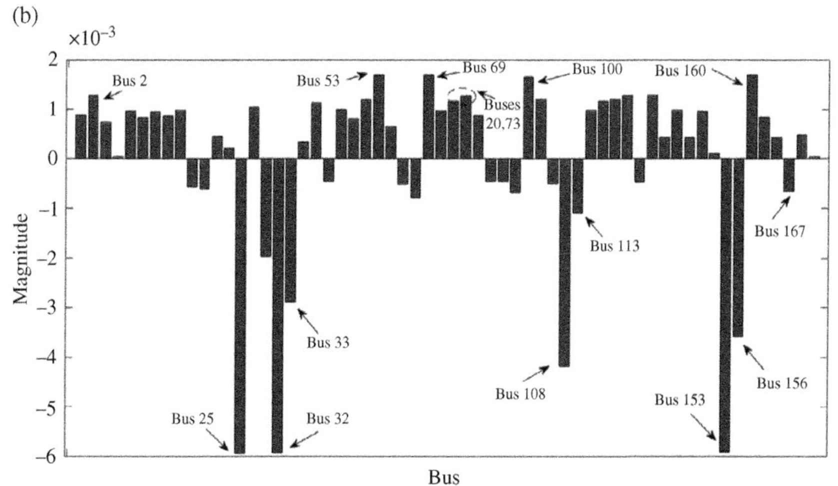

Figure 5.14 Leading eigenvector of the Markov transition matrix: (a) contingency scenario CE04 and (b) contingency scenario CE05 (unnormalized values).

In this section, basic correlations measures are extended to allow simultaneous analysis of multiple datasets within the framework of data fusion and diffusions maps. Three approaches to the analysis of correlation measures are discussed here:

1) Direct analysis of concatenated data;
2) The application of multiview DMs;
3) Consensus analysis of multivariate data.

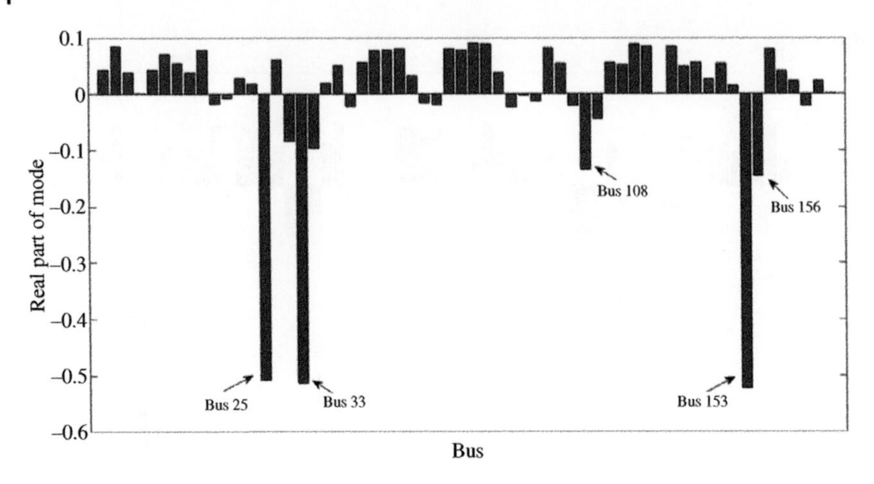

Figure 5.15 Dominant eigenvector extracted using the MCL for contingency scenario CE04.

5.5.5.1 Direct Analysis of Concatenated Data

A simple approach to enable a comparison of multitype data of complementary nature is the use of detrending techniques. Along the same vein as Ref. [33], the row-wise concatenated data are defined as:

$$X = \left[X_V \quad X_{Q_{gen}} \; X_{Q_{svc}}\right] \tag{5.18}$$

Because of the non-homogeneity of the data, data fusion techniques need to be utilized to reveal important relationships between pilot nodes and reactive power outputs. Alternatively, conventional normalization techniques can be combined with detrending techniques to allow a joint analysis of multimodal data. This is the approach adopted here.

In the first attempt to determine correlation measures, normalization techniques were applied to (5.18). Assume to this end, that the individual data types in (5.18) are independently normalized. In the second step, the overall data matrix $\hat{X}$ is detrended using linear or nonlinear detrending techniques.

The following three-step procedure is used to determine sensitivity relations between bus voltage deviations and reactive power outputs:

1) Normalize the individual data matrices in (5.18);
2) Detrend the data using wavelet shrinkage or a nonlinear detrending technique;
3) Determine a reduced-order model using data fusion. Compute the distance matrix $\boldsymbol{M}$ and determine the distance coefficients d_{ij} between generation buses, SVC buses, and wind farm buses, and bus voltage magnitudes using (5.13).

Table 5.10 The five largest interaction coefficients m_{ij} computed using (5.5) for the base operating case.

Bus	Bus (diffusion coefficients α_{ij})
153	25, 32, 33, 35, 156, 113
73	2, 811, 122, 100
136	26, 148, 81, 73, 2
167	17, 61, 87, 26, 136
61	61, 62, 94, 22, 125

Table 5.10 shows the five largest diffusion coefficients obtained using this approach. Comparisons with reactive power sensitivities show that DMs can provide an accurate characterization of the strength of interactions between bus voltage magnitudes and reactive power reserves and also have a strong physical interpretation. It is noted that the adopted approach identifies both SVCs and generators showing the largest strength of interaction with nearby buses.

The practical application of these concepts to complex power systems with increased wind penetration is deferred to Chapter 8.

5.5.5.2 Two-Way Correlation Analysis

With reference to Eq. (5.18), let $\boldsymbol{X} = \boldsymbol{X}_V$ and $\boldsymbol{Y} = \left[\boldsymbol{X}_{Q_{\text{wfs}}} \quad \boldsymbol{X}_{Q_{\text{gen}}} \ \boldsymbol{X}_{Q_{\text{svc}}}\right]$. Motivated by the notion of multiview DMs [42], the concatenated data are defined as:

$$\hat{\boldsymbol{X}} = \begin{bmatrix} \boldsymbol{O}_1 & \boldsymbol{X}^T\boldsymbol{Y} \\ \boldsymbol{Y}^T\boldsymbol{X} & \boldsymbol{O}_2 \end{bmatrix} \tag{5.19}$$

where $\boldsymbol{O}_1$ and $\boldsymbol{O}_2$ are matrices of zeros of appropriate dimensions. Observe that other formulations such as row-concatenated data can be used for the case of more than two data types.

5.5.5.3 Partial Least Squares Correlation

Given two data matrices $\boldsymbol{X} \in \Re^{N \times p}$ and $\boldsymbol{Y} \in \Re^{N \times m}$, this technique analyzes the correlation between matrices $\boldsymbol{X}$ and $\boldsymbol{Y}$. Following the formulation of Krishnan et al. [43] and Abdi and Williams [44], the cross-block correlation matrix, $\boldsymbol{R}$, is defined by:

$$\boldsymbol{R} = (\boldsymbol{Y})^T\boldsymbol{X} \tag{5.20}$$

Singular value decomposition (SVD) of $\boldsymbol{R}$ yields

$$R = U\Sigma V^T \tag{5.21}$$

Here, U and V are orthonormal matrices called saliences containing the left and right singular vectors of R, respectively, and Σ is a diagonal matrix of the nonzero singular values. The latent variables, L_x and L_y, which are linear combinations of the original values, are obtained by projecting the data matrices onto their respective saliences as:

$$L_x = XV$$

and

$$L_y = YU$$

Several observations are of interest here:

- Matrices L_x and L_y give the datasets spatial patterns or shapes;
- The correlation matrix R gives the correlation between a given output signal, Q_j, and a given input signal, V_k. Accordingly, the correlation coefficient R_{ij} between i*th* bus's voltage magnitude with respect to the i*th* reactive power output is obtained from the rows of the correlation matrix.

When coupled with real-time information, these techniques have also the potential to correlate reactive power reserves with bus voltage deviation and provide operational warnings.

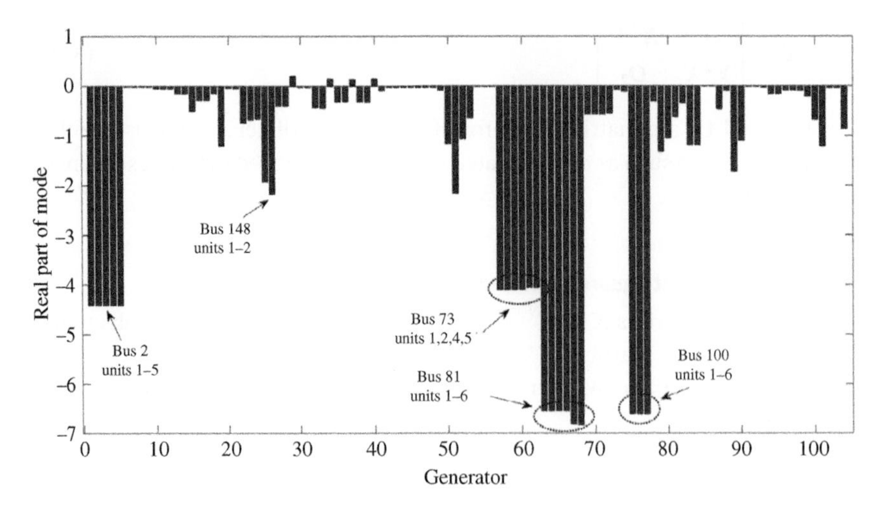

Figure 5.16 Partial least squares regression of datasets X and Y.

Figure 5.16 shows partial least squares regression results for data matrices let $\boldsymbol{X} = \boldsymbol{X}_V$ and $\boldsymbol{Y} = \left[\boldsymbol{X}_{Q_{wfs}} \quad \boldsymbol{X}_{Q_{gen}}\ \boldsymbol{X}_{Q_{svc}}\right]$. Results correlate well with previous results in Figures 5.14 and 5.15.

This example demonstrates that correlation analyzes may identify key relationships between observed bus voltage deviations and output reactive power from wind farms and synchronous generators, such as sensitivity relationships, or levels of strength associated with reactive power control areas. It is found that both simple correlation analysis and partial least squares regression perform well for the simulated records.

5.6 Summary

Over the last few decades, various forms of voltage monitoring and reactive power management systems have been developed. These architectures offer the possibility to calculate in near-real-time several aspects of voltage control and VAR management such as distances to voltage collapse and network reactive control, distances to instability, and sensitivity calculations. The (geographical) dispersion of dynamic recorders and the changing network structure and operating conditions, however, create wide-area voltage monitoring issues that must be addressed using advanced data-based analysis approaches and correlation techniques.

In this chapter, a new use of spectral analysis tools for dimensionality reduction and clustering of high-dimensional datasets, based on the normalized graph Laplacian has been introduced. The proposed technique combines the inherent abilities of graph-theoretical techniques with spectral clustering and visualization methods to identify loosely interconnected voltage control areas and reconstruct system behavior using selected measurements. Attention has been focused on three main aspects, namely the identification of critical system zones showing a coherent behavior and the associated reactive power sources, the use of spectral analysis to identify the critical buses, and the computation of reduced-order models.

Issues such as reserve capacity control and the efficient utilization and coordination of reactive power sources need to be addressed to achieve fast voltage monitoring and control and keep the reactive power output margin against system contingencies.

References

1. Coordinated Voltage Control in Transmission Networks. (2007). CIGRE Task Force C4.602 (February 2007).

2. Corsi, S., Possi, M., Sabelli, C., and Serrani, A. (2004). The coordinated automatic voltage control of the Italian transmission grid - Part I: reasons of the choice and overview of the consolidated hierarchical system. *IEEE Transactions on Power Systems* **19** (4): 1723–1732.
3. Cañizares, C.A., Cavallo, C., Pozzi, M., and Corsi, S. (2005). Comparing secondary voltage regulation and shunt compensation for improving voltage stability and transfer capability in the Italian power system. *Electric Power Systems Research* **73**: 67–76.
4. Lagonotte, P., Sabonnadiere, J.C., Léost, J.Y., and Paul, J.P. (1989). Structural analysis of the electrical system: application to secondary voltage control in France. *IEEE Transactions on Power Systems* **4** (2): 479–486.
5. Taylor, C.W. (1994). *Power System Voltage Stability*. McGraw-Hill.
6. Taylor, C.W. (1993). Survey of effective and practical solutions for longer-term voltage stability. *Electrical Power and Energy Systems* **15** (4): 217–220.
7. Taylor, C.W. (2000). The Future in on-line security assessment and wide-area stability control. Proceedings of the 2000 IEEE Power Engineering Society Winter Meeting.
8. AEMO. (2013). Australian energy market operator, wind turbines plant capabilities report. Wind Integration Studies. https://www.aemo.com.au/-/media/Files/PDF/Wind_Turbine_Plant_Capabilities_Report.pdf/.
9. Miller, N.W., Guru, D., and Clark, K. (2008). Wind generation applications for the cement industry. Proceedings of the IEEE Cement Industry Technical Conference Record.
10. Miller, N., MacDowell, J., Chmiel, G. et al. (2012). Coordinated voltage control for multiple wind plants in eastern wyoming: analysis and field experience.
11. Ullah, N.R. and Bhattacharya, K. (2008). Wind farms as reactive power ancillary service providers—technical and economic issues. *IEEE Transactions on Energy Conversion* **24** (3): 661–672.
12. Taranto, G.N., Martins, N., Falcao, D.M. et al. (2000). Benefits of applying secondary voltage control schemes to the Brazilian system. Proceedings of the 2000 Power Engineering Society Summer Meeting (July 2000).
13. Sancha, J.L., Fernandez, J.L., Cortez, A., and Abarca, J.T. (1996). Secondary voltage control: analysis, solutions and simulation results for the Spanish transmission system. *IEEE Transactions on Power Systems* **11** (2): 630–638.
14. Paul, J.P., Leost, J.Y., and Tesseron, J.M. (1987). Survey of the secondary voltage control in France: present realization and investigations. *IEEE Transactions on Power Systems* **PWRS-2** (2): 505–511.
15. Kundur, P. (1994). *Power System Stability and Control, The EPRI Power System Engineering Series*. New York, NY: McGraw-Hill.
16. Carpentier, J.L. (1987). "CRIC", a new active-reactive decoupling process in load flows, optimal power flows and system control. *IFAC Proceedings* **20** (6): 59–64.

17. Zhong, J., Nobile, E., and Bose, A. (2004). Localized reactive power markets using the concept of voltage control areas. *IEEE Transactions on Power Systems* **19** (3): 1555–1561.
18. Cotilla-Sanchez, E., Hines, P., and Barrows, C. (2013). Multi-attribute partitioning of power networks based on electrical distance. *IEEE Transactions on Power Systems* **28** (4): 4979–4987.
19. Tyuryukanov, I., Popov, M., van der Meijden, M.A.M.M., and Terzija, V. (2018). Discovering clusters in power networks from orthogonal structure of spectral embedding. *IEEE Transactions on Power Systems* **33** (6): 6441–6451.
20. Jiang, T., Bai, L., Ji, L., and Li, F. (2007). Spectral clustering-based partitioning of volt/VAR control areas in bulk power systems. *IET Generation, Transmission & Distribution* **11** (5): 1126–1133.
21. Saugata, S., Biswas, T., and Srivastava, A.K. (2014). Performance analysis of a new synchrophasor based real time voltage stability monitoring (RT-VSM) tool. Proceedings of the 2014 North American Power Symposium (NAPS).
22. Sun, H., Guo, Q., Zhang, B. et al. (May 2013). An adaptive zone division-based automatic voltage control system with applications in China. *IEEE Transactions on Power Systems* **28** (2): 1816–1828.
23. Morison, K., Wang, X., Moshref, A., and Edris, A. (2008). Identification of voltage control areas and reactive power reserve; an advancement in on-line voltage security assessment. Proceedings of the 2008 IEEE Power and Energy Society General Meeting – Conversion and Delivery of Electrical Energy in the 21st Century.
24. Roman-Messina, A. (2015). *Wide-area Monitoring of Interconnected Power Systems, IET, Power and Energy Series 77, Stevenage.* UK: IET, The Institute of Engineering and Technology.
25. Li, H., Li, F., Xu, Y., Y. et al. (2010). Adaptive voltage control with distributed energy resources: algorithm, theoretical analysis, simulation, and field test verification. *IEEE Transactions on Power Systems* **25** (3): 1638–1647.
26. Messina, A.R. and Vittal, V. (2007). Extraction of dynamic patterns from wide-area measurements using empirical orthogonal functions. *IEEE Transactions on Power Systems* **2** (2): 682–692.
27. Newman, M. (2010). *Networks*, 2e. Oxford, UK: Oxford University Press.
28. Dray, S., Legendre, P., and Peres-Neto, P.R. (2006). Spatial modelling: a comprehensive framework for principal coordinate analysis of neighbour matrices (PCNM). *Ecological Modelling* **196**: 483–493.
29. Arvizu, C.M. and Messina, A.R. (2016). Dimensionality reduction in transient simulations: a diffusion maps approach. *IEEE Transactions on Power Delivery* **31** (5): 2379–2389.

30. David, G. and Averbuch, A. (2012). Hierarchical data organization, clustering and denoising via localized diffusion folders. *Applied and Computational Harmonic Analysis* **33** (1): 1–23.
31. Griffith, D. (2000). Eigenfunction properties and approximations of selected incidence matrices employed in spatial analyzes. *Linear Algebra and Its Applications* **321** (1-3): 95–112.
32. Moghadas, S.M. and Jaberi-Douraki, M. (2019). *Mathematical Modelling*. Hoboken, NJ: Wiley.
33. Román-Messina, A. (2020). *Data Fusion and Data Mining for Power System Monitoring*. Boca Raton, FL: CRC Press.
34. S. Van Dongen, Graph Clustering by Flow Simulation. PhD thesis, University of Utrecht, May 2000.
35. Enright, A.J., Van Dongen, S., and Ouzounis, C.A. (2002). An efficient algorithm for large-scale detection of protein families. *Nucleic Acids Research* **30** (7): 1575–1584.
36. Malmstrom, R.D., Lee, C.T., Van Wart, A.T., and Amro, R.E. (2014). Application of molecular-dynamics based Markov state models to functional proteins. *Journal of Chemical Theory and Computation* **10**: 2648–2657.
37. Cazade, P.A., Zheng, W., Prada-Garcia, D. et al. (2015). A comparative analysis of clustering algorithms: O2 migration in truncated hemoglobin I from transition networks. *The Journal of Chemical Physics* **142**: 025103-1-15.
38. Bezdek, J.C. (1981). *Pattern Recognition with Fuzzy Objective Function Algorithms*. New York, NY: Plenum Press.
39. Bezdek, J.C., Ehrlich, R., and Full, W. (1984). FCM: The Fuzzy c-means clustering algorithm. *Computers and Geosciences* **10** (2–3): 191–203.
40. Zadeh, L.A. (1971). Similarity relations and fuzzy orderings. *Information Sciences* **3**: 177–200.
41. Barocio, E., Pal, B.C., Thornhill, N.F., and Roman-Messina, A. (2015). A dynamic mode decomposition framework for global power system oscillation analysis. *IEEE Transactions on Power Systems* **30** (6): 2902–2912.
42. Lindenbaum, O., Yeredor, A., Salhov, M., and Averbuch, A. (2020). Multi-view diffusion maps. *Information Fusion* **57**: 127–149.
43. Krishnan, A., Williams, L.J., b., McIntosh, A.R., and Abdi, H. (2011). Partial least squares (PLS) methods for neuroimaging: a tutorial and review. *Neuroimage* **5**: 455–475.
44. Abdi, H. and Williams, L.J. (2013). *Partial least squares methods: partial least squares correlation and partial least square regression*. In: *Computational Toxicology*, Methods in Molecular Biology, 930, vol. **II** (eds. B. Reisfeld and A.N. Mayeno), 543–579. New York, NY: Human Press-Springer.

6

Advanced Control Synthesis

Modern power grids, characterized by high penetration of inverter-interfaced generation sources, suffer from a lack of rotational inertia and governor control, which may cause faster frequency dynamics and larger frequency deviations and even instability problem. Over the last few years, several inertia-emulation-based control strategies have been developed and applied to support system frequency in power systems with significant renewable generation. Conceptually, those control strategies manipulate the converter power injections in response to local frequency deviations to support system stability.

In *Chapter 4*, the problems of inertia emulation and placement were discussed, based on the notion of virtual synchronous generator (VSG) control. In this chapter, advanced inertia-emulation-based control strategies to enable high penetration levels of microgrid (MG) are discussed. These approaches combine an adaptive energy storage system (ESS) dispatch strategy with an MG-controlled islanding scheme to provide frequency support and enhance small-signal stability and voltage regulation.

6.1 Introduction

Power grids worldwide are experiencing a significant transformation, arising from the increased penetration of distributed generations (DGs). As discussed in previous chapters, reduced rotational inertia in the grid may adversely affect frequency response and system control and lead to degrade performance of traditional control schemes. This, in turn, may result in large frequency and voltage deviations and, potentially, load shedding and instability [1–5]. Advanced control of grid-connected MGs, however, has the potential to offset the intermittent nature of distributed energy resources and provide stability support to the host utility during emergency conditions, which is emphasized in this chapter.

Renewable Integrated Power System Stability and Control, First Edition.
Hêmin Golpîra, Arturo Román-Messina, and Hassan Bevrani.

6.2 Frequency Dynamics Enhancement

6.2.1 Background: The Concept of Flexible Inertia

In *Chapter 2*, the concept of center of gravity (COG) was introduced to study the long-term power–frequency transients following large perturbations. Using the COG concept, the original power system can be represented by the simplified equivalent model of Figure 6.1 [6].

To develop an advanced ancillary control scheme for frequency support, consider an equilibrium mechanical system with equilibrium point δ_0. The equilibrium condition requires that the resultant torque in the COG vanishes, namely [7],

$$F_1 \cos \delta L_1 - F_2 \cos \delta L_2; \quad F_k = m_k g h_k, \quad k = 1, 2 \tag{6.1}$$

where F, L, and δ denote the applied force, the lever arm length, and the angle between the beam and the horizon, respectively. Figure 6.2 shows a schematic representation of this model.

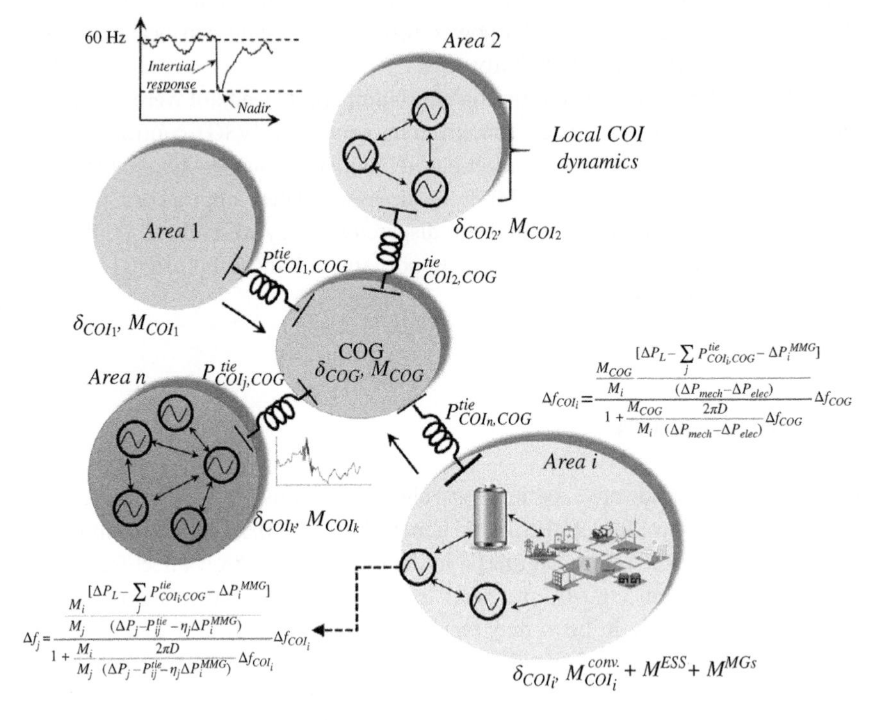

Figure 6.1 Conceptual overview of the adopted control scheme showing the grid-parallel configuration of interconnected MGs.

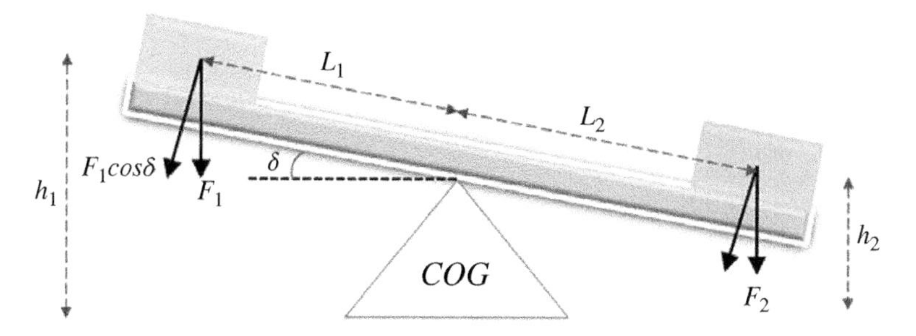

Figure 6.2 Mechanical system explaining the adopted control scheme.

Equation (6.1) establishes that any change in L should be compensated by F and vice versa to keep the equilibrium point δ_0.

The same reasoning used in (6.1) and Figure 6.2 may be applied to the equivalent model of Figure 6.1. In analogy to a mechanical system, the inertia M can be interpreted as the mass, the fictitious reactance X is the length of the lever arm, and the inertia-weighted disturbance $\Delta P_i/M_i$ is a force.

The size of the inertia-weighted perturbation will cause a frequency variation, given by the swing equation of the form [8]

$$\frac{df(t)}{dt} = \frac{T_m(t) - T_e(t)}{2M} \tag{6.2}$$

where $T_m(t)$ and $T_e(t)$ are the time-varying mechanical input torque and the electrical output torque, respectively. Incorporating the tie-line flows, (ΣP^{tie}), as well as the load dynamics (D), into (6.2) results in [3, 9]

$$\dot{f} = \frac{1}{2\pi M}\left[\Delta P - 2\pi D \Delta f - \sum P^{tie}\right] \tag{6.3}$$

where P^{tie} in the equivalent system of Figure 6.1 has the form [10]

$$P^{tie}_{COI_i,COG} = \frac{|V_{COI_i}||V_{COG}|}{X^{tie}_{COI_i,COG}} \sin\left(\delta_{COI_i} - \delta_{COG}\right) \tag{6.4}$$

Furthermore, the fictitious reactance X^{tie} can be calculated using (2.28), i.e.

$$\min_{X^{tie}_{COI_i},\, COG} \textstyle\sum_{S=1}^{S} p_S \sum \xi_S^- - \xi_S^+$$

$$s.t.$$

$$\sum_{i=1}^{n} \frac{1}{\omega_{COG}} \frac{|V_{COI_i}||V_{COG}|}{X^{tie}_{COI_i,COG}} \sin\left(\delta_{COI_i} - \delta_{COG}\right) + \xi_S^- - \xi_S^+ = \sum_{i=1}^{n} \Delta P^{MMG}_{i,s}$$

$$\xi_S^-, \xi_S^+ \geq 0$$

$$0 < X^{tie}_{COI_i,COG} \leq 1 \tag{6.5}$$

According to (6.3) and (6.5), the Eq. (6.1) for the equivalent system of Figure 6.1 can be reorganized as

$$\frac{\Delta P_1}{M_1}X_1 + \frac{\Delta P_2}{M_2}X_2 + \frac{\Delta P_3}{M_3}X_3 + \dots + \frac{\Delta P_n}{M_n}X_n = 0 \tag{6.6}$$

The equilibrium condition (6.6), as the basic principle of the inertia-based control scheme for frequency support, relies on the following simple yet effective premise [11]: "Any variation in the equivalent reactance, X in Figure 6.1, following a disturbance, will be compensated by a corresponding change in $\Delta P_i/M_i$ so that the equilibrium condition is satisfied." Area i in Figure 6.1 schematically describes the variation of F, through manipulation of M, in the equivalent system. Referring to Figure 6.1 and taking the contribution of multi-MGs (MMGs) into account, the swing Eq. (6.3), relative to the COG, can be expressed in the form of (6.7) [11].

$$\dot{f}_{COI_i} = \frac{1}{2\pi M_{COI_i}}\left[\Delta P_i - 2\pi D_i \Delta f_{COI_i} - P^{tie}_{COI_i,COG} - \Delta P_i^{MMG}\right] \tag{6.7}$$

Equation (6.7) establishes that frequency deviations exceeding a predetermined threshold can be limited by suitable modifying the center of inertia (COI) inertia, M_{COIi}. This, in turn, results in a change in $P^{tie}{}_{COIi,COG}$ according to (6.3), which further affects frequency dynamics (see Eq. (2.4) in *Chapter 2*). Moreover, (6.7) suggests that the proper transition of MMGs from the grid-connected mode to the islanded mode can be used to aid frequency support. Generally stated, for area i, the total inertia can be explicitly decomposed into three constituent components, as:

$$M_i = M_{COI_i}^{Conv.} + M^{ESS} + M^{MMG_s} \tag{6.8}$$

where $M_{COI_i}^{Conv.}$, M^{ESS}, and M^{MMGs} represent, respectively, the conventional synchronous inertia and the inertia provided by the ESS and MMGs.

Key parameters of interest to be controlled via manipulation of (6.8) include the frequency nadir, the rate of change of frequency (*RoCoF*), and the frequency deviation during a given time interval of interest. Using this framework, the ESS output is manipulated to reduce the *RoCoF* and the frequency nadir through the emulated virtual inertia. Disconnecting the MMGs from the network (islanding) in a time horizon greater than the response time of ESS, on the other hand, increases the frequency evolution to satisfy the rolling window criterion. The effects of the MMGs islanding strategy on the frequency response are examined based on (6.8) for two different operating conditions of the MGs, i.e. power import from the grid and power export to the grid. These cases are discussed separately as follows [11]:

1) Power import operating mode: In the inertial response period where $T_m < T_e$, the MG islanding causes the torque T_e to decrease, which in turn decreases the *RoCoF* and nadir. On the other hand, for the time interval beyond the nadir, where $T_m > T_e$, islanding leads to a greater acceleration torque and thus a faster frequency recovery.
2) Power export operating mode: In this case, the MG islanding, reinterpreted as loss of inertia according to (6.8), renders frequency dynamics faster. The inertia-based control strategy imposes the islanding of the MMGs beyond the frequency arrest period.

6.2.2 Frequency Dynamics Propagation

Inertia manipulation via coordinated control action of ESS and MMGs relies on the propagation of frequency dynamics in the reduced equivalent system, represented by Figure 6.1 and (6.5) [see Equations (2.15) and (2.17) in *Chapter 2*]. Using this notion, it is assumed that any disturbance may be aggregated and applied to the COG. As discussed in *Chapter 2* and [12], any change in the instantaneous system frequency following a disturbance can be presented in the COG formulation, given by (6.9).

$$2M_{COG}\frac{df_{COG}(t)}{dt} = \Delta R(t) - \Delta P_L \tag{6.9}$$

Note that (6.9) incorporates the additional power delivered through frequency response, i.e. R [MW], into the formulation [13]. By integrating (6.9) and assuming that the frequency response is delivered by linearly increasing the active power with a fixed slope (R/T_D) during inertial response [14], one could write

$$\Delta f_{COG}(t) = \frac{\frac{1}{2}\frac{R}{T_D}t^2 - \Delta P_L t}{2M} \tag{6.10}$$

where Δf_{COG}, ΔP_L, and T_D are the COG frequency deviation, imbalance power, and the delivery time of primary frequency response, respectively. The frequency nadir for the COG can be calculated by setting the derivative of (6.10) with respect to t equal to zero, namely

$$\frac{\partial \mid \Delta f_{COG}(t) \mid}{\partial t} = 0 \tag{6.11}$$

which leads to

$$\frac{\frac{R}{T_D}t - \Delta P_L}{2M} = 0 \tag{6.12}$$

Equation (6.12) is satisfied when

$$t = \frac{\Delta P_L}{\frac{R}{T_D}} \tag{6.13}$$

By substituting (6.13) in (6.10), one obtains:

$$\Delta f_{nadir}^{COG} = f_{nadir}^{COG} - f_0 = \frac{\Delta P_L^2 T_D}{4MR} \tag{6.14}$$

Further substitution of (6.2) into (6.14), and noting that

$$\frac{d\Delta f(t)}{dt} = RoCoF \tag{6.15}$$

results in

$$\Delta f_{nadir}^{COG} = \frac{4M^2 \left(\frac{d\Delta f}{dt}\right)^2 T_D}{4MR} = \frac{M(RoCof)^2 T_D}{R} \tag{6.16}$$

All the independent parameters in (6.16) are system-dependent except for R/T_D. As a common assumption, the dependent variable Δf_{nadir} is set to the maximum allowable value to calculate the worst-case R/T_D. Accordingly, the COG frequency dynamics may be specified by (6.16). The problem of interest, however, is to calculate the local dynamics of specified buses exhibiting undesired behavior. Alternatively, local behavior in (6.16) can be obtained using sensitivity relations in the reduced equivalent system of the form (6.5). Dividing (6.7) by the classical form of (6.9) gives

$$\frac{df_{COI_i}}{df_{COG}} = \frac{M_{COG}}{M_{COI_i}} \frac{\left[\Delta P_i - 2\pi D \Delta f_{COI_i} - P_{COI_i,COG}^{tie} - \Delta P_i^{MMG}\right]}{(\Delta P_{mech} - \Delta P_{elec})} \tag{6.17}$$

where ΔP^{MMG} defines the difference between the total generation and the local MMG load. By knowing the COG dynamics of interest in (6.9) and (6.16), and the frequency sensitivity relation in (6.17), one can write

$$\Delta f_{COI_i} = \frac{M_{COG}}{M_{COI_i}} \frac{\left[\Delta P_i - 2\pi D \Delta f_{COI_i} - P_{COI_i,COG}^{tie} - \Delta P_i^{MMG}\right]}{(\Delta P_{mech} - \Delta P_{elec})} \Delta f_{COG} \tag{6.18}$$

or equivalently,

$$\Delta f_{COI_i} = \frac{\frac{M_{COG}}{M_{COI_i}} \frac{\left[\Delta P_L - \sum_j P_{COI_i,COG}^{tie} - \Delta P_i^{MMG}\right]}{(\Delta P_{mech} - \Delta P_{elec})}}{1 + \frac{M_{COG}}{M_{COI_i}} \frac{2\pi D}{(\Delta P_{mech} - \Delta P_{elec})} \Delta f_{COG}} \Delta f_{COG} \tag{6.19}$$

As is apparent in (6.19), the MMGs capacity ΔP^{MMG} affects the COI dynamics through its effect on the power imbalance and the associated sensitivity factor.

Let now the swing Eq. (6.3) be rewritten as

$$\dot{f}_j = \frac{1}{2\pi M_j}\left[\Delta P_j - P_{ij}^{tie} - 2\pi D_j \Delta f_j - \eta_j \Delta P_i^{MMG}\right] \tag{6.20}$$

for bus j, where

$$\sum_j \eta_j = 1; \quad j \in i; \quad \forall i \tag{6.21}$$

and η_j is defined as the ratio of the total MMGs capacity at bus j to the total MMGs capacity in the associated area i. Following the same procedure as that in (6.19), it can be shown that

$$\Delta f_j = \frac{\dfrac{M_{COI_i}}{M_j} \dfrac{\left[\Delta P_L - \sum_j P_{COI_i,COG}^{tie} - \Delta P_i^{MMG}\right]}{\left(\Delta P_j - P_{ij}^{tie} - \eta_j \Delta P_i^{MMG}\right)}}{1 + \dfrac{M_{COI_i}}{M_j} \dfrac{2\pi D}{\left(\Delta P_j - P_{ij}^{tie} - \eta_j \Delta P_i^{MMG}\right)} \Delta f_{COI_i}} \Delta f_{COI_i} \tag{6.22}$$

where M_j and M_{COG} have the same interpretation as in (6.8).

6.2.3 Inertia-Based Control Scheme

For each *generator bus*, the frequency nadir, the *RoCoF,* and the frequency evolution, calculated by (6.19) and (6.22), may be compared with those of the standard values. For buses that frequency trends away from the acceptable values, dispatching of the ESSs and tripping of the MMGs may be used to change $M_{i(j)}$ and η_j to mitigate undesired frequency variations. Figure 6.3 shows a flowchart representation of the advanced inertia-based control scheme.

To build some intuition about the flowchart of Figure 6.3, assume that the decline of frequency response of area i COI (generator j) deviates from the acceptable value. Referring back to Figure 6.2, and by properly dispatching the ESSs and/or the MMG islanding, the position of the COI equivalent generator relative to the COG may be changed to oppose variations in frequency response. An illustration of the inertia-based control strategy is shown in Figure 6.4. It should be emphasized that the transfer function of the phase-locked loop (PLL), $K_f(s)$, used in the voltage source converter control strategies, takes the SOC and the frequency as inputs and represents the internal inverter control strategy behavior. Using the approach, the virtual inertia loop would be activated upon the RoCoF and the frequency nadir exceeds the normal values, following the inception of a fault.

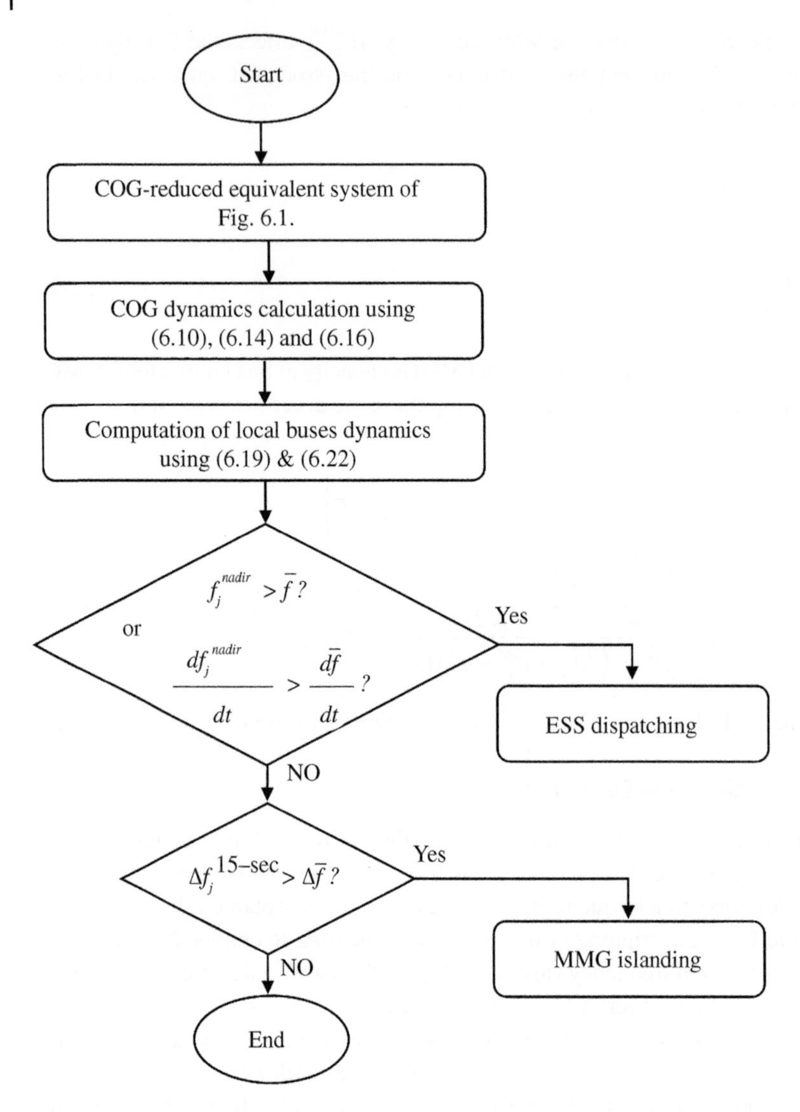

Figure 6.3 Flowchart representation of the proposed control strategy; $\bar{f}$ represents an upper acceptable frequency value.

6.2.4 Flexible Inertia: Practical Considerations

It is mandatory to reconnect the islanded MMGs to the host grid in steady state after the fault. Some considerations should be made regarding the synchronization of the MMGs and the host grid to ensure a smooth voltage and current transition

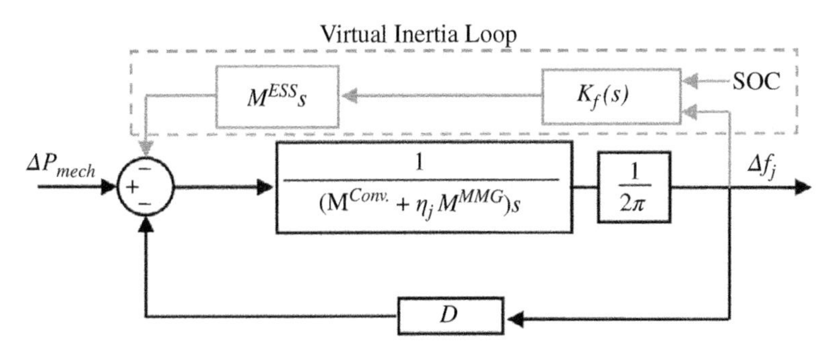

Figure 6.4 Frequency response model with virtual inertia.

and safe reconnection of the two grids [15, 16]. The early static switch closing causes low-frequency beat voltage across the switch. This beat voltage can cause a significant beat current, which leads to adverse effects such as possible resonance with mechanical structure and additional power loss. Figure 6.5 shows the beat voltage and the current flow from the switch for the early closing of the switch.

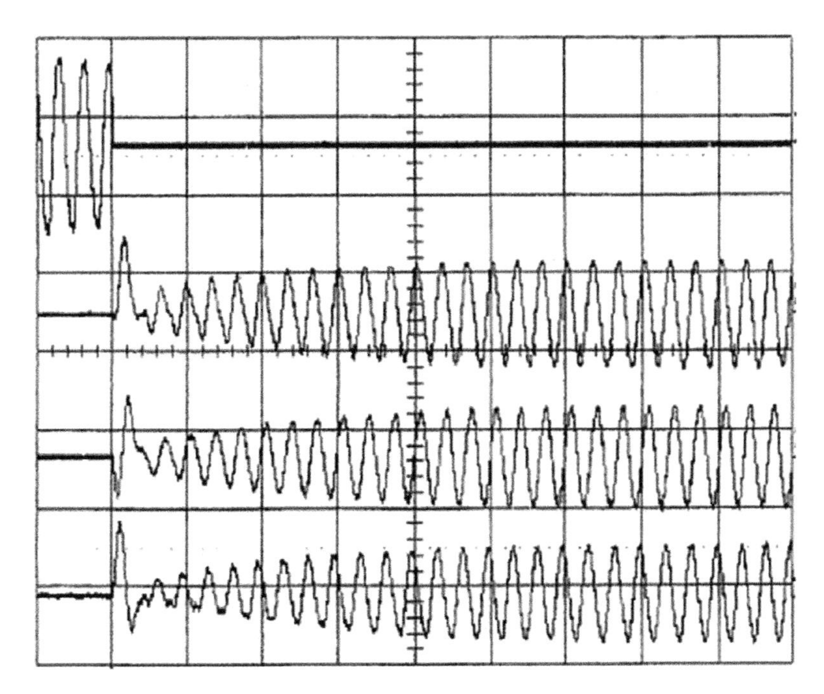

Figure 6.5 Beat voltage for early closing of the switch; upper plot: beat voltage across the switch, second plot: current of phase A, third plot: current of phase B, fourth plot: current of phase C.

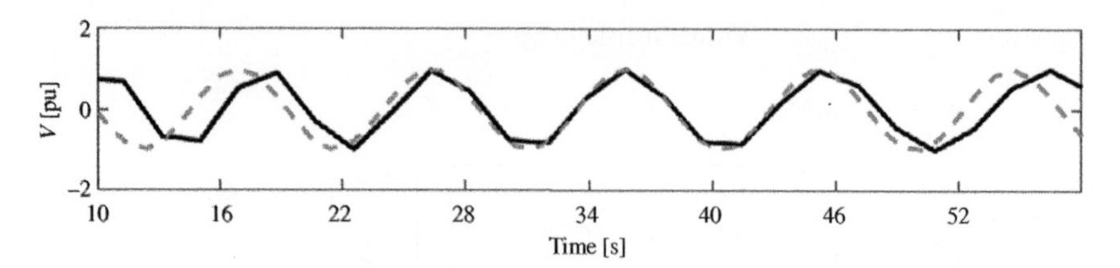

Figure 6.6 Voltages on either side of the static switch [15].

Monitoring the voltage across the static switch reveals that the MMGs and the utility should be synchronized when the phase difference between voltages of the two grids is approximately zero. Figure 6.6 shows voltages on either side of the static switch. From these results, the interval time between 33 and 42 seconds is an appropriate synchronization time to close the switch [17].

6.2.5 Results and Discussions

Exploratory studies to assess the impact of the ESSs dispatching and the MMGs islanding on the system inertia and frequency dynamics control are conducted on two test systems: (a) a simple two-area, four-machine test system and (b) a 16-machine, five-area 68-bus test model of the New York New England (NYNE) test system. Without loss of generality, uniform distribution for the uncertain parameter ΔP^{MMGs} is considered in (6.5).

1) *Two-area system:*
   The single-line diagram of the system is shown in Figure 6.7. The disturbance considered is the trip of 1400 MW generation, i.e. $\Delta P_L = 14$ pu, in Area 2. The

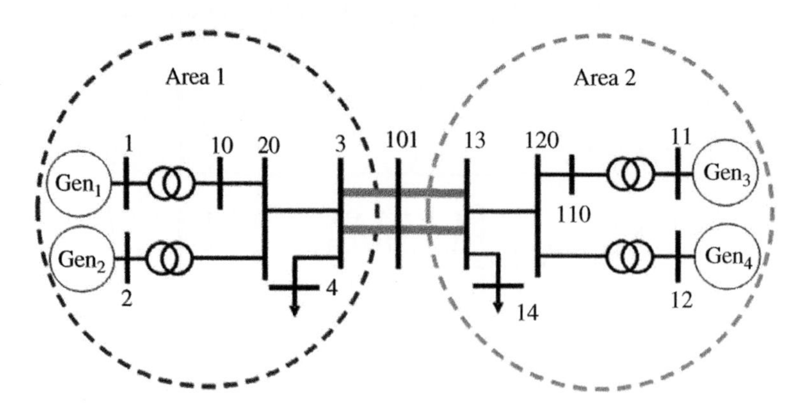

Figure 6.7 Single-line diagram of two-area system.

reduced equivalent model of the system is derived based on (6.5) to gain insight into the nature of the frequency behavior. For this purpose, 10 random scenarios of equal probability, i.e. $P_s = 0.1$ in (6.5), are generated. Minimization of (6.5) reveals that the reduced equivalent model can be characterized by

$$X^{tie}_{COI_1,COG} = 0.2134; \quad X^{tie}_{COI_2,COG} = 0.3172 \tag{6.23}$$

Bus 1 frequency behavior in the conventional system, obtained by setting $\eta = 0$ in (6.22), can be calculated by substituting (6.23) in (6.19) and (6.22) as

$$\Delta f_1 = \frac{1.67}{1 + 0.97\Delta f_{COI_1}} \Delta f_{COG} \tag{6.24}$$

Similarly, for bus 12, in Area 2, one can write:

$$\Delta f_{12} = \frac{1.16}{1 + 0.08\Delta f_{COI_2}} \Delta f_{COG} \tag{6.25}$$

Equations (6.24) and (6.25) suggest that the frequency nadir, for *generator buses* in Area 2, limits the ability of the host grid to accommodate high levels of MGs generation. A related problem of interest is that of determining the number of ESSs as well as the generation capacity of the MMGs that can be integrated into the grid to efficiently support system frequency. From (6.2) and for the base system, one can write

$$\frac{df_1}{dt} = \frac{14pu}{M_1} \tag{6.26}$$

and, for the system with high penetration of MGs generation and installed ESSs, we have

$$\frac{df_2}{dt} = \frac{14pu}{M_2 + M^{ESS}} \tag{6.27}$$

where $M_2 = M_1 + M^{MMG}$. Dividing (6.27) by (6.26) gives

$$\frac{df_2}{df_1} = \frac{M_1}{M_2 + M^{ESS}} \tag{6.28}$$

Multiplying both sides of (6.28) in the term Δf_1, on the other hand, gives

$$\frac{df_2}{df_1} \times \Delta f_1 = \frac{M_1}{M_2 + M^{ESS}} \times \Delta f_1 \tag{6.29}$$

Equation (6.29) can be rewritten in the compact form as

$$\Delta f_2 = \frac{M_1}{M_2 + M^{ESS}} \Delta f_1 \tag{6.30}$$

in which M^{ESS} and M^{MMG} (M_2) are unknown parameters to be determined. A similar expression for the frequency evolution during a time interval of interest may be derived to complete the set of two equations with two unknown parameters. As the frequency dynamics beyond the inertial response horizon is predominantly affected by the MMGs islanding, a 15-second window starting from the frequency nadir time is used to characterize frequency dynamics as [18]

$$f(T_n + 15) - f(T_n) = \frac{\frac{1}{2}R(T_n + 15) - \Delta P_L}{M_1 + M^{MMG}} - \frac{\frac{1}{2}R(T_n) - \Delta P_L}{M_1 + M^{MMG} + M^{ESS}} \quad (6.31)$$

where T_n is the time at which the frequency nadir occurs.

Substituting the actual values ($\Delta f_1 = 1.1$, $M_1 = 2 \times 6.5$) and the marginal quantities [17] ($\Delta f_2 = 0.8$, $\Delta f^{15\text{-}sec} = 0.96$) in (6.30) and (6.31) reveals that the ESSs should provide 0.9% of the base system inertia and that the emulated inertia is $M^{MMG} = 1.4$.

Experimental results for the UOK-MG show that

$$M^{MMG} = 1.89\frac{P^{Genset}}{P^{MMG}} = 1.89\frac{5}{22} = 0.43\ [s] \quad (6.32)$$

where P^{Genset} and P^{MMG} define the generated power of the synchronous-based DG and the overall MG generation, respectively (details on how to calculate (6.32) are given in *Chapter 2*), and the inertia constant in (6.32) is on a 15-kVA base. Accordingly, $M^{MMG}=1.4$ would be realized in the base of the system, i.e. 100 MW, by penetration of

$$P^{MMG} = \frac{1.4 \times 100}{0.43 \times 0.015} \times 0.022 = 477.52[MW] \quad (6.33)$$

which allows a 17.5% penetration level of MMGs. On the other hand, providing 0.9% of the base system inertia by ESSs results in [11]

$$M^{ESS} = 0.009 \times M_2 = \frac{0.5J_{VI}\omega^2 - VAh^{ESS}}{VA_{rated}} \rightarrow 0.009 \times 6.5 = \frac{VAh^{ESS}}{900}$$
$$\rightarrow VAh^{ESS} = 52.65 \quad (6.34)$$

It then follows that the number of ESS units, which could provide such an energy, is

$$52.65 = N \times V^{ESS} \times Ah^{ESS} = N \times 12 \times 480 \rightarrow N = 9140 \quad (6.35)$$

This means that 9140 ESS units, rated 12 V and 480 Ah, are required to maximize the MG penetration level. Following the same procedure as that of (6.25) for the MG-penetrated system including 9140 ESS units gives

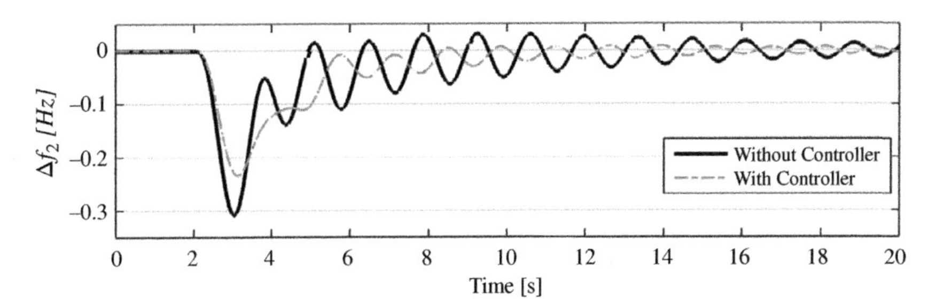

Figure 6.8 ESS effect on frequency response.

$$\Delta f_{12} = \frac{1.015}{1 + 0.01\Delta f_{COI_2}} \Delta f_{COG} \tag{6.36}$$

Comparing (6.25) with (6.36) suggests that the inertia-based control scheme can effectively reduce the frequency nadir. Figure 6.8 exhibits the appropriateness of the adopted control strategy, assuming that, in the online control system center, the frequency event is detected after receiving five samples with increasing or decreasing rate [17]; in our simulations, the ESSs are triggered 0.1 second following the inception of the fault.

2) *NYNE system:*
NYNE test system of Figure 6.9 is used to further illustrate the efficiency of the inertia-based control scheme. Five different contingency scenarios, including the trip of major generating units and load shedding at selected buses, are considered for the studies. Results in Table 6.1, comparing the capability of the host grid to accommodate MGs with and without the advanced control scheme, show the effectiveness of the inertia-based control scheme.

Table 6.1 reveals that the penetration level of MMGs into the system is limited by the frequency nadir at bus 60, in Area 1, for contingency 3. This contingency causes the COG and COI to experience the frequency nadirs of 49.78 and 49.89 Hz, respectively. Using (6.19) and (6.22), the frequency dynamics of bus 60 can be expressed in terms of the COG frequency behavior by the sensitivity relation

$$f_{60}^{nadir} = \frac{1.346}{1 + 0.363 f_{COI_1}} f_{COG} \tag{6.37}$$

then

$$f_{60}^{nadir} = \frac{1.346}{1 + 0.363(49.89/50)} 49.78 = 49.19 \tag{6.38}$$

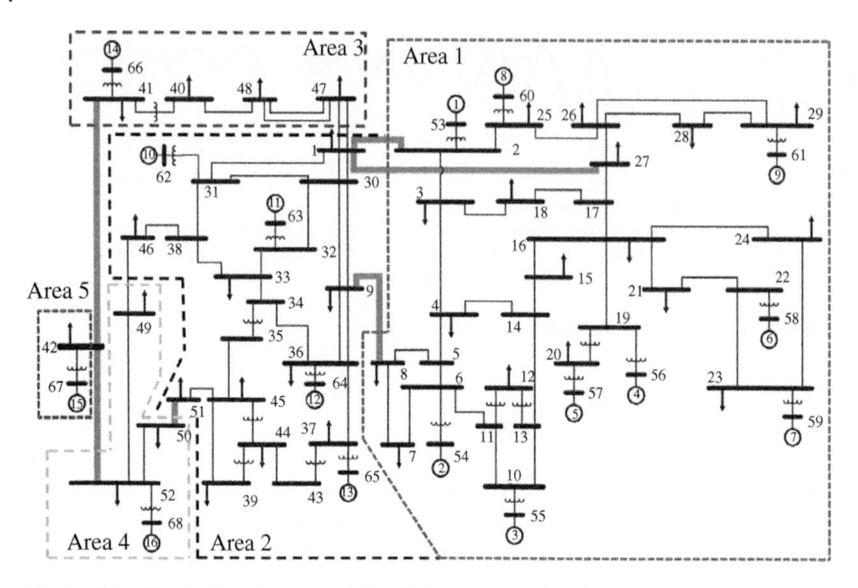

Figure 6.9 Single-line diagram of the 68-bus system showing coherent areas and their interconnections.

Table 6.1 Comparison of maximum penetration level for NYNE test system with and without the controller.

Scenario	Fault	MP without controller (%)	MP with the proposed controller (%)	ESS/ load (%)
1	G_{12}	14.23	22.37	2.83
2	L_{14}	11.34	21.78	3.01
3	G_{16}	10.87	28.11	2.98
4	L_{37}	10.56	26.06	3.54
5	L_{42}	12.01	29.73	3.16

MP: maximum penetration level.

Inspection of (6.38) shows that when the COG frequency nadir reaches 49.78 Hz, the frequency at bus 60 drops to 49.19 Hz, thus limiting the penetration of MG generations into the system. Installing 11 MW of ESSs at bus 60 increases the first swing frequency amplitude to 49.42 Hz, and the penetration level of the system increases by 17%. Table 6.2 reports the required ESSs to increase the penetration level of MMGs. It should be noted that as generators G_{14}–G_{16} represent the aggregated behavior of areas 3–5 connected to the New York power system, the

Table 6.2 Required ESS capacity to enable high penetration of MG.

Scenario	Area 1 (%)	Area 2 (%)	ESS/load (%)
1	5.71	3.89	2.99
2	3.89	2.03	1.82
3	5.91	4.21	3.16
4	6.33	1.70	2.39
5	3.51	4.53	2.60

specification of the ESSs capacity for such equivalent areas is far from reality. While the reported results in columns 2 and 3 of Table 6.2 are expressed in percentage based on each area generation capacity, the last column is expressed in percentage based on the total system load.

It should be emphasized that using this approach, each area in the system is characterized by two equations of the form (6.30) and (6.31). Given a system with n areas, solving the set of $2 \times n$ equations allows to determine the number of ESSs and MMGs ratings in the overall system.

Figure 6.10 compares the penetration levels as well as the frequency nadirs for the system with and without the inertia-based controller. For completeness, the

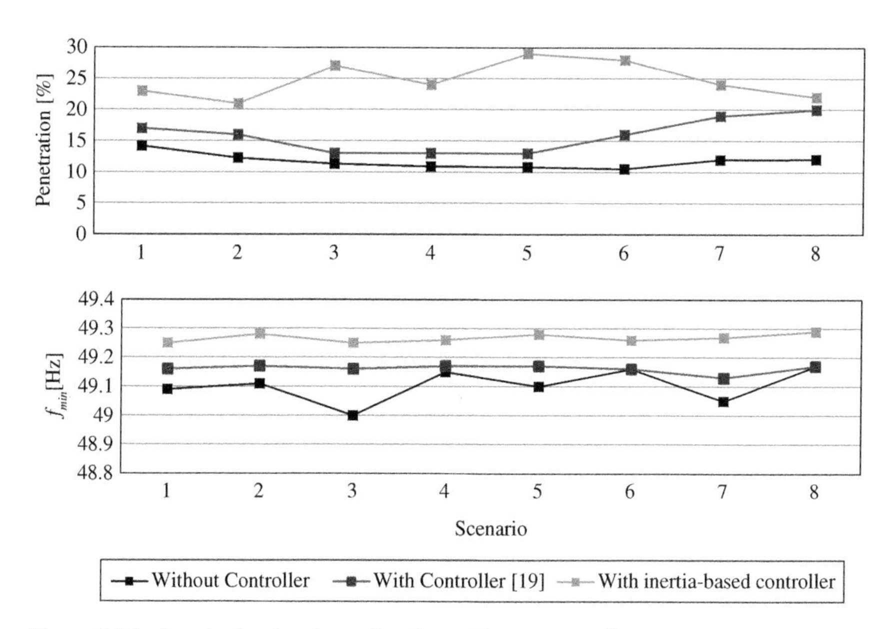

Figure 6.10 Penetration level as a function of frequency nadir.

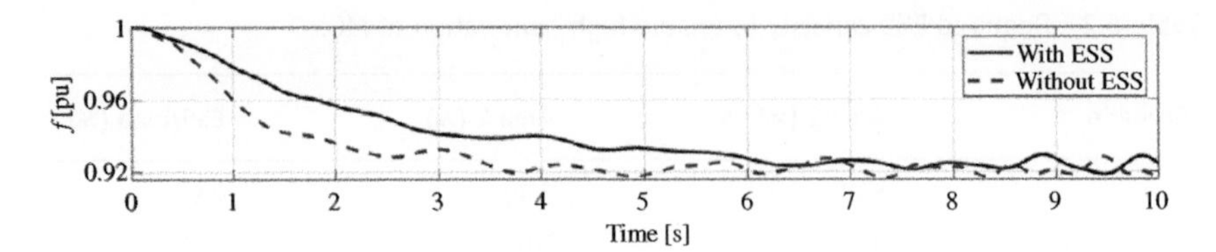

Figure 6.11 ESS effect on frequency response.

effectiveness of the inertia-based control scheme is compared with those strategies introduced in [19] where synchronous condensers in addition to the inertial response from wind farms are employed to enhance frequency dynamics.

Furthermore, Figure 6.11 shows the time-domain simulation results for scenario 3. It indicates that the *RoCoF* and the frequency nadir are much less pronounced in the case with the ESS and that without the ESS. It should be noted that frequency nadir occurs before the fourth second of simulation, as the delivery time of primary frequency in the base system is 3.5 seconds, according to Ref. [20].

6.3 Small Signal Stability Enhancement

6.3.1 Key Concept

In this section, an advanced control scheme based on the flexible inertia concept is proposed to simultaneously improve voltage regulation and small-signal stability performance. For this purpose, phase difference versus voltage deviation ($\Delta\delta - \Delta V$) graphs are introduced as a powerful tool for the realization of a flexible inertia-based control scheme. In conventional systems, automatic voltage regulator (AVR) and power system stabilizer (PSS) are assumed to act on the voltage and the rotor angle to improve voltage regulation/transient stability and small-signal stability performance. The AVRs and PSSs produce torques in phase with the rotor angle and the speed variations, respectively. However, both the AVR and PSS employ the field voltage to produce these torques, which are not in phase. More generally stated, a single control signal is employed to satisfy two conflict control actions [21]. Accordingly, it seems that using the $\Delta\delta - \Delta V$ graphs helps to make a trade-off between the voltage regulation and small-signal stability performance in the control synthesis process [22, 23]. In what follows, the possibility of dispatching the ESSs in some *generator buses* in addition to disconnecting the MMGs in some others based on the ($\Delta\delta - \Delta V$) indices is investigated.

6.3.2 Control Scheme Design

Following a system perturbation, a graph for each *generator bus* is plotted in the plane of phase difference versus voltage deviation ($\Delta\delta - \Delta V$). In this plane, the variables are defined as:

$$\Delta\delta_i(t) = \delta_i(t) - \delta_{0i} \tag{6.39}$$

and

$$\Delta V_i(t) = V_i(t) - V_{0i} \tag{6.40}$$

where δ_{0i} and V_{0i} are the initial values of rotor angle and terminal voltage related to ith generator, respectively. To establish the effective criteria for triggering of the inertia-based control scheme, parameters normalization is suggested. For this purpose, consider the maximum variable variations as:

$$\Delta\delta_{\max}(t) = \max\{|\ \Delta\delta_i(t)\ |\} \tag{6.41}$$

and

$$\Delta V_{\max}(t) = \max\{|\ \Delta V_i(t)\ |\} \tag{6.42}$$

Accordingly, one could rewrite (6.39) and (6.40) in the normalized form of

$$\Delta\delta_i(t) = \frac{\Delta\delta_i(t)}{\Delta\delta_{\max}(t)} \tag{6.43}$$

and

$$\Delta V_i(t) = \frac{\Delta V_i(t)}{\Delta V_{\max}(t)} \tag{6.44}$$

Economic reasons in addition to environmental constraints cause transmission lines to operate close to their limits. Therefore, the desired operation of a power system after being subjected to a disturbance could be theoretically achieved when the system returns to the planned operating point, characterized by voltage profile, nominal frequency, and transmitted power. Considering the power flow between buses i and j, following a system perturbation, in the form of one could define the following conditions to guarantee the desired performance:

$$P_{ij} = \frac{|\ V_{0i} + \Delta V_i(t)|V_{0j} + \Delta V_j(t)\ |}{X_{ij}} \sin\left[(\delta_{0i} + \Delta\delta_i(t)) - (\delta_{0j} + \Delta\delta_j(t))\right] \tag{6.45}$$

1) Voltage deviations ΔV_i for each generator converge to zero.
2) The difference between the angle deviations, i.e. $\Delta\delta_i - \Delta\delta_j$, converges to zero.

Ideally, all the connected generators satisfy the aforementioned conditions, and hence, they are characterized by (1, 1) in the normalized $(\Delta\delta - \Delta V)$ plane. However, in a real power grid, the system returns to a new operating point different from the initial one, and hence, the voltage and the difference between the rotor angle deviations of the generators differ from each other. Therefore, control actions should be taken to force the terminal voltage deviations and the difference between the rotor angle deviations into the desired value, i.e. zero. In other words, for the stable, secure, and reliable system operation, all the generators' operating points must be located in the minimum distance from (1, 0) in the normalized $(\Delta\delta - \Delta V)$ plane. This could be mathematically represented by the minimization problem of

$$\min\{|\Delta\delta_i + \Delta V_i| - |(1,0)|\} \tag{6.46}$$

which could be rewritten as

$$\min\left\{\sqrt{(\Delta\delta_i)^2 + (\Delta V)^2} - 1\right\} \tag{6.47}$$

in the Cartesian system.

The minimum of (6.47) can be calculated by setting the derivative of the equation with respect to ΔV or $\Delta\delta$ (both give the same results) to zero, namely

$$\frac{2\Delta\delta_i + 2\Delta V_i \frac{d(\Delta V_i)}{d(\Delta\delta_i)}}{2\sqrt{(\Delta\delta_i)^2 + (\Delta V_i)^2}} = 0 \tag{6.48}$$

Hence,

$$2\Delta\delta_i + 2\Delta V_i \frac{d(\Delta V_i)}{d(\Delta\delta_i)} = 0 \tag{6.49}$$

from which it follows that

$$|\Delta\delta_i| = |\Delta V_i| = 0.707 \tag{6.50}$$

As stated, the inertia-based control scheme tries to conduct all the committed generators to the prefault condition, i.e. (1,1) in the normalized plane. Therefore, one could define the secure operation region in the plane as:

$$\|\Delta\delta_i\| \geq 0.707 \tag{6.51}$$

and

$$\|\Delta V_i\| \geq 0.707 \tag{6.52}$$

Moreover, parameters normalization imposes

$$\|\Delta\delta_i\| \leq 1 \tag{6.53}$$

and

$$\|\Delta V_i\| \leq 1 \tag{6.54}$$

The sign of the acceleration torque in (6.2) specifies the sign of phase and voltage variations in (6.51) through (6.54). According to (6.51)–(6.54), the stable region of the system in the normalized plane of phase difference versus voltage deviation can be represented in Figure 6.12.

For the generators located in the *CDEFC* area, the conventional controllers are well-tuned on which the synchronizing and damping torques are positive. However, for the generators located out of the desired region, the conventional controller parameters, including AVR and PSS, should be retuned to return the generators to the *CDEFC* area. However, the main aim is to perform the same job as AVR–PSS via the realization of flexible inertia.

Assume that some generating units, for a given time of interest, are located in the desired region *CDEFC* while others are out of the region. The aim is to conduct all the generating units to the desired region *CDEFC*. The logic of the inertia-based control scheme is:

> While for the generating units located in the desired region CDEFC, dispatching of the ESSs and consequentially increasing the effective grid

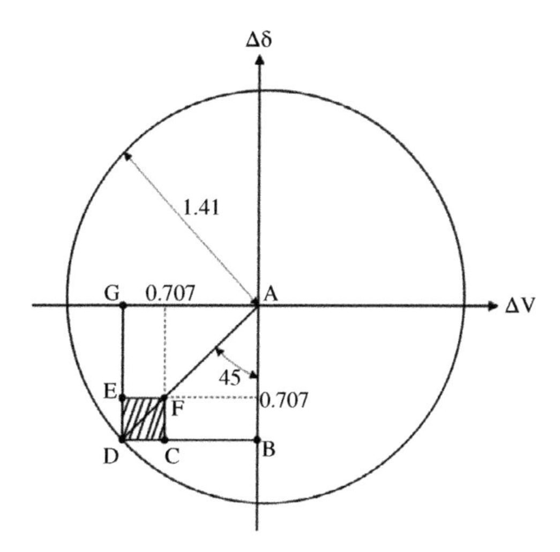

Figure 6.12 Stable region of power system

inertia can decrease the parameter variations for the machines out of the desired region, disconnecting the MMGs decreases effective grid inertia which in turn increases the parameter variations.

Accordingly, the inertia-based control scheme changes $\Delta\delta_i$, ΔV_i and the maximum variable variations in (6.43) and (6.44) and hence, causes the generators to move toward the desired region. Of note that the control scheme would be realized in several steps in such a way that in each step, and for generators in the region of interest, only 10% of the installed ESSs would be dispatched. On the other hand, only 10% of the synchronized MMGs are disconnected in each step of the algorithm.

6.3.3 Simulation and Results

The effectiveness of the inertia-based control scheme is investigated on the NYNE test system of Figure 6.7. In each *generator bus* of the system, the ESSs are placed according to the optimal results of Figure 6.13. Details on how to calculate Figure 6.13 are given in *Chapter 4* using Figure 4.21.

Figure 6.14 demonstrates the capability of the inertia-based control scheme to enhance the system stability in compliance with the voltage regulation as well as small-signal stability performance. Figure 6.14a shows the position of the

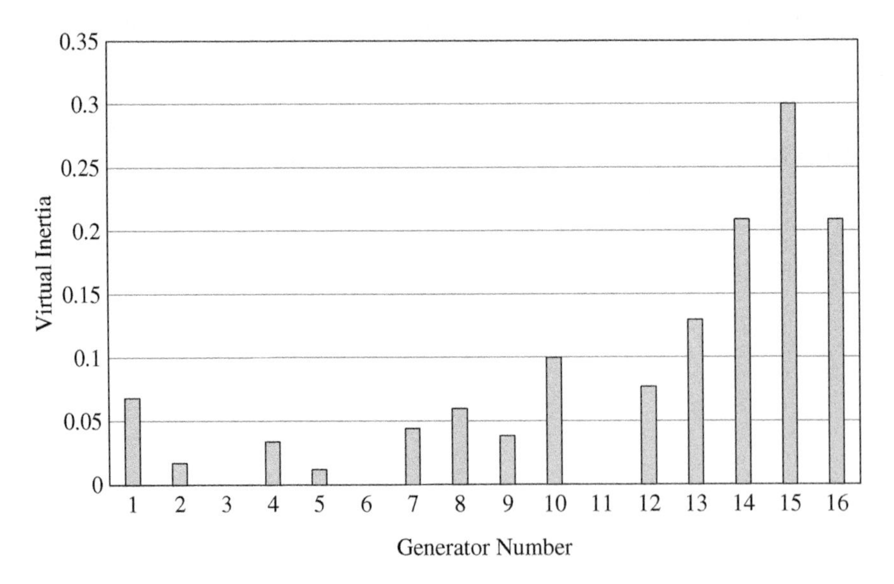

Figure 6.13 Virtual inertia allocation for NYNE test system.

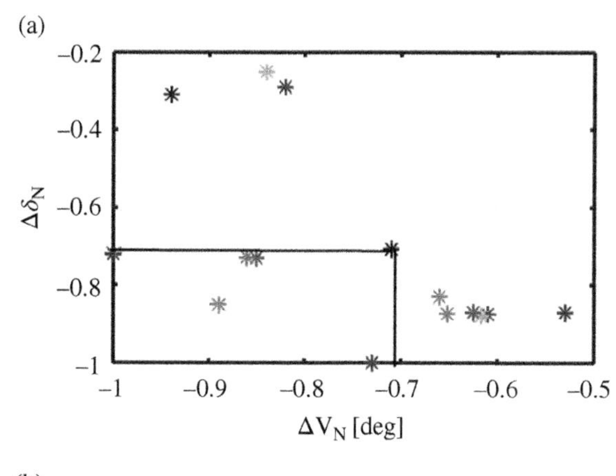

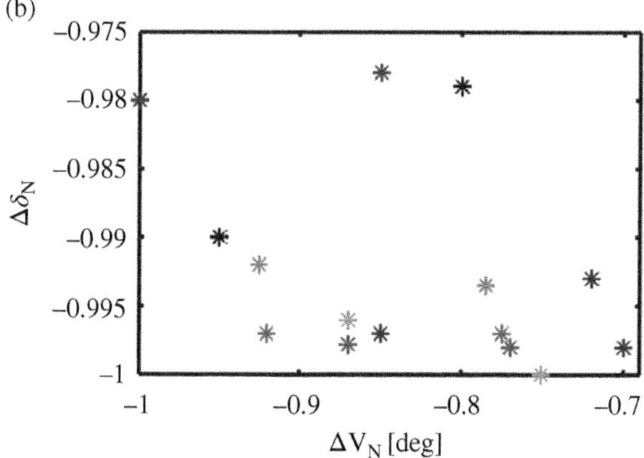

Figure 6.14 Trace of generating units in the $\Delta\delta - \Delta V$ plane in response to the visualization of flexible inertia after the outage of generator 2.

committed generators in the normalized $\Delta\delta - \Delta V$ plane just 1 second after the outage of generator 2. It can be seen that except for six generators, others are out of the desired region. As discussed earlier, dispatching the ESSs in the buses associated with the generators in the desired region and disconnecting the MMGs in other buses improve the system performance. Figure 6.14b suggests that the flexible-inertia-based control scheme can return all the generating units to the desired region.

Further, the effectiveness of the flexible inertia concept to improve the voltage regulation as well as the small-signal stability performance is assessed considering

(a)

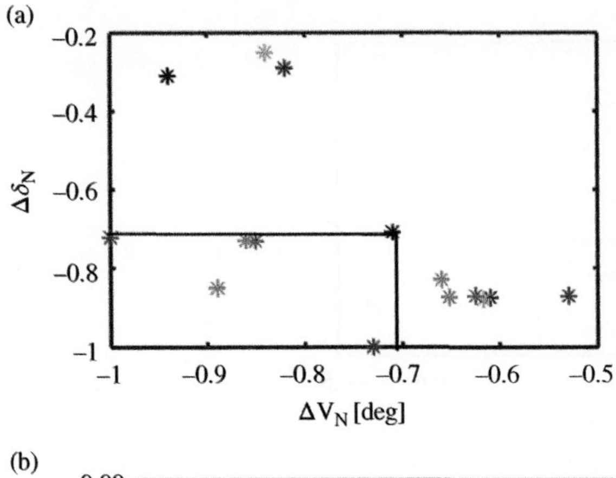

(b)

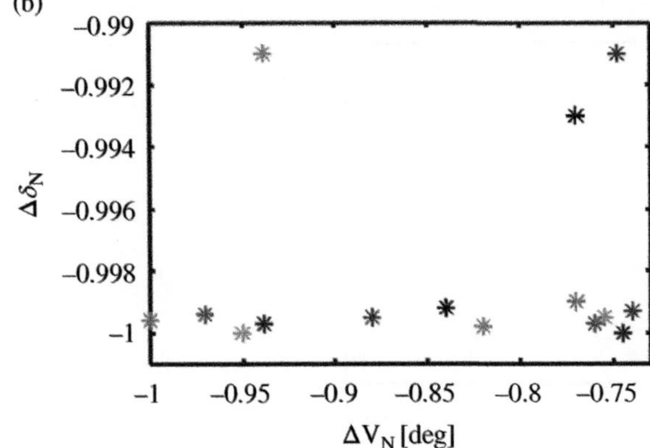

Figure 6.15 Trace of generating units in the $\Delta\delta$ – ΔV plane in response to visualization of flexible inertia after outage of generator 16.

the outage of generator 16. Figure 6.15 reveals that all the generating units are located in the desired region in about 20 seconds after the inception of the fault. The time-domain simulation result for the applied fault further demonstrates the high capability of the proposed control scheme (Figure 6.16).

Regarding the time horizon of voltage instability (>2 seconds) and transient instability (about 2 seconds), the sampling time interval to dispatch the ESSs and disconnect the MMGs can be set to 2 seconds. However, the proposed control strategy employs the first sample at a smaller time (less than 2 seconds) to improve its reliability.

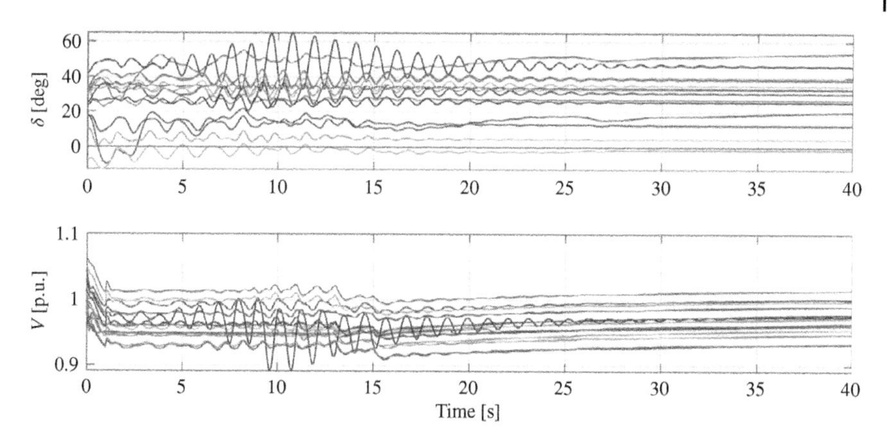

Figure 6.16 Effects of flexible inertia on voltage regulation and small-signal stability.

6.4 Summary

This chapter deals with the designing of advanced control schemes to improve power system stability. The discussed advanced control schemes rely on the inertia manipulation in the system to mitigate undesired frequency, voltage, and rotor angle dynamics. The developed control schemes depend on the stochastic equivalent model of the power system to enable high penetration levels of MGs. The approaches combine the adaptive ESS dispatch strategy with the MG controlled islanding scheme to provide a stability support for the host grid.

References

1. Hatziargyriou, N. et al. (2020). Stability definitions and characterization of dynamic behavior in systems with high penetration of power electronic interfaced technologies. IEEE, Tech. Rep.
2. Hatziargyriou, N., Asano, H., Iravani, R., and Marnay, C. (2007). Microgrids. *IEEE Power and Energy Magazine* **5** (4): 78–94.
3. Ulbig, A., Borsche, T.S., and Andersson, G. (2013). Impact of low rotational inertia on power system stability and operation. arXiv preprint arXiv:1312.6435.
4. Hatziargyriou, N. et al. (2017). Contribution to bulk system control and stability by distributed energy resources connected at distribution network, IEEE, PES-TR22.
5. Lasseter, R. et al. (2002). *Integration of Distributed Energy Resources. The CERTS Microgrid Concept.* Berkeley, CA: Lawrence Berkeley National Labaratory (LBNL).

6. Golpîra, H. and Messina, A.R. (2018). A center-of-gravity-based approach to estimate slow power and frequency variations. *IEEE Transactions on Power Systems* **33** (1): 1026–1035.
7. Meriam, J.L. and Kraige, L.G. (2012). *Engineering Mechanics: Dynamics*. Wiley.
8. Bevrani, H. (2009). *Robust Power System Frequency Control*. Springer.
9. Golpîra, H. (2019). Bulk power system frequency stability assessment in presence of microgrids. *Electric Power Systems Research* **174**: 105863.
10. Golpira, H. and Bevrani, H. (2011). Application of GA optimization for automatic generation control design in an interconnected power system. *Energy Conversion and Management* **52** (5): 2247–2255.
11. Golpîra, H., Messina, A.R., and Bevrani, H. (2019). Emulation of virtual inertia to accommodate higher penetration levels of distributed generation in power grids. *IEEE Transactions on Power Systems* **34** (5): 3384–3394.
12. Golpira, H., Atarodi, A., Amini, S. et al. (2020). Optimal energy storage system-based virtual inertia placement: a frequency stability point of view. *IEEE Transactions on Power Systems.*
13. Teng, F. and Strbac, G. (2017). Full stochastic scheduling for low-carbon electricity systems. *IEEE Transactions on Automation Science and Engineering* **14** (2): 461–470.
14. Teng, F., Trovato, V., and Strbac, G. (2016). Stochastic scheduling with inertia-dependent fast frequency response requirements. *IEEE Transactions on Power Systems* **31** (2): 1557–1566.
15. Lasseter, R.H. (2006). Control and design of microgrid components. *PSERC Publication* **06–03**.
16. Chen, C.-L., Wang, Y., Lai, J.-S. et al. (2009). Design of parallel inverters for smooth mode transfer microgrid applications. *IEEE Transactions on Power Electronics* **25** (1): 6–15.
17. Standard BAL-003-1 (2013). *Frequency Response and Frequency Bias Setting*. NERC.
18. Golpîra, H., Seifi, H., Messina, A.R., and Haghifam, M.-R. (2016). Maximum penetration level of micro-grids in large-scale power systems: frequency stability viewpoint. *IEEE Transactions on Power Systems* **31** (6): 5163–5171.
19. Ahmadyar, A.S., Riaz, S., Verbič, G. et al. (2018). A framework for assessing renewable integration limits with respect to frequency performance. *IEEE Transactions on Power Systems* **33** (4): 4444–4453.
20. GE (2010). *Frequency Response: Basics*. USA: GE Energy.
21. Golpîra, H., Bevrani, H., and Naghshbandy, A.H. (2010). A survey on coordinated design of automatic voltage regulator and power system stabilizer. *International Review of Automatic Control* **3** (2): 172–182.
22. Golpira, H., Bevrani, H., and Naghshbandy, A.H. (2012). An approach for coordinated automatic voltage regulator–power system stabiliser design in large-

scale interconnected power systems considering wind power penetration. *IET Generation, Transmission & Distribution* **6** (1): 39–49.

23. Batmani, Y. and Golpîra, H. (2019). Automatic voltage regulator design using a modified adaptive optimal approach. *International Journal of Electrical Power & Energy Systems* **104**: 349–357.

7

Small-Signal and Transient Stability Assessment Using Data-Driven Approaches

In the past decades, a variety of approaches have been introduced to assess small and large systems performance, from linear, model-based techniques to data-driven approaches. This includes efforts to extract spatial shapes, determine modal properties, and assess the energy exchange between interconnected systems, among other issues. A common approach is to simultaneously analyze the data sequences using techniques such as the multichannel Prony method or Koopman mode decomposition. Despite the importance of these methods, more general, global multiscale methods that rely on the joint analysis of large data sets are needed to cope with the ever-increasing utilization of distributed generators (DGs) and the quantity and complexity of the recorded system behavior.

As the number and distribution of inverter-based generation resources grow, it becomes increasingly challenging to extract modal properties from heterogeneous and complex, measured responses following system perturbations. Knowledge of the fundamental characteristics of global oscillation modes provides valuable information about the stability of oscillatory phenomena and may help to identify machines and wind and solar photovoltaic (PV) farms and their control systems involved in the exchange of oscillating energy and the design or modification of controllers.

This chapter identifies basic issues important in understanding the oscillatory performance of wind and solar PV penetrated power systems. A fundamental study of the characterization of the dynamic behavior of power systems with increased renewable generation is presented. The study is motivated by the need to further clarify the participation of wind and PV farms in interarea oscillations. A second goal is to assess the performance of modern data-driven modal tools and analysis techniques.

The use of modal characterization techniques is illustrated on two test systems: a 6-machine, 10-bus test system, and a 5449-bus, 635-generator test system. By using a simple example, the effect of renewable generation on system dynamic

Renewable Integrated Power System Stability and Control, First Edition.
Hêmin Golpîra, Arturo Román-Messina, and Hassan Bevrani.

performance is first investigated. The challenges that need to be confronted for such approaches to be more widely applied by the power system community are then discussed in a large-scale power system. Analytical criteria to describe the energy relationships in the observed oscillations are derived, and a physical interpretation of the system modes is suggested. Future research topics are summarized and described.

7.1 Background and Motivation

The rapid growth of inverter-based distributed renewable generation and the increased availability of measured data have made the application of data-driven techniques to the analysis of system dynamic behavior imperative. Conceptually, the study of large systems characterized by a complex mix of generating resources, often involving multiple contingency scenarios and more sophisticated (and non-linear) control strategies, poses multiple challenges. Critical aspects include the identification of wind and solar PV locations that participate significantly in critical oscillatory behavior [1–3], the extraction of modal information (spatial and temporal modes) of the system [4–6], and the analysis of control interactions [7].

Power system oscillatory behavior can be analyzed in terms of modes, expressed as exponentially modulated sinusoids, exhibited in signals measured on the system [8]. These signals are driven by the behavior of a large, nonlinear, time-variant system and may contain responses involving disparate timescales. Typically, system oscillatory behavior is commonly analyzed using small-signal (linear) analysis and time-domain simulations. The modal properties (i.e. modal damping, frequency, and shape) of these modes can be affected by various factors including flow patterns, control actions, market conditions, and the increased penetration of renewable energy sources (RESs) [9]. Moreover, measured data from distributed energy resources may exhibit nonlinear trends, transient ramps, or random fluctuations that may obscure or make difficult physical interpretation. In addition, the use of converter interfaces precludes a direct interpretation of the effect of renewable generation on modal behavior [10].

More advanced algorithms are required to extract the spatiotemporal structures concealed in the data that represent the relevant system dynamics as well as to assess the impact of operational and planning strategies on system performance. Using small-signal analysis to assess the impact of DGs/RESs on system dynamic performance can be a difficult and challenging task. First, power system models are becoming more complex and nonlinear. Further, current renewable generation models are designed to capture, essentially, linear, balanced, fundamental frequency behavior and may not be suited for analyzing control interactions or other phenomena. Second, the analysis and interpretation of oscillatory behavior in large power systems with significant penetration of distributed generation may be difficult or provide partial information on various aspects of system performance.

This motivates the need to develop simplified models that significantly reduce the number of degrees of freedom by neglecting less relevant details of the involved physics to remain computationally tractable. Algorithms for multiple unknown signal extraction from selected system measurements are well suited to analyze complex dynamic behavior and can provide complementary information to the small-signal analysis. Using data-driven approaches, dynamic trends and phase relationships between key system signals from measured data can be obtained efficiently [11, 12]. In addition, energy relationships of interest in the study and classification of global modes can be readily determined.

Modern data-driven methods rely increasingly on advanced algorithms and analysis methods that help to overcome some of the size-related limitations of measured data sets. This chapter investigates the application of modal extraction techniques to examine oscillatory behavior in the presence of high wind and solar PV penetration.

Techniques to extract modal parameters directly from observational data are critically examined and compared with well-established modal analysis techniques. Factors that affect the performance of the method are also discussed, including the effects of nonlinear trends, data quality, and sampling design. Connections with other modal identification methods are also investigated.

7.2 Modal Characterization Using Data-Driven Approaches

Measured data sets are often relatively large and may exhibit nonlinearities and other artifacts. In this section, several data-driven approaches are examined to various aspects associated with the integration of RESs. First, a conventional mass–spring system is used to introduce the nature of oscillatory phenomena. The notation is also introduced.

7.2.1 Modal Decomposition

A great deal of physical insight about the nature of oscillatory phenomena can be gained by viewing a multimachine system as a lossless mass–spring system. Without loss of generality, consider an n-degree-of-freedom lossless mechanical system, $\boldsymbol{M}\ddot{\boldsymbol{q}} + \boldsymbol{K}\boldsymbol{q} = \boldsymbol{0}$, where $\boldsymbol{q} \in \Re^{n \times 1}$ is the coordinate vector, and the mass and spring constant matrices $\boldsymbol{M} \in \Re^{n \times n}$ and $\boldsymbol{K} \in \Re^{n \times n}$ are symmetric matrices [13, 14], and damping effects are neglected.

The equation of motion can also be written as a pair of first-order equations as [14]

$$\begin{bmatrix} \boldsymbol{M} & \boldsymbol{0} \\ \boldsymbol{0} & \boldsymbol{M} \end{bmatrix} \begin{bmatrix} \dot{\boldsymbol{q}} \\ \dot{\boldsymbol{p}} \end{bmatrix} + \begin{bmatrix} \boldsymbol{0} & -\boldsymbol{M} \\ \boldsymbol{K} & \boldsymbol{0} \end{bmatrix} \begin{bmatrix} \boldsymbol{q} \\ \boldsymbol{p} \end{bmatrix} = \begin{bmatrix} \boldsymbol{0} \\ \boldsymbol{0} \end{bmatrix}$$

Use of the transformation $\boldsymbol{q} = \boldsymbol{M}^{-1/2}\boldsymbol{x}$ in the second-order model yields the second-order system $\ddot{\boldsymbol{x}} + \boldsymbol{A}\boldsymbol{x} = \boldsymbol{0}$, with $\boldsymbol{A} = \boldsymbol{M}^{-1/2}\boldsymbol{K}\boldsymbol{M}^{-1/2}$. A similar interpretation can be given for the first-order set of linear differential equations.

The free natural response is given by

$$\boldsymbol{x}(t) = \sum_{j=1}^{n} \underbrace{A_j \sin\left(\omega_j t + \theta_j\right)}_{a_j(t)} \boldsymbol{v}_j \tag{7.1}$$

where $\boldsymbol{n}$ is the number of modes, the $\boldsymbol{v}_j's$ are the natural modes, and the terms $a_j(t) = A_j \sin(\omega_j t + \theta_j)$ represent the time modulation of the natural modes, in which A_j, ω_j, and θ_j are the modal amplitude, frequency, and phase, associated with the jth mode respectively.

Let now $\boldsymbol{a}(t) = [a_1(t)\ a_2(t) ...\ a_n(t)]^T \in \Re^{n \times 1}$ be the vector containing the time evolution of the modes, $a_j(t) = A_j \sin(\omega_j t + \theta_j)$, $j = 1, ..., n$, and $\boldsymbol{V} = [\boldsymbol{v}_1\ \ \boldsymbol{v}_2 ...\ \ \boldsymbol{v}_n] \in \Re^{nxn}$ be the modal matrix of modal coordinates. It follows that:

$$\boldsymbol{x}(t) = \boldsymbol{V}\boldsymbol{a}(t) \tag{7.2}$$

By evaluating (7.2) for a given time interval, $t_1, t_2, ..., t_N$, the system response can be written in terms of the modal matrix and the matrix of modal coordinates as

$$\boldsymbol{X}(t) = [\boldsymbol{x}(t_1)\ \boldsymbol{x}(t_2) ...\ \boldsymbol{x}(t_N)] = \begin{bmatrix} x_1(t_1) & x_1(t_2) & \cdots & x_1(t_N) \\ x_2(t_1) & x_2(t_2) & \cdots & x_2(t_N) \\ \vdots & & \ddots & \\ x_n(t_1) & x_n(t_2) & \cdots & x_n(t_N) \end{bmatrix} = \boldsymbol{V}\boldsymbol{A}(t) \tag{7.3}$$

where, $\boldsymbol{x}_k(t_j) = [x_k(t_j)\quad x_k(t_j) ...\quad x_k(t_j)], j = 1, ..., N$, and

$$\boldsymbol{A} = \begin{bmatrix} a_1(t_1) & a_1(t_2) & \cdots & a_1(t_N) \\ a_2(t_1) & a_2(t_2) & \cdots & a_2(t_N) \\ \vdots & \vdots & \ddots & \\ a_n(t_1) & a_n(t_2) & \cdots & a_n(t_N) \end{bmatrix}$$

is the matrix of time-dependent coefficients.

Alternatively, one can rewrite (7.3) as

$$X(t) = \sum_{j=1}^{n} a_j(t)\boldsymbol{v}_j = \boldsymbol{V}\boldsymbol{a}(t) \tag{7.4}$$

where the $a_j(t)$ are time-dependent coefficients.

Equation (7.4) shows that the time evolution of the system states can be expressed as a linear combination of modal components and that this representation can be used to analyze mode–state relationships. In the literature, several measures of energy and coupling between modes and states have been proposed in both analytical models [15, 16] and data-driven techniques [12].

To introduce the more general ideas that follow, let the matrix $\boldsymbol{A}$ have a set of n distinct eigenvalues $(\lambda_1, \lambda_2, ..., \lambda_n)$ and a corresponding set of right and left eigenvectors be $\boldsymbol{U} = (\boldsymbol{u}_1, \boldsymbol{u}_2, ..., \boldsymbol{u}_n)$ and $\boldsymbol{V} = (\boldsymbol{v}_1, \boldsymbol{v}_2, ..., \boldsymbol{v}_n)$. Defined in this way, the modal components, $\boldsymbol{U}$, form a set of orthogonal basis vectors [16]. In Refs. [15, 16], it is shown that participation factors (PFs), $P_{ki} = u_{ki}v_{ki}$ can be considered as the participation of the kth state when only the ith mode is excited. Alternatively, the PFs can be shown to be modal energies of the unforced system $\dot{\boldsymbol{x}} = \boldsymbol{A}\boldsymbol{x}$ or be associated with more general mode–state relationships as discussed in later sections of this chapter.

In what follows, the applicability of time-domain methods to characterize system dynamic behavior in power systems with large penetration of distributed renewable generation is investigated. These methods provide a basis for the modal decomposition of an ensemble of measured data obtained from measurements or experiments that can be related to aforementioned linear analyses.

Among the various multichannel modal extraction methods, attention is focused on three methods: Koopman mode decomposition analysis, singular value decomposition (SVD)-based Prony analysis, and dynamic mode decomposition (DMD).

7.2.2 Multisignal Prony Analysis

The Prony methods have been described in detail in several papers [17–20]. For completeness, a concise review of these methods is presented as follows.

7.2.2.1 Standard Prony Analysis

Following the nomenclature of Hauer [17], consider a linear time-invariant system represented by the model $\dot{\boldsymbol{x}} = \boldsymbol{A}\boldsymbol{x}, \hat{y} = \boldsymbol{C}\boldsymbol{x},$ with $\boldsymbol{x}(t_o) = \boldsymbol{x}_o$. Let now the noisy measured signal $\hat{y}(t), t = 0, ..., N-1$ be expressed as a superposition of the true signal, $y(t)$, and noise $\varepsilon(t)$, which for the sake of convenience is assumed to be white, as [21]

$$\hat{y}(t) = y(t) + \varepsilon(t), \quad t = 0, \ldots, N-1 \tag{7.5}$$

where N is the number of samples.

Assume further that the ringdown waveform $y(t)$ can be written as a superposition of Q exponentials as [18, 22]

$$y(t) = \sum_{i=1}^{Q} A_i e^{\lambda_i t} \cos(2\pi f_i t + \phi_i) \tag{7.6}$$

Defining

$$B_i = A_i e^{\,j\phi_i}$$
$$z_i = e^{\lambda_i \Delta t} = e^{(\sigma_i + j\omega_i)\Delta t}$$

Equation (7.6) can be rewritten in the discrete domain ($t = t_k = k\Delta t$), in the form

$$y(k) = \sum_{i=1}^{n} B_i z_i^k \tag{7.7}$$

where $y(k) = y(t_k)$, $k = 0, 1, \ldots, N-1$, the $B_i \in C$ are the signal residues, $\lambda_i \neq \lambda_j$ for $i \neq j$, $z_i = e^{\lambda_i t_k}$, Δt is the sampling space, and the parameters $\{B_i, z_i\}$, $i = 1, \ldots, n$ are unknown complex parameters to be determined; n denotes the order of the model.

Neglecting noise in (7.5), and assuming that $N = 2n$, it is possible to fit $\hat{y}(t)$ to a model of the form (7.7) [23]. Formally, for $0 \leq k \leq N-1$, it follows from this expression that

$$\underbrace{\begin{bmatrix} z_1^0 & z_2^0 & \cdots & z_n^0 \\ z_1^1 & z_2^1 & \cdots & z_n^1 \\ \vdots & \vdots & \ddots & \vdots \\ z_1^{N-1} & z_2^{N-1} & \cdots & z_n^{N-1} \end{bmatrix}}_{Z \in C^{n\times n}} \underbrace{\begin{bmatrix} B_1 \\ B_2 \\ \vdots \\ B_n \end{bmatrix}}_{B \in C^{n\times 1}} = \underbrace{\begin{bmatrix} y(0) \\ y(1) \\ \vdots \\ y(N-1) \end{bmatrix}}_{y \in \Re^{n\times 1}}, \quad k = 0, \ldots, N-1 \tag{7.8}$$

or, in compact form, $\boldsymbol{ZB} = \boldsymbol{y}$, where $\boldsymbol{Z}$ is a Vandermonde matrix [24].

Associated with matrix $\boldsymbol{Z}$ is the characteristic equation (the Prony polynomial),

$$p(z) = \prod_{i=1}^{n}(z - z_i) = \sum_{m=0}^{n} a_m z^{n-m} = z^n - \sum_{m=0}^{n-1} a_m z^m, \quad a_o = 1 \tag{7.9}$$

and the companion or Frobenius matrix [24]

$$\boldsymbol{C}_n = \begin{bmatrix} 0 & 0 & \cdots & 0 & a_o \\ 1 & 0 & \cdots & 0 & a_1 \\ 0 & 1 & \cdots & 0 & a_2 \\ \vdots & \vdots & \ddots & \vdots & \vdots \\ 0 & 0 & \cdots & 1 & a_{n-1} \end{bmatrix} \in C^{n \times n}$$

where the a_js are unknown coefficients to be determined and $det(z\boldsymbol{I}_n - \boldsymbol{C}_n(a)) = p(z)$, in which $\boldsymbol{I}_n$ is the $n \times n$ identity matrix. It is noted that the eigenvalues of $\boldsymbol{C}_n$ coincide with the zeros of the Prony polynomial in (7.9).

It can be verified after some algebra that (7.9) satisfies a forward linear predictor model (LPM) of the form [25, 26]

$$\sum_{m=0}^{n} a_m y(k-m) = 0$$

Using the known data values, $y(k)$, this equation can be rewritten in matrix form as

$$\underbrace{\begin{bmatrix} y(n) & y(n-1) & \cdots & y(1) \\ y(n+1) & y(n) & \cdots & y(2) \\ y(n+2) & y(n+1) & \cdots & y(3) \\ \vdots & \vdots & \ddots & \vdots \\ y(N-1) & y(N-2) & \cdots & y(N-n) \end{bmatrix}}_{Y} \underbrace{\begin{bmatrix} a_1 \\ a_2 \\ \vdots \\ a_n \end{bmatrix}}_{a} = - \underbrace{\begin{bmatrix} y(n+1) \\ y(n+2) \\ y(n+3) \\ \vdots \\ y(N) \end{bmatrix}}_{y}, \quad n+1 \leq k \leq N \tag{7.10a}$$

or

$$\underbrace{\begin{bmatrix} y(n-1) & y(n-2) & \cdots & y(0) \\ y(n) & y(n-1) & \cdots & y(1) \\ y(n+1) & y(n) & \cdots & y(2) \\ \vdots & \vdots & \ddots & \vdots \\ y(N-1) & y(N-2) & \cdots & y(N-n-1) \end{bmatrix}}_{Y} \underbrace{\begin{bmatrix} a_1 \\ a_2 \\ \vdots \\ a_n \end{bmatrix}}_{a} = - \underbrace{\begin{bmatrix} y(n) \\ y(n+1) \\ y(n+2) \\ \vdots \\ y(N-1) \end{bmatrix}}_{y}, \quad n \leq k \leq N-1 \tag{7.10b}$$

in which it is assumed that $a_o = 1$.

Equations (7.10a) and (7.10b) can be rewritten in the form $\boldsymbol{Ya} = \boldsymbol{y}$, where $\boldsymbol{Y}$ is a matrix of dimension $(N-n) \times n$. This system of equations can be solved directly for the a_js if $N = 2n$, as $\boldsymbol{a} = \boldsymbol{Y}^{-1}\boldsymbol{y}$ [23].

As noted in Ref. [21], the Prony method is essentially a procedure to determine the z_is in (7.7) without resorting to nonlinear optimization. This can be achieved using the following three-step approach:

1) Determine vector $\boldsymbol{a}$ by solving (7.10),
2) Compute the roots of the polynomial $p(z) = 0$ by solving (7.9),
3) Solve (7.8) for the B_i as $\boldsymbol{B} = \boldsymbol{Z}^{-1}\boldsymbol{y}$.

This is often referred to as the standard or classical Prony method.

Having determined the complex amplitudes and phases, the modal parameters in (7.6) can be computed as

$$\sigma_i = \log \frac{|z_i|}{T}$$
$$\omega_i = tg^{-1}(Im(z_i) \mid Re(z_i))/T$$

and

$$A_i = |z_i|$$
$$\varphi_i = tg^{-1}(Im(z_i)| Re(z_i))$$

in which $T = \Delta t$ is the sampling interval.

The reader is referred to Refs. [17, 19, 20, 27] for further details about the practical implementation of this method and its potential applications.

7.2.2.2 Modified Least-Squares Algorithm

In practical applications, there are more data points than parameters $N > 2n$. This results in an overdetermined equation of the form (7.10b)

$$\underbrace{\begin{bmatrix} y(n-1) & y(n-2) & \cdots & y(0) \\ y(n) & y(n-1) & \cdots & y(1) \\ y(n+1) & y(n) & \cdots & y(2) \\ \vdots & \vdots & \ddots & \vdots \\ y(N-1) & y(N-3) & \cdots & y(N-n-1) \end{bmatrix}}_{\boldsymbol{Y}} \underbrace{\begin{bmatrix} a_1 \\ a_2 \\ \vdots \\ a_n \end{bmatrix}}_{\boldsymbol{a}} = - \underbrace{\begin{bmatrix} y(n) \\ y(n+1) \\ y(n+2) \\ \vdots \\ y(N-1) \end{bmatrix}}_{\boldsymbol{b}}, N > 2n \tag{7.11}$$

which has a noniterative, least-squares solution of the form

$$\boldsymbol{a} = \boldsymbol{Y}^{\dagger}\boldsymbol{b} = -\left(\boldsymbol{Y}^T\boldsymbol{Y}\right)^{-1}\boldsymbol{Y}^T\boldsymbol{b}$$

where $\boldsymbol{Y}^{\dagger}$ denotes the Moore–Penrose pseudoinverse of $\boldsymbol{Y}$.

Similarly, one can write

$$\begin{bmatrix} z_1^0 & z_2^0 & \cdots & z_n^0 \\ z_1^1 & z_2^1 & \cdots & z_2^1 \\ \vdots & \vdots & \ddots & \vdots \\ z_1^{N-1} & z_2^{N-1} & \cdots & z_2^{N-1} \end{bmatrix} \begin{bmatrix} B_1 \\ B_2 \\ \vdots \\ B_n \end{bmatrix} = \begin{bmatrix} y(0) \\ y(1) \\ \vdots \\ y(N-1) \end{bmatrix} \tag{7.12}$$

The aforementioned results extend readily to an arbitrary number of measured data as discussed further.

7.2.2.3 Multichannel Prony Analysis

Extension of the aforementioned procedures to the multichannel case has been considered by Trudnowski and coworkers [19]. Let y_m, $m = 1, \ldots, M$ denote a set of measurements (channels) with N data points each, $k = 0, 1, \ldots, N-1$. In this case, the LPM in (7.10b) can be rewritten as

$$\underbrace{\begin{bmatrix} y_1(n-1) & y_1(n-2) & \cdots & y_1(0) \\ y_1(n) & y_1(n-1) & \cdots & y_1(1) \\ \vdots & \vdots & \vdots & \vdots \\ y_1(N-1) & y_1(N-2) & \cdots & y_1(N-n-1) \\ \vdots & \vdots & \vdots & \vdots \\ y_M(n-1) & y_M(n-2) & \cdots & y_M(0) \\ y_M(n) & y_M(n-1) & \ddots & y_M(1) \\ \vdots & \vdots & \ddots & \vdots \\ y_M(N-1) & y_M(N-2) & \ldots & y_M(N-n-1)] \end{bmatrix}}_{\hat{Y}_m} \underbrace{\begin{bmatrix} \hat{a}_1 \\ \hat{a}_2 \\ \vdots \\ \hat{a}_n \end{bmatrix}}_{\hat{a}_m} = - \underbrace{\begin{bmatrix} \text{Block } 1 \begin{cases} y_1(n) \\ y_1(n+1) \\ \vdots \\ y_1(N-1) \end{cases} \\ \vdots \\ \text{Block } M \begin{cases} y_M(n) \\ y_M(n+1) \\ \vdots \\ y_M(N-1)] \end{cases} \end{bmatrix}}_{\hat{y}_m} \tag{7.13}$$

where the $\hat{a}_j's$ are the modified signals' amplitudes, and each signal is defined by a vector, $\boldsymbol{y}_j = \begin{bmatrix} y_j(0) & y_j(1) & \cdots & y_j(N-1) \end{bmatrix}^T, j = 1, \ldots, M$.

Given this result, the following three-step procedure is used to compute the Prony parameters

1) Build M data blocks and construct (7.13) as a column-wise concatenation of data blocks as

$$\hat{\boldsymbol{Y}}_m = \begin{bmatrix} \hat{\boldsymbol{Y}}_1 \\ \hat{\boldsymbol{Y}}_2 \\ \vdots \\ \hat{\boldsymbol{Y}}_M \end{bmatrix}$$

with

$$\hat{\boldsymbol{Y}}_j = \begin{bmatrix} y_j(n-1) & y_j(n-2) & \cdots & y_j(0) \\ y_j(n) & y_j(n-1) & \cdots & y_j(1) \\ \vdots & \vdots & \ddots & \vdots \\ y_j(N-1) & y_j(N-2) & \cdots & y_j(N-n-1) \end{bmatrix}, j = 1, \ldots, M$$

2) Solve (7.13) for the unknown vector $\hat{\boldsymbol{a}}_m$,
3) Obtain the roots of the Prony polynomial (7.9),
4) Calculate the residues $\boldsymbol{B}_i$, from the linear system,

$$\underbrace{\begin{bmatrix} y_1(0) & y_2(0) & \cdots & y_M(0) \\ y_1(1) & y_2(1) & \cdots & y_M(1) \\ y_1(2) & y_2(2) & \cdots & y_M(2) \\ \vdots & \vdots & \ddots & \vdots \\ y_1(N-1) & y_2(N-1) & \cdots & y_M(N-1) \end{bmatrix}} = \hat{\boldsymbol{V}}_{vand}\hat{\boldsymbol{B}}$$

with

$$\hat{\boldsymbol{V}}_{vand} = \begin{bmatrix} 1 & 1 & \cdots & 1 \\ \hat{z}_1 & \hat{z}_2 & \cdots & \hat{z}_n \\ \hat{z}_1^1 & \hat{z}_2^1 & \cdots & \hat{z}_n^1 \\ \vdots & \vdots & \ddots & \vdots \\ \hat{z}_1^{N-1} & \hat{z}_2^{N-1} & \cdots & \hat{z}_n^{N-1} \end{bmatrix}$$

$$\hat{\boldsymbol{B}} = \begin{bmatrix} \hat{B}_1^1 & \hat{B}_1^2 & \cdots & \hat{B}_1^M \\ \hat{B}_2^1 & \hat{B}_2^2 & \cdots & \hat{B}_2^M \\ \hat{B}_3^1 & \hat{B}_3^2 & \cdots & \hat{B}_3^M \\ \vdots & \vdots & \ddots & \vdots \\ \hat{B}_n^1 & \hat{B}_n^2 & \cdots & \hat{B}_n^M \end{bmatrix}$$

Robust, more efficient interpretations of this model are discussed in subsequent sections.

7.2.2.4 Hankel-SVD Methods

These methods take advantage of the Hankel structure of the data matrix and can be used to efficiently exploit the application of advanced analytical methods. Other related techniques based on SVD analysis of a Hankel matrix are the MUSIC (multiple signal classification) methods [28] described in Chapter 6 and the approximate Prony method.

Given a data matrix y_m, $m = 1, \ldots, M$, one can define a Hankel matrix, $\hat{\boldsymbol{Y}}_H$, of the form [29, 30],

$$\hat{\boldsymbol{Y}}_H = \left[\hat{Y}_{ij}\right] = \begin{bmatrix} y(2) & y(3) & y(4) & \cdots & y(L) \\ y(3) & y(4) & y(5) & \cdots & y(L+1) \\ y(4) & y(5) & y(6) & \cdots & y(L+2) \\ \vdots & \vdots & \ddots & & \vdots \\ y(L) & y(L+1) & y(L+2) & \cdots & y(N) \end{bmatrix}$$

$$\in \Re^{L\times L}, \textit{a Hankel matrix}$$

where $L = N - M + 1, M = N/2$, and the elements $\hat{Y}_{ij}$ depend only on the sum of the indices, i.e. $\hat{Y}_{i+1,j+1} = y(i+j),\ \ i+j = 0, 1, 2, \ldots, L.$

An algorithm is now suggested for extracting Prony modes from multichannel data based on Hankel SVD analysis. As outlined in previous sections, the minimum norm solution to the linear prediction equation $\hat{\boldsymbol{Y}}\hat{\boldsymbol{a}} = \boldsymbol{b}$ is given by $\hat{\boldsymbol{a}} = \hat{\boldsymbol{Y}}^{\dagger}\boldsymbol{b}$. A more efficient solution for noisy measurements can be obtained from SVD analysis of matrix $\hat{\boldsymbol{Y}}$.

Suppose the SVD of the Hankel matrix $\hat{\boldsymbol{Y}}_H$ is given by

$$\hat{\boldsymbol{Y}}_H = \boldsymbol{U}\boldsymbol{\Lambda}\boldsymbol{V}^T \tag{7.14}$$

where $\boldsymbol{\Lambda}$ is a diagonal matrix with positive, real values in the upper left and zeroes elsewhere, and $\boldsymbol{U},\boldsymbol{V}$ are the left and right singular vector matrices, respectively.

In Refs. [31, 32], it is shown that the pseudo inverse $\left(\hat{\boldsymbol{Y}}_H\right)^{\dagger}$ of $\hat{\boldsymbol{Y}}_H$ is related to the SVD of $\hat{\boldsymbol{Y}}_H$ by the relation

$$\left(\hat{\boldsymbol{Y}}_H\right)^{\dagger} = \boldsymbol{V}\boldsymbol{\Sigma}^{\dagger}\boldsymbol{U}^T \tag{7.15}$$

in which $\boldsymbol{\Sigma}^{\dagger}$is obtained from $\boldsymbol{\Sigma}$ by replacing each positive diagonal entry by its reciprocal. Related approaches are introduced in Refs. [22, 32], where connections are made with the Matrix Pencil method and the eigensystem realization algorithm.

From the SVD of the Hankel matrix in Eq. (7.15), it is then possible to construct a least-squares estimate of the Hankel matrix for the ideal, noiseless signal. Once the model (7.15) has been computed and the LPM is solved, the roots, z, of the characteristic polynomial, are obtained; the associated Ritz eigenvalues [12], and their modal amplitudes are computed from

$$\begin{aligned} \lambda_k &= conj(-\log(z_k)/\Delta t) \\ z_{n_k} &= e^{\lambda_k \Delta t} \\ Z_{jk} &= (z_{n_k})^{j-1} \end{aligned} \tag{7.16}$$

An overview of the *developed* Kumaresan–Tufts (KT) procedure used in the sections that follow to extract modal parameters is given in Table 7.1. Some comments on the details of this algorithm are given as follows.

The extension to the multichannel case is immediate; in this case, m row-concatenated Hankel matrices $\hat{\boldsymbol{Y}}_H = \left[\hat{\boldsymbol{Y}}_{H_1} \quad \hat{\boldsymbol{Y}}_{H_2} \cdots \quad \hat{\boldsymbol{Y}}_{H_M}\right]$ are needed for efficient computation of the model. It should be observed that the method analyzes all data blocks or channels simultaneously; as a result, the correlation between them can be fully exploited, thereby improving the signal-to-noise ratio (SNR) and robustness.

As the number of measurements grows, as in the case of distributed generation, and measurements are increasingly nonlinear and sparse, it becomes increasingly challenging to extract modal properties from large data sets. The following subsections explore the application of other dynamic mode representations.

7.2.3 Koopman and Dynamic Mode Decomposition Representations

An alternative method to the modal analysis of nonlinear complex systems is based on the notion of the Koopman operator [33–36]. A concise review of this method is given in this section.

Table 7.1 Hankel-based Prony analysis algorithm.

Given a record of data sequences y_m, $m = 1, ..., N$:
1) Arrange the data into a Hankel matrix $\hat{\boldsymbol{Y}}_{Hm_{i+1,j+1}} = y(i+j),\ \ i+j = 0, 1, 2, ..., L$.
2) Compute the SVD of the Hankel matrix $[\boldsymbol{U\Sigma V}] = svd(\hat{\boldsymbol{Y}}_H);\ \ \hat{\boldsymbol{Y}}_H = \boldsymbol{V\Sigma}^T\boldsymbol{U}$ and arrange the singular values in order of decreasing magnitude $\sigma_1 \geq \sigma_2 \geq ... \geq \sigma_n$,
3) Solve the LPM of the system model and determine a least-squares estimate of the measurement data,
4) Calculate the complex amplitudes $\boldsymbol{z}_i = e^{\lambda_i t}, \boldsymbol{Z}(s, r) = \boldsymbol{z}_i^{s-1}$
5) Determine the residues of the signals y_m.

7.2.3.1 The Koopman Operator

Consider a discrete-time system evolving on an M-dimensional manifold

$$\boldsymbol{x}_{k+1} = \boldsymbol{f}(\boldsymbol{x}_k), \boldsymbol{x} \in \Re^M \tag{7.17}$$

$k = 0, 1, 2, ..., N-1$; where k is an integer index.

Let now $g(\boldsymbol{z}) : M - \Re$ be any scalar-valued function (a measurement of the state or observable). The Koopman operator, U, is a linear operator that maps g into a new function:

$$Ug(\boldsymbol{x}) = g(\boldsymbol{z})$$

The key idea behind Koopman analysis is to study the system dynamics in equation (7.17), from measured data using the eigenspectrum of U. Assume to this end that φ_j and λ_j denote the eigenfunctions and eigenvalues (Koopman modes) of the Koopman operator, respectively, given by

$$U\boldsymbol{\varphi}_j(x) = \lambda_j \boldsymbol{\varphi}_j(x), \quad j = 1, 2, ... \tag{7.18}$$

where for a sufficiently long N, the Koopman eigenfunctions form an orthonormal expansion basis.

In practical applications, one is interested in functions $\mathbf{g}(\boldsymbol{x}) = \begin{bmatrix} g_1(\boldsymbol{x}) & g_2(\boldsymbol{x}) & \dots & g_p(\boldsymbol{x}) \end{bmatrix} : M \rightarrow p$, with $p < N$. If each of the components of g lies within the span of the eigenfunctions, the time evolution of the functions can be expanded as

$$\mathbf{g}(\boldsymbol{x}) = \sum_{j=1}^{\infty} \varphi_j(\boldsymbol{x}) \boldsymbol{v}_j \tag{7.19}$$

From (7.18) and (7.19) one has that

$$\boldsymbol{x}_{k+1} = g(\boldsymbol{x}_k) = \sum_{j=1}^{\infty} U^k \boldsymbol{\varphi}_j(\boldsymbol{x}_o) \boldsymbol{v}_j = \sum_{j=1}^{\infty} \lambda_j^k \boldsymbol{\varphi}_j(\boldsymbol{x}_o) \boldsymbol{v}_j \tag{7.20}$$

where use has been made of (7.18). Physically, (7.20) indicates that the observable $g(\boldsymbol{x}_k)$ is decomposed into vector coefficients, $\boldsymbol{v}_j$, called Koopman modes whose temporal behavior is given by the associated eigenvalues λ_j.

As discussed further, the phase of the eigenvalues determines its frequency, while its modulus determines the growth rate. The magnitude $\boldsymbol{\varphi}_j(\boldsymbol{x}_o)\boldsymbol{v}_j$ is used as a measure of the relative participation of a mode to the modal decomposition. Refer to [12, 34] for the detailed implementation of the method. Applications to other power system dynamic phenomena are discussed in [37].

7.2.4 Dynamic Mode Decomposition

Recently, the DMD method and its variants have been developed to approximate a few functions $\boldsymbol{\varphi}_j$, using two sets of time-ordered sequences of data snapshots.

To introduce this method, assume that $x_j(t_k)$ denotes an element of observation, where x_j is the jth grid or measurement point (a sensor), and t_k, $k = 1, 2, ..., N$ is the time at which the observations are made.

Define the data matrix $\boldsymbol{X}$ as

$$\boldsymbol{X} = \boldsymbol{X}_1^N = [\boldsymbol{x}_1 \quad \boldsymbol{x}_2 \quad \cdots \quad \boldsymbol{x}_N] = \begin{bmatrix} x_1(t_1) & x_1(t_2) & \cdots & x_1(t_N) \\ x_2(t_1) & x_2(t_2) & \cdots & x_2(t_N) \\ \vdots & \vdots & \ddots & \vdots \\ \underbrace{x_m(t_1)}_{x_1} & x_2(t_2) & \cdots & \underbrace{x_m(t_N)}_{x_N} \end{bmatrix} \in \Re^{m \times N} \tag{7.21}$$

where $\boldsymbol{x}_j = [x_1(t_j) \quad x_2(t_j) ... \quad x_m(t_j)]^T, j = 1, ..., N,$ and the superscript, N, denotes end time; $\boldsymbol{m}$ represents the number of sensors or measurement locations.

The method assumes that the data sequences or snapshots, $\boldsymbol{x}_j$, in (7.21) are generated by a discrete-time linear dynamical system whose evolution is governed by the linear mapping. Formally, these methods postprocess the sequence of snapshots (7.21) and relate two consecutive data fields through a linear mapping of the form

$$\boldsymbol{x}_{k+1} = \boldsymbol{A}\boldsymbol{x}_k + \boldsymbol{\eta}_k, \qquad k = 0, 1, ..., N-1 \tag{7.22}$$

where $\boldsymbol{A}$ is an unknown operator matrix of dimension $\boldsymbol{m}\times\boldsymbol{m}$ for a time step $\boldsymbol{\Delta}\, t$, and matrix $\boldsymbol{A}$ is real, asymmetrical, and high-dimensional; $\boldsymbol{\eta}_k$ is some noise process.

This equation defines a Krylov sequence [38]

$$\begin{aligned} \boldsymbol{x}_1 &= \boldsymbol{A}\boldsymbol{x}_o \\ \boldsymbol{x}_2 &= \boldsymbol{A}\boldsymbol{x}_1 = \boldsymbol{A}^2\boldsymbol{x}_o \\ \boldsymbol{x}_3 &= \boldsymbol{A}\boldsymbol{x}_2 = \boldsymbol{A}^3\boldsymbol{x}_o \\ &\vdots \\ \boldsymbol{x}_N &= \boldsymbol{A}\boldsymbol{x}_{N-1} = \boldsymbol{A}^N\boldsymbol{x}_o \end{aligned} \tag{7.23}$$

or, in a compact form

$$\boldsymbol{X}_2^N \in \Re^{m \times N} = \underbrace{[\boldsymbol{x}_2 \quad \boldsymbol{x}_3 \quad \cdots \quad \boldsymbol{x}_N]}_{\boldsymbol{X}_2^N} = \boldsymbol{A}\underbrace{[\boldsymbol{x}_1 \quad \boldsymbol{x}_2 \quad \cdots \quad \boldsymbol{x}_{N-1}]}_{\boldsymbol{X}_1^{N-1}} = \boldsymbol{A}\boldsymbol{X}_1^{N-1} \tag{7.24}$$

with

$$\boldsymbol{X}_2^N \in \Re^{m \times N} = [\boldsymbol{x}_2 \quad \boldsymbol{x}_3 \quad \cdots \quad \boldsymbol{x}_N]$$
$$\boldsymbol{X}_1^{N-1} \in \Re^{m \times N} = [\boldsymbol{x}_1 \quad \boldsymbol{x}_2 \quad \cdots \quad \boldsymbol{x}_{N-1}]$$

It immediately follows that the linear operator $\boldsymbol{A}$ can be obtained as follows

$$\boldsymbol{A} = \boldsymbol{X}_2^N \left(\boldsymbol{X}_1^{N-1}\right)^{\dagger} \tag{7.25}$$

The process of computing $\boldsymbol{A}$ directly from (7.25) may not be feasible, especially when the size of the data set increases or when the model is ill-conditioned. Several formulations to estimate the eigenvalues of matrix $\boldsymbol{A}$ have been developed and tested. For completeness, a brief review of these methods is presented further. Connections to other approaches are reviewed.

7.2.4.1 SVD-Based Methods

The DMD algorithm used in this section is adapted from [12, 38, 39], which is briefly summarized further.

Following the derivation in [38], consider the data sequence (7.23). The Krylov method relies on the fact that the vectors of the sequence $\boldsymbol{X}_1^{N-1} = [\boldsymbol{x}_1 \quad \boldsymbol{x}_2 \quad \cdots \quad \boldsymbol{x}_{N-1}]$ sooner or later become linearly dependent. When this condition is satisfied, one has that,

$$\boldsymbol{x}_N = c_1\boldsymbol{x}_1 + c_2\boldsymbol{x}_2 + \cdots + c_{N-1}\boldsymbol{x}_{N-1} \tag{7.26}$$

where the $c_j s$ are the unknown expansion coefficients.

Equation (7.26) can be rewritten in compact form as

$$\boldsymbol{x}_N \approx c_1\boldsymbol{x}_1 + c_2\boldsymbol{x}_2 + \cdots + c_{N-1}\boldsymbol{x}_{N-1} = \boldsymbol{X}_1^{N-1}\boldsymbol{c} \tag{7.27}$$

where $\boldsymbol{c} = [c_1 \quad c_2 \quad \ldots \quad c_{N-1}]^T$ is a vector of unknown coefficients.

Now, recalling that $\boldsymbol{X}_2^N = [\boldsymbol{x}_2 \quad \boldsymbol{x}_3 \quad \cdots \quad \boldsymbol{x}_N]$ and making use of (7.27) results in

$$\boldsymbol{X}_2^N = [\boldsymbol{x}_2 \quad \boldsymbol{x}_3 \quad \cdots \quad \boldsymbol{x}_N] = \left[\boldsymbol{x}_2 \quad \boldsymbol{x}_3 \quad \cdots \quad \boldsymbol{X}_1^{N-1}\boldsymbol{c}\right] \tag{7.28}$$

and hence

$$\boldsymbol{X}_2^N = \boldsymbol{A}\boldsymbol{X}_1^{N-1} = \boldsymbol{X}_1^{N-1}\boldsymbol{S} \tag{7.29}$$

where $\boldsymbol{S}$ is the Frobenius or companion matrix

$$\boldsymbol{S} = \begin{bmatrix} 0 & 0 & \cdots & 0 & c_o \\ 1 & 0 & \cdots & 0 & c_1 \\ 0 & 1 & \cdots & 0 & c_2 \\ \vdots & \vdots & \ddots & \vdots & \vdots \\ 0 & 0 & \cdots & 1 & c_{N-1} \end{bmatrix} \in \Re^{N-1 \times N-1} \tag{7.30}$$

As discussed earlier, the eigenvalues of $\boldsymbol{S}$ are a subset of the eigenvalues of matrix $\boldsymbol{A}$ and are obtained from the characteristic polynomial

$$p_{\boldsymbol{S}}(\lambda) = \lambda^N - c_{N-1}\lambda^{N-1} - c_{N-2}\lambda^{N-2} - \ldots c_1 = 0$$

In practical application, linear dependence in (7.27) occurs gradually. As a result, a residual is obtained from (7.29) of the form [38]

$$\begin{cases} \boldsymbol{r} = \boldsymbol{X}_2^N - \boldsymbol{X}_1^{N-1}\boldsymbol{S} \\ \boldsymbol{A}\boldsymbol{X}_1^{N-1} - \boldsymbol{X}_1^{N-1}\boldsymbol{S} = \boldsymbol{r}\boldsymbol{e}_{\boldsymbol{p}}^T \end{cases}$$

where

$$\boldsymbol{r}\boldsymbol{e}_{\boldsymbol{p}}^T = \begin{bmatrix} 0 & 0 & \cdots & \rho_1 \\ \vdots & \vdots & \ddots & \vdots \\ 0 & 0 & \cdots & \rho_n \end{bmatrix}$$

The problem is now how to find the vector **c**, which determines the residues, $\boldsymbol{r}$, and therefore matrix $\boldsymbol{S}$. From the aforementioned results, the eigenvalues of $\boldsymbol{A}$ approximate those of $\boldsymbol{S}$ when $\|\boldsymbol{r}\|_2 \to 0$.

Following the earlier discussion, it can be seen that matrix **S** can be determined by

$$\underbrace{\text{argmin}}_{\boldsymbol{S}} \left\|\boldsymbol{X}_2^N - \boldsymbol{X}_1^{N-1}\boldsymbol{S}\right\|_2 \tag{7.31}$$

It is straightforward to show that a solution to the aforementioned optimization problem is given by $\boldsymbol{S} = \left(\boldsymbol{X}_1^{N-1}\right)^{\dagger}\boldsymbol{X}_2^N$. In practical applications, however, an alternate algorithm to compute matrix $\boldsymbol{S}$ and its associated eigenvalues and eigenvectors, based on SVD of the companion matrix, is favored when the matrix is rank deficient ($M > N$) [40].

Let to this end the SVD of matrix $\boldsymbol{X}_1^{N-1}$ be expressed as

$$\boldsymbol{X}_1^{N-1} = \boldsymbol{U\Sigma W}^T \tag{7.32}$$

where $\boldsymbol{U}$ is an orthogonal matrix of right singular values, $\boldsymbol{\Sigma}$ is the diagonal matrix of singular values, and $\boldsymbol{W}$ is the matrix of left singular values. Noting now that $\boldsymbol{X}_2^N = \boldsymbol{A}\boldsymbol{X}_1^{N-1}$, and substituting into (7.32), yields

$$\boldsymbol{X}_2^N = \boldsymbol{A}\underbrace{\boldsymbol{U}\boldsymbol{\Sigma}_m\boldsymbol{W}^T}_{\boldsymbol{X}_1^{N-1}} \tag{7.33}$$

Inserting (7.29) in this expression, one has that

$$\boldsymbol{X}_2^N = \boldsymbol{X}_1^{N-1}\boldsymbol{S} = \boldsymbol{U}\boldsymbol{\Sigma}\boldsymbol{W}^T\boldsymbol{S} = \boldsymbol{A}\boldsymbol{U}\boldsymbol{\Sigma}_m\boldsymbol{W}^T \tag{7.34}$$

Multiplying (7.34) on the left by $\boldsymbol{U}^T$results in

$$\boldsymbol{\Sigma}\boldsymbol{W}^T\boldsymbol{S} = \boldsymbol{U}^T\boldsymbol{A}\left(\boldsymbol{U}\boldsymbol{\Sigma}_m\boldsymbol{W}^T\right)$$

from which it follows that

$$\hat{\boldsymbol{S}} = \boldsymbol{U}^T\boldsymbol{A}\boldsymbol{U} \tag{7.35}$$

Then, solving (7.33) for $\boldsymbol{A}$ and inserting in (7.33) yields

$$\hat{\boldsymbol{S}} = \boldsymbol{U}^T\boldsymbol{A}\boldsymbol{U} = \boldsymbol{U}^T\boldsymbol{X}_2^N\boldsymbol{W}^T\boldsymbol{\Sigma}_m^{-1} \tag{7.36}$$

where $\boldsymbol{\Lambda} = \text{diag}[\lambda_1 \quad \lambda_2 \cdots \quad \lambda_p]$ is the diagonal matrix of eigenvalues, and $\boldsymbol{\Phi} = [\boldsymbol{\phi}_1 \quad \boldsymbol{\phi}_2 \cdots \quad \boldsymbol{\phi}_r]$ is the matrix of eigenvectors, with $\boldsymbol{\phi}_j = \left[\phi_{1j} \quad \phi_{2j} \ldots \quad \phi_{mj}\right]^T$.

Suppose now that $\hat{\boldsymbol{S}}$ is diagonalizable with an eigenvalue decomposition

$$\hat{\boldsymbol{S}} = \boldsymbol{Y}\boldsymbol{\Lambda}\boldsymbol{Y}^{-1} \tag{7.37}$$

where $\boldsymbol{\Lambda} = diag[\lambda_1 \quad \lambda_2 \ldots \quad \lambda_m] \in C^{m\times m}$ is a diagonal matrix consisting of empirical Ritz eigenvalues, λ_j, and $\mathbf{Y} = [y_1 \quad y_2 \ldots \quad y_m] \in C^{m\times m}$ is the matrix of right eigenvectors, respectively.

Use of (7.36) into (7.37), yields

$$\hat{\boldsymbol{S}} = \boldsymbol{Y}\boldsymbol{\Lambda}\boldsymbol{Y}^{-1} = \boldsymbol{U}^T\boldsymbol{X}_2^N\boldsymbol{W}^T\boldsymbol{\Sigma}_m^{-1} \tag{7.38}$$

and hence,

$$\boldsymbol{X}_2^N = \boldsymbol{U}\hat{\boldsymbol{S}}\boldsymbol{\Sigma}_m\boldsymbol{W}_m = \boldsymbol{U}\left(\boldsymbol{Y}\boldsymbol{\Lambda}\boldsymbol{Y}^{-1}\right)\boldsymbol{\Sigma}_m\mathbf{W}_m^T \tag{7.39}$$

or

$$\boldsymbol{X}_2^N = \boldsymbol{U}\hat{\boldsymbol{S}}\boldsymbol{\Sigma}_m\boldsymbol{W}_m = \underbrace{\boldsymbol{U}\boldsymbol{Y}}_{\substack{spatial\\ sructure}}\underbrace{\left(\boldsymbol{\Lambda}\boldsymbol{Y}^{-1}\boldsymbol{\Sigma}_m\boldsymbol{W}_m\right)}_{\substack{Temporal\\ structure}} = \boldsymbol{\Phi}\boldsymbol{\Lambda}\boldsymbol{\Gamma}_m(t) \tag{7.40}$$

where $\mathbf{\Gamma}_m(t) = \mathbf{Y}^{-1}\mathbf{\Sigma}_m \boldsymbol{W}_m$, and $\mathbf{\Phi} = \boldsymbol{UY}$. A physical interpretation of this model is provided in Barocio et al. [12].

The computation of the modal decomposition can be summarized in the following five steps:

1) Given a data set $\boldsymbol{X} = [\boldsymbol{x}_1 \quad \boldsymbol{x}_2 \quad \cdots \quad \boldsymbol{x}_N] \in \Re^{m \times N}$, build matrices $\boldsymbol{X}_1^N$ and $\boldsymbol{X}_1^{N-1}$,
2) Decompose the data matrix $\boldsymbol{X}$ using SVD,
3) Compute

$$\hat{\mathbf{A}} = \boldsymbol{U}^* \boldsymbol{X}^T \boldsymbol{W} \mathbf{\Sigma}^{-1},$$

4) Compute the eigenvalues and eigenvectors of $\hat{\mathbf{A}}$ from

$$\left(\hat{\mathbf{A}} - \mathbf{\Lambda}\right)\boldsymbol{W} = \mathbf{0},$$

5) Calculate the DMD modes as

$$\mathbf{\Psi} = \boldsymbol{X}^T \boldsymbol{V} \mathbf{\Sigma}^{-1} \boldsymbol{W}$$

7.2.4.2 The Companion Matrix Approach

As discussed earlier, the eigenvectors of $\hat{\mathbf{S}}$ form a basis for the span of $\boldsymbol{A}$. Now, referring back to Eq. (7.23), one can write

$$\begin{aligned} \boldsymbol{x}_1 &= \boldsymbol{A}\boldsymbol{x}_o \\ \boldsymbol{x}_2 &= \boldsymbol{A}\boldsymbol{x}_1 \\ \boldsymbol{x}_3 &= \boldsymbol{A}\boldsymbol{x}_2 = \boldsymbol{A}^2\boldsymbol{x}_1 \\ \boldsymbol{x}_4 &= \boldsymbol{A}\boldsymbol{x}_3 = \boldsymbol{A}^3\boldsymbol{x}_1 \\ &\vdots \\ \boldsymbol{x}_N &= \boldsymbol{A}\boldsymbol{x}_{N-1} = \boldsymbol{A}^{N-1}\boldsymbol{x}_1 \end{aligned}$$

$$\text{or} \quad \boldsymbol{x}_i = \boldsymbol{A}^{i-1}\boldsymbol{x}_1, i = 2, 3, \ldots, N.$$

From linear system theory, each data vector can be expanded in the form (refer to Eq. (7.1)). Following [40] let

$$\boldsymbol{x}_1(t) = \sum_{j=1}^{m} a_j \boldsymbol{\phi}_j \tag{7.41}$$

and

$$\boldsymbol{x}_i(t) = \sum_{j=1}^{m} \boldsymbol{A}^{i-1} a_j \boldsymbol{\phi}_j , \quad i = 1, 2, \ldots$$

It is left to the reader to prove that the time evolution of the ith state can be expressed as

$$\boldsymbol{x}_i(t,\boldsymbol{x}) = \sum_{j=1}^{m} \lambda_j^{i-1} a_j \boldsymbol{\phi}_j, \quad i = 1, 2, \ldots \tag{7.42}$$

where the $a_j s$ are the time-dependent amplitudes, the $\boldsymbol{\varphi}_j s$ are the eigenvectors of $\hat{\mathbf{S}}$, and $\lambda_j = \sigma_j + i\omega_j$ is the complex frequency of the associated mode.

Equation (7.42) can be conveniently rewritten in the vector-matrix form

$$\mathbf{X}(t) = \boldsymbol{\Phi}\boldsymbol{V}_{vand}\mathbf{c} \tag{7.43}$$

where $\boldsymbol{\Phi} = [\boldsymbol{\phi}_1 \quad \boldsymbol{\phi}_2 \cdots \quad \boldsymbol{\phi}_m]$, is the matrix of spatial structures (mode shapes), and

$$\boldsymbol{V}_{vand} = \begin{bmatrix} 1 & \lambda_i^1 & \lambda_i^k \cdots & \lambda_i^k \\ 1 & \lambda_i^k & \lambda_i^k \cdots & \lambda_i^k \\ \vdots & \vdots & \ddots & \vdots \\ 1 & \lambda_i^k & \lambda_i^k \cdots & \lambda_i^k \end{bmatrix} \tag{7.44}$$

is the Vandermonde matrix.

Several approaches to compute the vector of modal amplitudes, $\mathbf{c}$, have been suggested in the literature. References [41, 42] are cited to this end.

Table 7.2 gives a simplified procedure for this approach.

Table 7.2 Krylov-based DMD algorithm.

Dynamic mode decomposition algorithm

Given a data set $\boldsymbol{X} = [\boldsymbol{x}_1 \quad \boldsymbol{x}_2 \quad \cdots \quad \boldsymbol{x}_N] \in \Re^{m \times N}$:

1) Build matrices $\boldsymbol{X}_2^N$ and $\boldsymbol{X}_1^{N-1}$,
2) Compute the SVD of $\boldsymbol{X}_o^{N-1}$, $[\boldsymbol{U}, \boldsymbol{\Sigma}, \boldsymbol{V}] = svd[\boldsymbol{X}_o^{N-1}]$,
3) Compute the rank, r, of $\boldsymbol{\Sigma}$ and compute a truncated approximation of ***U, S, V,***
4) Compute matrix $\boldsymbol{X}$ as

   $$\boldsymbol{X} = \boldsymbol{U}^T \boldsymbol{X}_1^N \boldsymbol{V} \boldsymbol{\Sigma}^{-1},$$

5) Calculate the eigenvalues and eigenvectors, $\boldsymbol{\Lambda}$, $\boldsymbol{U}$ of $\boldsymbol{X}$, with $\boldsymbol{\Lambda} = diag[\lambda_1 \quad \lambda_2 \ldots \quad \lambda_r]$,
6) Compute the Vandermode matrix using (7.44),
7) Determine the Ritz eigenvalues and their associated frequency and damping factors σ and ω.
8) Compute mode shapes and mode amplitudes in (7.40),
9) If desired, reconstruct the original measurements using (7.43) as
   $\hat{\boldsymbol{X}} = \boldsymbol{\Phi}\mathbf{c}\boldsymbol{V}_{vand}$,

7.2.4.3 Energy Criteria

In pursuing the analogy with linear modal analysis in Section 7.2.1, assume that the time evolution of a measured signal, $\boldsymbol{x}(t)$, can be represented by a decomposition of the form (7.21). In discrete time

$$\boldsymbol{x}(t) = \sum_{j=0}^{r-1} \lambda_j^{t/\Delta t} a_j \tilde{\boldsymbol{\phi}}_j \tag{7.45}$$

where $k = t/\Delta t$.

In what follows, analytical criteria to describe the energy relationships in the observed oscillations are derived and a physical interpretation of the system modes is suggested. For simplicity and clarity of exposition, assume that the eigenvectors $\boldsymbol{\varphi}_j$ are normalized to unity, so that

$$\hat{\boldsymbol{\phi}}_j = \frac{\boldsymbol{\phi}_j}{\|\boldsymbol{\phi}_j\|_2}$$

It follows from basic considerations that the total averaged energy, $E(T)$, over a time interval $[0, T]$ can be expressed as

$$E(T) = \frac{1}{T}\int_0^T |\boldsymbol{x}(t)|^2 dt = \frac{1}{T}\int_0^T \left| \sum_{j=0}^{r-1} \lambda_j^{t/\Delta t} c_j \tilde{\boldsymbol{\phi}}_j \right|^2 dt$$

or

$$E(T) = \frac{1}{T}\int_0^T \left| \sum_{j=0}^{r-1} \left(\lambda_j^{t/\Delta t} c_j \tilde{\boldsymbol{\phi}}_j\right)^T \left(\lambda_j^{t/\Delta t} c_j \tilde{\boldsymbol{\phi}}_j\right) \right|^2 dt \tag{7.46}$$

Performing the product on the right-hand side of (7.46) yields

$$E(T) = \frac{1}{T}\int_0^T \sum_{j=0}^{r-1} \left(\tilde{\boldsymbol{\phi}}_j^T \tilde{\boldsymbol{\phi}}_j \lambda_j^{t/\Delta t} c_j\right) dt$$

Hence, the averaged energy for the jth mode is

$$E_j(T) = \frac{1}{T}\int_0^T \left(\tilde{\boldsymbol{\phi}}_j^T \tilde{\boldsymbol{\phi}}_j \lambda_j^{t/\Delta t} c_j\right) dt \tag{7.47}$$

Noting, finally, that $e^{\log x} = x$, and integrating (7.47) with respect to t gives

$$E_j = \frac{1}{T}\|\phi_j\|^2 \int_0^T \lambda_j^{2\sigma_j t} dt = \|\phi_j\|^2 \frac{e^{2\sigma_j T} - 1}{2\sigma_j T} \tag{7.48}$$

which is used in Ref. [43] as a metric for ranking system modes.

As shown in (7.48), modal energy is proportional to the square of the mode shape amplitude, the selected time interval, T, and the damping decay, σ_j. This makes this criterion useful to identify and isolate specific dynamic behavior associated with a given mode and timescale.

The next sections describe the application of the aforementioned techniques in the context of the oscillatory system response in more detail.

7.3 Studies of a Small-Scale Power System Model

As a first motivating example, a six-machine test system is used in this section to investigate the effect of inverter-based generation on system dynamic performance [44]. The test system consists of 10 buses, five generators, and a synchronous compensator; the total system load is 430 MW.

Figure 7.1 depicts a single-line line diagram of the test system showing generation and transmission resources. In this representation, Bus 1 is an infinite bus and is modeled by a classical representation with a large inertia value. The rest of the generators (buses 1, 3, 4, 6, 11, and 12) are represented by a fourth-order model equipped with an IEEE type 1 excitation system. In this model, the machine at bus 4 is a synchronous condenser (SC).

7.3.1 System Data and Operating Scenarios

Relevant load flow and transmission parameters for this system are given in Tables 7.3–7.5. Full details of the model are provided in Refs. [44, 45], which are briefly summarized here.

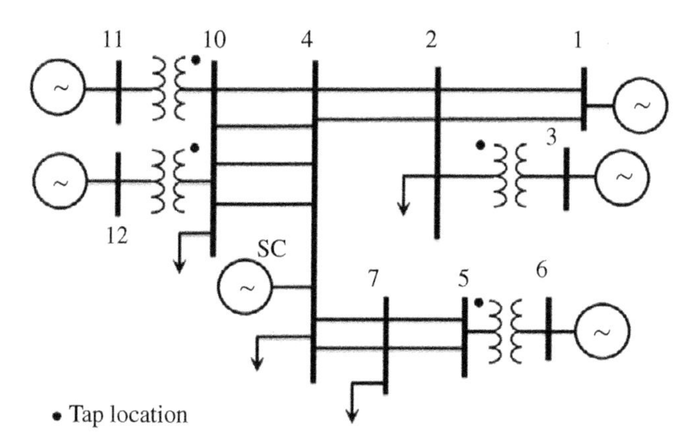

Figure 7.1 Ten-bus, six-machine test system.

Table 7.3 Load-flow data.

			Load		Generation	
Bus	**Voltage (pu)**	**Angle (°)**	**MW**	**MVAr**	**MW**	**MVAr**
1	1.0000	0.0000	0.00	0.00	52.69	−2.04
2	1.0388	−8.2649	100.00	25.00	0.00	0.00
3	1.0500	−1.9973	0.00	0.00	100.00	16.28
4	1.0500	13.7510	100.00	25.00	0.00	26.85
6	1.0683	1.4698	0.00	0.00	0.00	0.00
7	1.0504	−7.7070	30.00	15.00	0.00	0.00
10	1.0360	−13.8039	200.00	50.00	0.00	0.00
12	1.0500	−7.5449	0.00	0.00	100.00	18.65

Table 7.4 Transmission line data.

From bus	To bus	Number of circuits in parallel	*R* in pu	*X* in pu	Line charging in MVAR at 1 pu voltage
1	2	2	0.08	0.6	2.5
2	4	2	0.08	0.4	10.0
4	7	2	0.08	0.4	10.0
4	10	4	0.08	0.4	10.0
5	7	2	0.08	0.4	10.0

Table 7.5 Transformer data.

From bus	To bus	*R* in pu	*X* in pu[a]	Transformer tap ratio[b]
2	3	0.02	0.12	1.02
5	6	0.01	0.06	1.02
10	11	0.02	0.12	1.02
10	12	0.02	0.12	1.02

a Impedances are in pu on a 100 MVA base.
b The tapped side corresponds to the bus listed on the left.

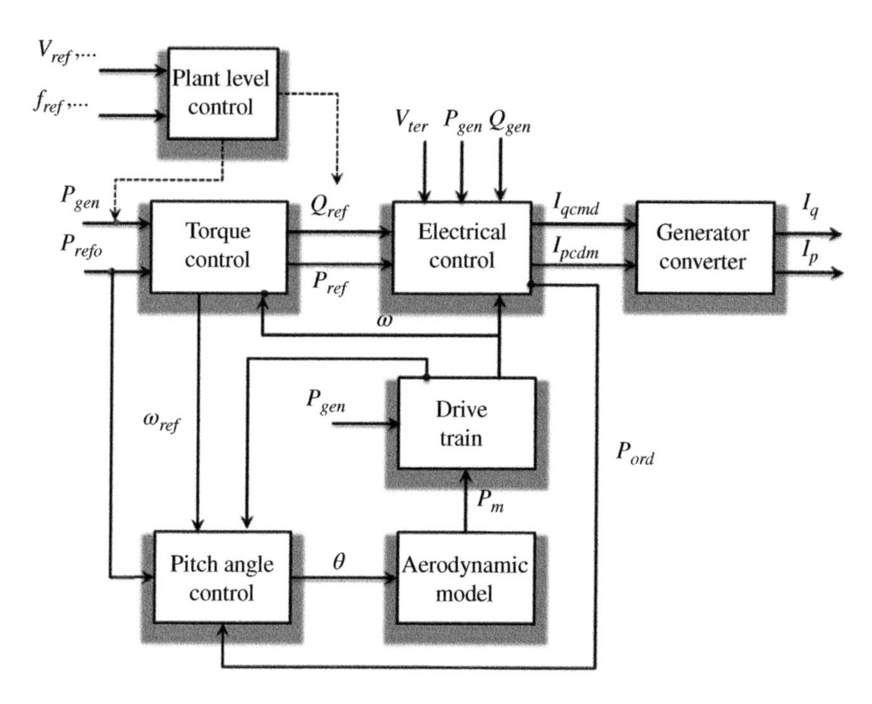

Figure 7.2 Generic wind farm/solar photovoltaic model.

In the studies described further, wind turbines of the wind farms (WFs) are modeled to represent typical (Type 3) doubly fed induction generator (DFIG) configurations, similar to the GE WFs' models [46, 47]. Three main control loops are considered in the analysis (see Figure 7.2 adapted from [46, 47]):

1) The electric control system model
2) The mechanical control system model
3) The generator model (DFIG or voltage source converter model)

Chapter 8 discusses the particular implementations of these models.

Four different operating scenarios were used in the simulations described as follows:

Case A. A base case with an infinite bus representation at bus 1,

Case B. A modified case where the infinite bus at bus 1 is replaced by a wind generator,

Case C. A modified case in which the infinite bus at bus 6 is replaced by an equivalent wind generator and bus 1 is an infinie bus,

Case D. A modified case in which the infinite bus 1 and the thermal generator at bus 6 are replaced by equivalent wind turbine generators.

Exploratory studies assessing the impact of renewable generation on system dynamic performance are described in the following subsections.

7.3.2 Exploratory Small-Signal Analysis

The base-case system (*Case A*) has five modes of oscillation at approximately 0.5, 0.89, 1.47, 1.72, and 2.24 Hz associated with synchronizing power flows among the five generators and the synchronous compensator [48]. Table 7.6 lists the eigenvalues, frequencies, damping ratios of the electromechanical modes of interest. Column 5 in this table summarizes the oscillation patterns.

For reference, the speed-based mode shapes are given in Figure 7.3 for each mode in Table 7.6, computed used a small-signal analysis software. Values are normalized with respect to the largest entry. Note that, because of the infinite bus representation, the magnitude of the speed (angle) entry for all modes is zero. Readers can find a more detailed discussion on the nature of these modes in Ref. [49].

Eigenvalue analysis of *Case A* identifies two low-damped low-frequency electromechanical modes at about 0.50 and 0.89 Hz and three higher frequency modes at about 1.47, 1.75, and 2.34 Hz associated with more localized behavior. From a physical perspective, the right eigenvector entries have large magnitudes at machines having strong participation in the oscillations.

Tables 7.7–7.9 give eigenvalue results for *Cases B–D*.

Table 7.6 Electromechanical modes of the system.

Mode	Eigenvalue	Frequency (Hz)	Damping (%)	Oscillation pattern
1	$-0.0055 \pm j3.1503$	0.501	0.180	Buses 6, 12, 11, 4, 3 swing coherently against bus 1
2	$-0.1105 \pm j5.6202$	0.894	1.97	Buses 12, 3, 11, 4 vs. bus 6
3	$-0.2461 \pm j9.2564$	1.473	2.66	Buses 3, 4, 11 vs. 6 and 12
4	$-0.0918 \pm j11.0113$	1.752	0.83	Buses 4, 11 vs. 6, 12, 3
5	$-0.4009 \pm j14.6170$	2.326	2.74	Bus 11 vs. 6, 12, 4, 3

Case A with an infinite bus at bus 1.

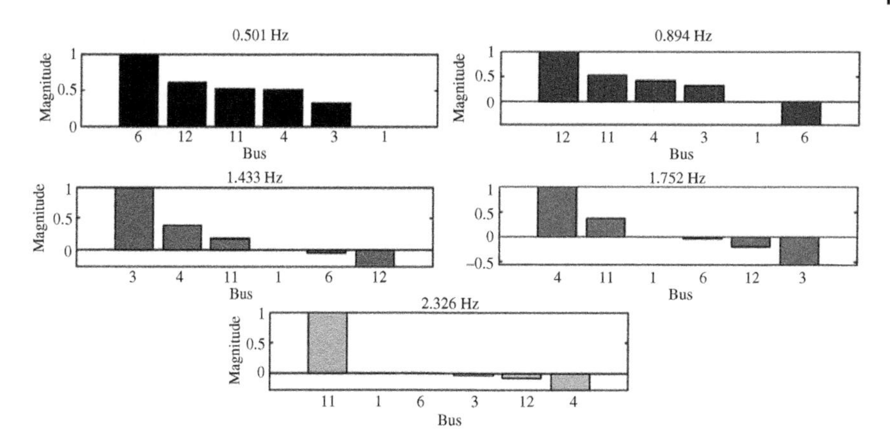

Figure 7.3 Normalized speed-based mode shapes of the electromechanical eigenvalues in Table 7.6 (base operating case).

Table 7.7 Electromechanical modes of the system (Case B with a wind farm at bus 1).

Mode	Frequency (Hz)	Damping (%)	Oscillation pattern
1	0.827	2.12	Buses 12, 3, 11, 4 vs. bus 6
2	1.335	3.81	Buses 3, 4, 11 vs. 6 and 12
3	1.752	0.63	Buses 4, 11 vs. 6, 12, 3
4	2.324	2.74	Bus 11 vs. 6, 3, 12, 4

Table 7.8 Electromechanical modes of the system (Case C with a WF farm at bus 6).

Mode	Frequency (Hz)	Damping (%)	Oscillation pattern
1	0.728	3.49	Buses 12,11,4,3 vs. 1
2	1.463	2.81	Buses 3, 4, 11 vs. 12
3	1.738	1.23	Buses 4, 11 vs. 12, 3
4	2.341	2.75	Bus 11 vs. 3, 12, 4

In all analyzed cases, damping improves with high wind generation, but as discussed earlier the effect of wind generation on system dynamic performance depends on several interacting factors.

In the following subsections, data-driven approaches that supplement information on conventional small-signal performance are investigated.

Table 7.9 Electromechanical modes of the system (Case D with wind farms at buses 1 and 6).

Mode	Frequency (Hz)	Damping (%)	Oscillation pattern
1	1.318	4.41	Buses 3, 4, 11 vs. 12
2	1.737	1.14	Buses 4, 11 vs. 12, 3
3	2.322	2.87	Bus 11 vs. 3, 12, 4

7.3.3 Large System Performance

Detailed system simulations have been carried out to characterize transient behavior as well as to examine the accuracy of the base case model. Emphasis is placed on the analysis of *Cases B* through *D*, which are assumed to represent distributed wind generation.

7.3.3.1 Cases B–C

Selected system simulations for cases B and C are shown in Figures 7.4–7.7. These solutions are thought to be representative of other operating scenarios described further. Based on these results, time-domain identification methods described in Section 7.2.2 are used to extract the modal characteristics of the system.

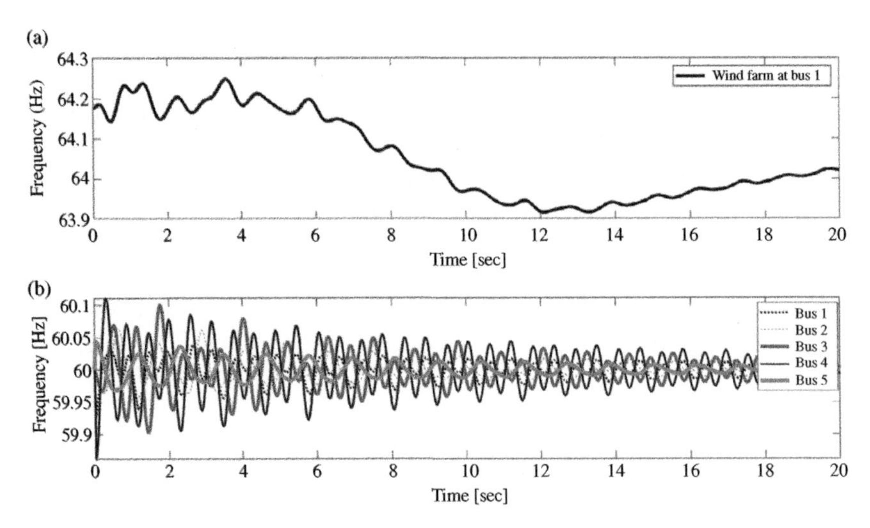

Figure 7.4 Speed deviations response following a three-phase fault at bus 7 (Case B); (a) bus 1, and (b) buses 1–5.

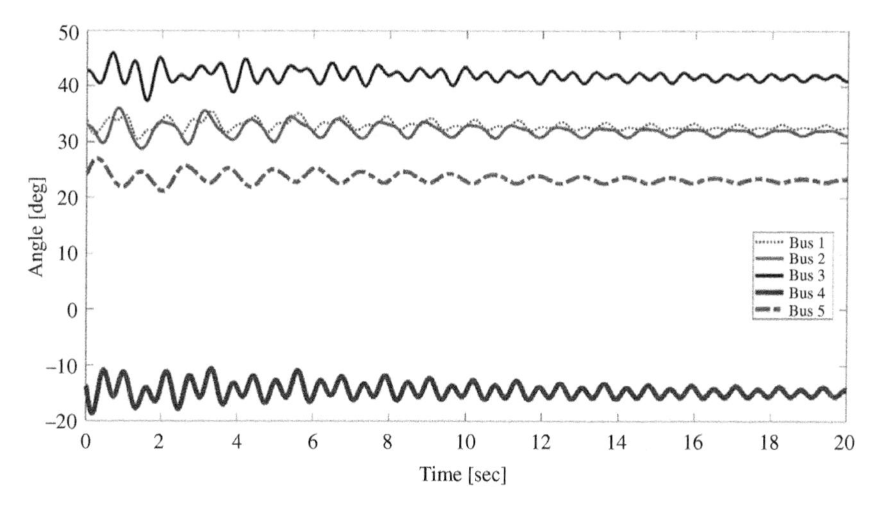

Figure 7.5 System rotor angle deviations following a three-phase fault at bus 7 (Case C).

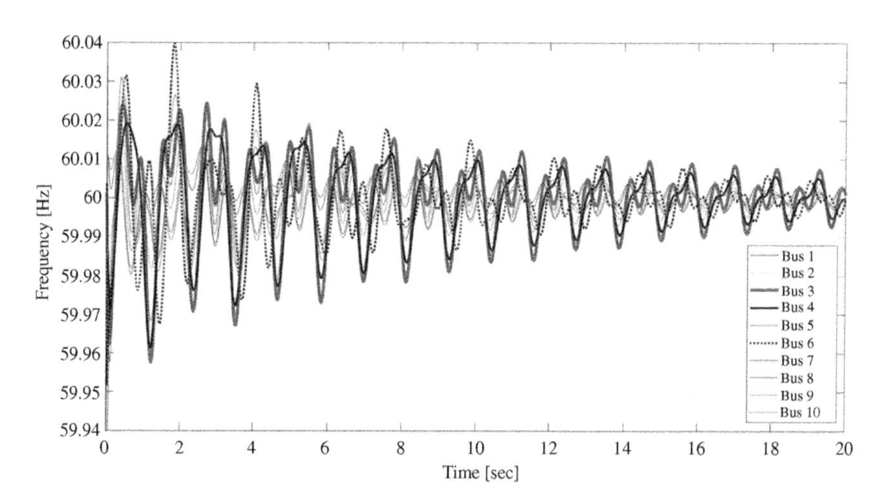

Figure 7.6 System bus frequency deviations following a three-phase fault at bus 7 (Case B).

Due to the small number of measurements, multisignal Prony analysis was used to identify the modes observed in the transient stability simulations [19]. A discussion of other modal extraction techniques is postponed until Section 7.4.

Table 7.10 shows Prony results for the generator speed deviations and generators' active output power (Case B) following a three-phase fault at bus 7. The good agreement between modal and data-driven approaches in Tables 7.5 and 7.8

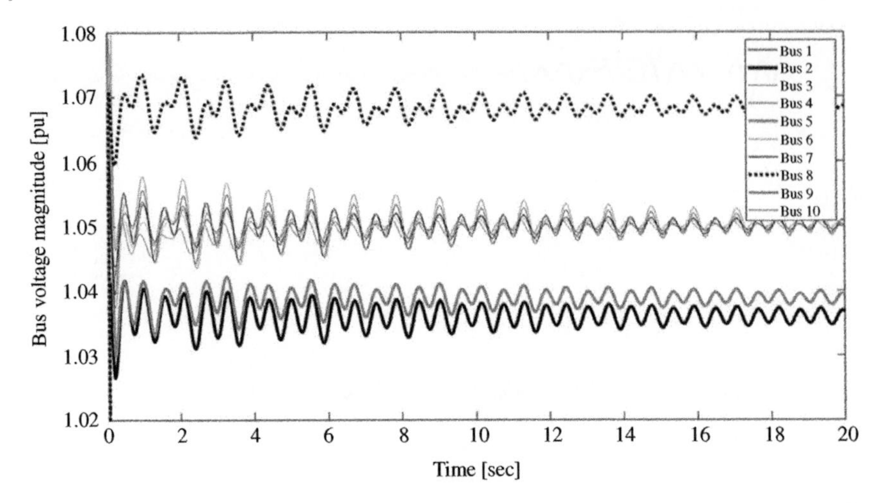

Figure 7.7 System bus voltage deviations following a three-phase fault at bus 7 (Case B).

Table 7.10 Multisignal Prony analysis for speed deviations and generator active power at buses 1, 2, 4, 6, 11, and 12 (Case B).

	Generator speeds		Generator active power	
Mode	**Frequency (Hz)**	**Damping (%)**	**Frequency (Hz)**	**Damping (%)**
1	0.863	1.798	0.862	1.740
2	1.381	2.311	1.381	2.271
3	1.741	0.632	1.741	0.628
4	1.879	9.191	1.879	9.191

ensures the correctness of the model. Results are found to be consistent, although generator active power is seen to provide a better characterization of the damping characteristics of mode 1 in Table 7.6, especially for the damping of mode 1.

A comparison of analytical results in Tables 7.6 with data-based results allows us to validate the small-signal analysis results.

7.3.3.2 Case D

In the discussion on the effect of renewable generation on the system dynamic performance, and to highlight the impact of inertia reduction, a two-WF case is now considered. In this analysis, generators at buses 1 and 6 were replaced by

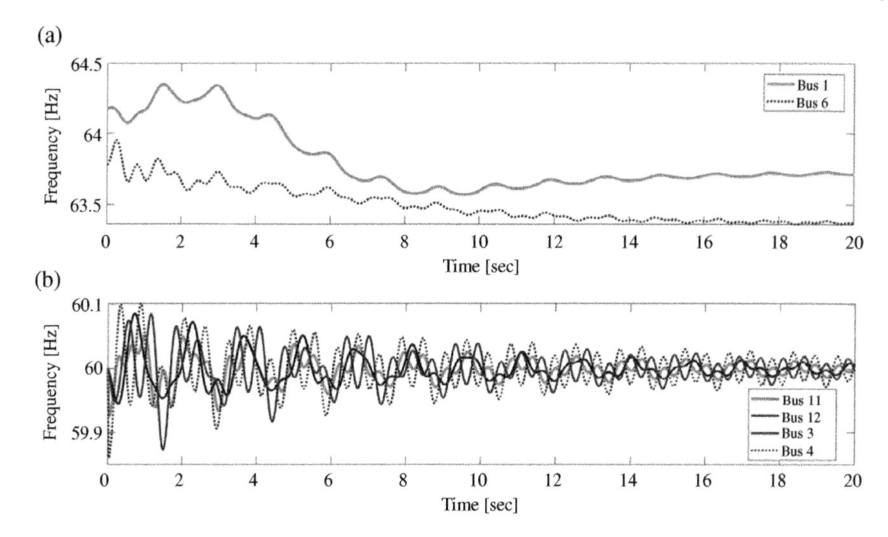

Figure 7.8 Speed deviations of system machines and wind farms following a three-phase fault at bus 7 (Case D); (a) buses 1 and 6, and (b) buses 3, 4, 11, and 12.

equivalent type 3 WFs. For simplicity and clarity of exposition, the same set of parameters for the control systems was utilized.

Figure 7.8 shows the system response to a three-phase fault at bus 7. Of interest in this plot are correlated motions of WFs at buses 1 and 6, which suggest the importance of identifying dynamic patterns and utilizing dynamic equivalencing techniques.

Further insight into the participation of WFs into system oscillatory behavior can be obtained from the analysis of PFs for the 1.318 Hz mode in Table 7.8. Similar results are obtained for other modes and are not discussed here.

As shown in Figure 7.9, the participation of the WFs to the modes of interest can be easily identified. The phase of PFs, however, does not have a direct interpretation, and other measures may be needed to supplement information on energy exchange, as well as the interaction paths.

Based on this analysis, data-driven approaches are used to supplement information on the effect of renewable generation on system dynamic behavior. To demonstrate the application of the aforementioned formulation, consider the time evolution of speed deviations following a three-phase stub fault at bus 7 in Figure 7.8. The time window selected for analysis is 20 seconds.

Figure 7.10 shows the time evolution of the extracted modes using Koopman analysis. Similar plots and results are obtained using DMD and other approaches. From Figure 7.10, system dynamic behavior is essentially dominated by two well-damped modes: a 1.21 Hz mode, and a 1.35 Hz mode.

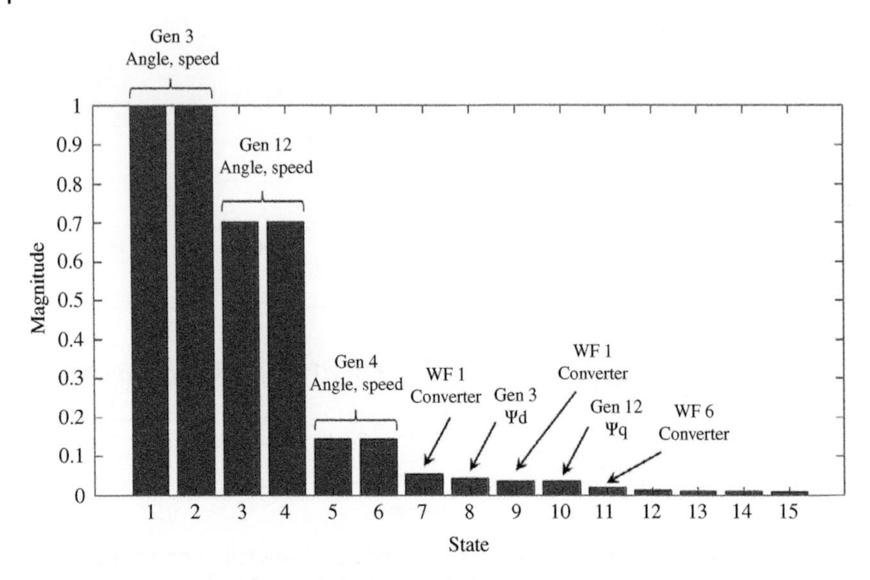

Figure 7.9 Participation factors for mode 1 in Table 7.5.

Table 7.11 shows the captured dominant modes. Multichannel Prony analysis indicates two modes with frequencies near 1.17 and 1.35 Hz that capture over 99% of the observed system response. Mode selection (column 3 in Table 7.11) was based on the transient energy ratio described in Refs. [43, 49].

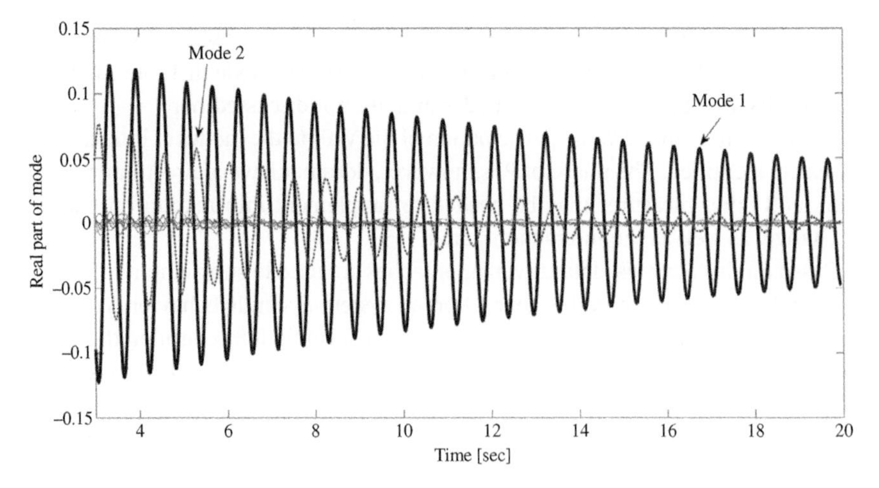

Figure 7.10 Dominant DMD modes from the decomposition $\sum_{j=0}^{d} \boldsymbol{a}_j(t)\boldsymbol{\psi}_j^T$ following a three-phase fault at bus 7.

Table 7.11 DMD analysis of generator bus frequency deviations.

Mode	Frequency (Hz)	Damping (%)	Participation (%)
1	1.717	0.504	66.88
2	1.355	2.401	22.76

Three-phase fault at bus 7.

The importance of the modal approximation in Eq. (7.20) or Eq. (7.28) is now obvious. Simulation results suggest that the time evolution of the system states can be approximated by the reduced-order model $\boldsymbol{X}(t) = \psi_1(t)\boldsymbol{a}_1^T(t) + \boldsymbol{a}_2(t)\psi_2^T$, where, as discussed earlier, the temporal coefficients a_js provide the temporal (amplitude) evolution of the modes, and the vector ψ give the approximate mode shape.

7.3.4 Mode Shape Identification

In the context of modal characterization, a key issue in the integration of wind and solar PV farms is the extraction of spatial relationships. Mode shapes play increasingly central roles in assessing energy transfer paths, assessing system damage, and complementing time-domain simulations [5]. DMD analysis is employed here to extract mode shapes from time-domain simulations.

Information on the nature of these modes can be gleaned from the mode shapes. To illustrate this issue, the shape of mode 1 for Case D in Table 7.12 was determined using conventional Prony analysis. In this case, the phase for mode 1 was determined by applying Prony analysis to each speed deviation in Figure 7.7.

In this analysis, each speed deviation was decomposed in the form

Table 7.12 Voltage-based mode shape of the interarea mode at 1.35 Hz extracted using Prony analysis for bus voltage magnitudes in Table 7.7 (Case D).

Machine/wind farm	Frequency, f_i (Hz)	Phase, φ_i (°)	Amplitude, A_i
WF at bus 1	1.355	−358.18	0.0211
Gen at bus 11	1.356	−176.55	0.0076
Gen at bus 12	1.356	−346.75	0.0233
Gen at bus 3	1.356	−171.77	0.0521
Gen at bus 4	1.356	−163.38	0.0205
WF at bus 6	1.355	−355.15	0.0086

$$\Delta\omega_j(t) = \sum_{i=1}^{Q} A_i e^{\lambda_i t} \cos(2\pi f_i t + \phi_i), j = 1, \ldots, 5 \tag{7.49}$$

Table 7.12 shows Prony results (PRS) for the speed-deviation signals. Several conclusions can be drawn for this analysis.

- The WFs at buses 1 and 12 and the WF at bus 6 swing nearly in phase and in opposition to generators at buses 11, 3, and 4.
- Bus 3 has the largest participation in the mode, followed by the generators at buses 12, 4, and the WF at bus 1.

Depending on the chosen algorithm, spatial structures can be real or complex. Insight into the applicability of these techniques can be gleaned from Figure 7.11 that shows the extracted voltage-based mode shape from simultaneous DMD analysis of the voltage measurements at generation buses 1, 3, 4, 6, 11, and 12.

Comparison of Prony results in Table 7.10 with DMD analysis in Figure 7.11 shows the correctness of the formulations.

7.3.5 Temporal Clustering

Clustering and dimensionality reduction play a critical role in the study of inter-system oscillations. Paramount to this approach is the ability to compress the complicated variability of the original data set into the smallest possible number of basic modes and their associated centroids.

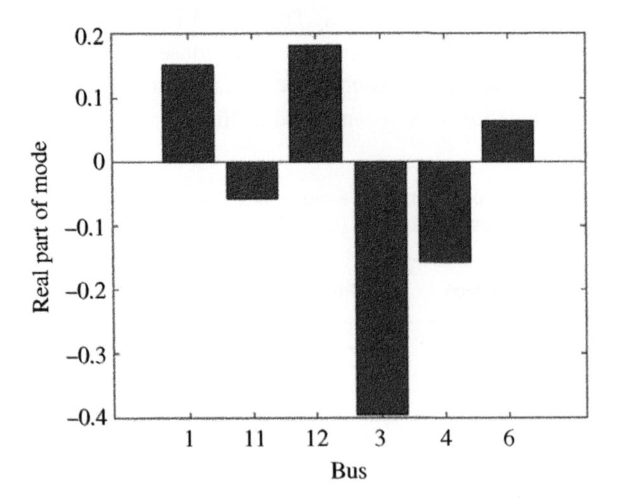

Figure 7.11 Speed-based mode shape for mode 1 in Table 7.3 computed using DMD analysis (Case D).

In Chapter 5 [49], an exploratory clustering technique based on fuzzy c-means that identifies groups of trajectories with similar compositions was implemented to divide the data into clusters and study power system coherent behavior.

Fuzzy clustering can be posed as the solution of the energy function:

$$f(U,v) = \sum_{k=1}^{K} \sum_{i=1}^{n} u_{ki}^{m} \|x_i - m_k\|^2 \tag{7.50}$$

where m_k are the cluster centers, the u_{ki}^{m} represent associated membership likelihoods (the degree of membership), $0 \leq u_{ki} \leq 1$, of the trajectory x_i being associated with the cluster center m_k.

The optimization problem is subject to the constraints

$$\begin{aligned} &\sum_{k=1}^{K} u_{ki} = 1, i = 1, \ldots, n \\ &u_{ki} \geq 0 \quad , k = 1, \ldots, K, i = 1, \ldots, n \end{aligned} \tag{7.51}$$

Using this approach (see Table 7.13), the system is divided into four areas and a centroid is associated with each area. In practice [49], it has been noted that, in the case of systems with high renewable generation, the centroids determined using the aforementioned techniques are a natural selection to monitor aggregated behavior.

In turn, the analysis of speed deviations in Figure 7.8 results in four clusters ($m = 4$ in [7.50]):

Cluster 1, which includes buses 11 and 4,
Cluster 2, which includes buses 12 and 3,
Cluster 3, which consists of bus 1, and
Cluster 4, which consists of bus 6.

Figure 7.12 shows the time evolution of the centroids for clusters 1 and 2 along with the buses included in the clusters.

Table 7.13 Clusters determined from c-means analysis of generator bus frequency data.

Cluster	Buses
1	12
2	4, 6, 11
3	3
4	1

Three-phase fault at bus 7.

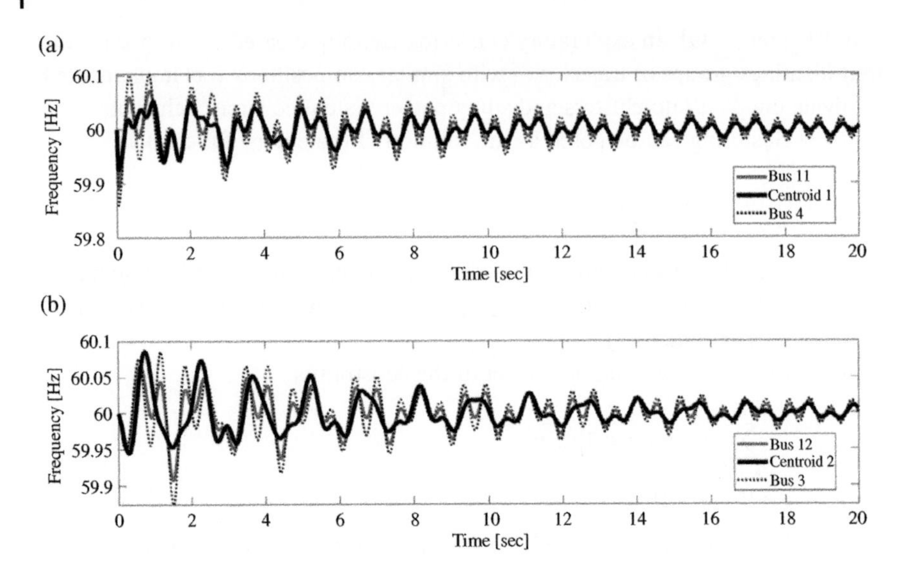

Figure 7.12 Speed signals following a three-phase fault at bus 7; (a) Cluster 1, and (b) Cluster 2.

Note that, by applying this technique, the center of angle (frequency) can be determined without simplifying assumptions.

7.4 Large-Scale System Study

In the system studied thus far, attention has been confined to basic active and reactive control strategies. The purpose now is to introduce some specific control alternatives in the study of a practical test system.

7.4.1 Case Study Description

The test system is based on a seven-area, 5449-bus representation of a large interconnected system. It includes several large-scale wind and solar PV farms, and SVCs. A single-line diagram of this system showing main buses and transmission lines and the location of selected WFs and solar PV plants is shown in Figure 7.13.

The system model consists of 635 fully represented generators, 5449 buses, 5292 transmission lines, and the detailed representation of 26 WFs and 3 equivalent solar PV farms; the total interconnected system load is 35 000 MW.

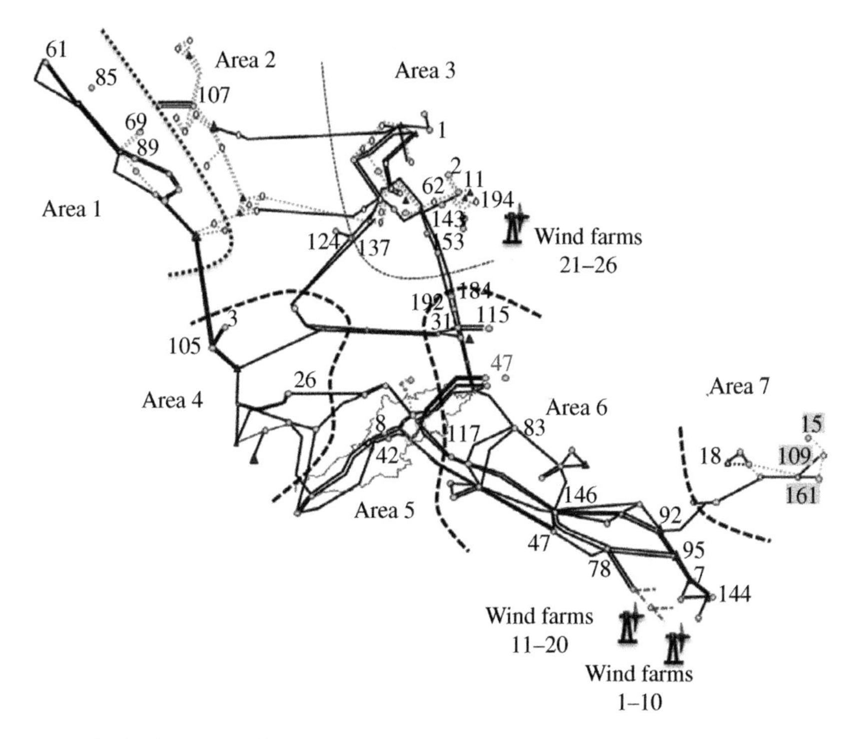

Figure 7.13 Schematic illustration of the seven-area test system showing major buses and transmission lines.

7.4.2 Renewable Generator Modeling

In the system model, all WFs are represented using detailed user models similar to those described in Section 7.2. Table 7.14 summarizes the main characteristics of these WFs. The block diagrams of the control systems, which are adapted from [46, 47], are given in Figure 7.14, while Figure 7.15 shows the two-mass model used in the simulation studies.

For this study, the following assumptions are made:

- All WFs are modeled by type 3 (DFIG) generic wind turbine models.
- Wind plant reactive power control is enabled.
- Voltage-dependent load characteristics are represented in detail.

7.4.3 Effect of Inverter-Based DGs on Oscillatory Stability

Small-signal stability analyses were conducted to examine the impact of wind generation on modal stability. Table 7.15 synthesizes the main characteristics of the slowest interarea modes in the system.

Table 7.14 Characteristics of DFIGs used in the study.

Wind farm	Type	Operating voltage (kV)	Voltage control	Active power control
1–2, 5–10, 14–15, 17–19	Type 3 DFIG	0.69	Yes	Yes
3, 11–13	Type 3 DFIG	12	Yes	Yes
4, 16	Type 3 DFIG	13.2	Yes	Yes

7.4.4 Large System Performance

Early transient stability studies have indicated that distributed nonsynchronous generation in Areas 3, 4, and 6 have a noticeable impact on small-signal stability. To excite the slowest electromechanical modes and investigate the impact of renewable generation on the oscillatory stability, several contingency scenarios were developed. These include large-loss-of-generation events, load shedding, double line outages, and short circuits at major buses.

Selected contingency scenarios are listed in Table 7.16.

The following measurements are selected for study:

- 935 bus frequency signals,
- 200 bus voltage magnitudes,
- 138 generator speed deviations, and
- 26 WF speed deviations.

Figures 7.16 and 7.17 show plots of frequency deviations of WFs 1–26 for three critical contingency scenarios in Table 7.16. In all cases, a nonlinear trend is observed, especially in the first few seconds of the simulations.

Studies conducted to extract dominant system behavior are discussed further. Table 7.17 shows the dominant system modes extracted using DMD applied to the generator speed deviation signals in Figure 7.16.

7.4.5 Model Validation

Extensive studies have been conducted to validate the system model using multi-signal modal extraction techniques. From the original simultaneous results, two data sets were initially selected to compare system results

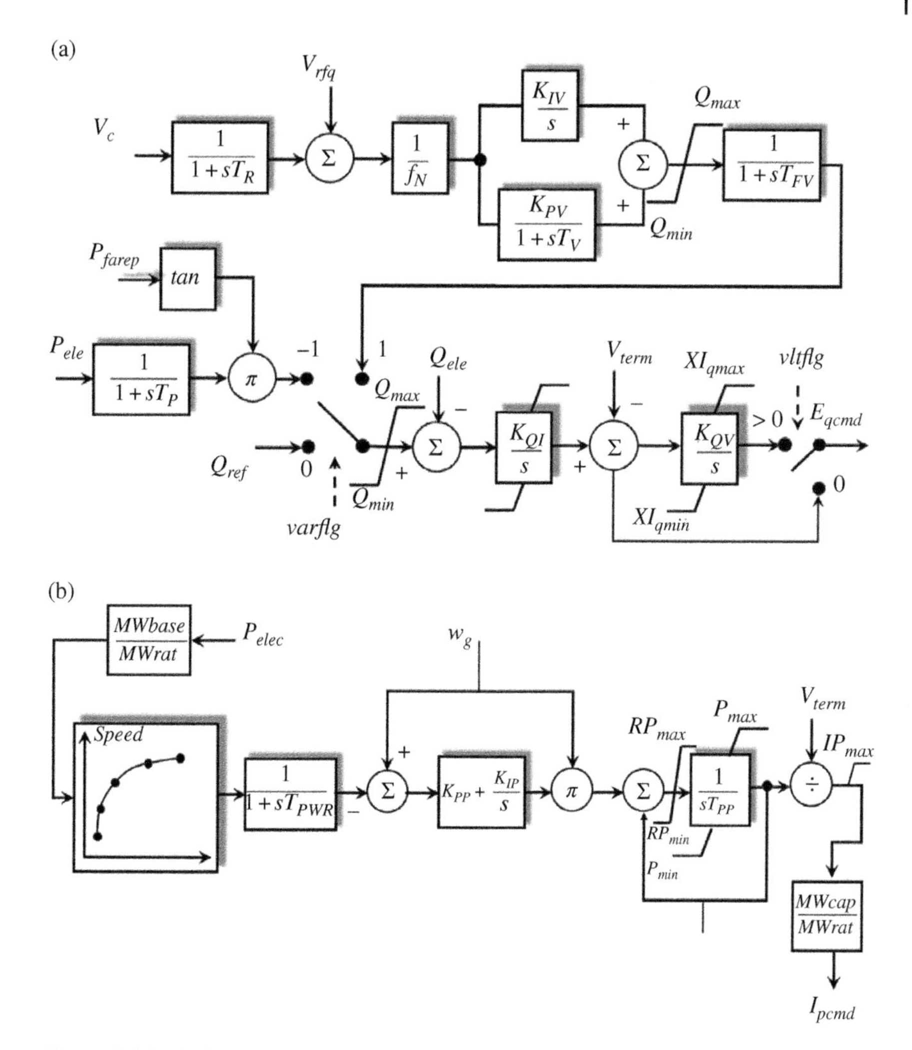

Figure 7.14 Active and reactive control characteristics: (a) reactive and (b) active control characteristics. Adapted from [46, 47].

1) *Data set 1* including 40-speed deviations at major selected generators and 26 WFs. This scenario allows for Prony analysis of the data,
2) *Data set 2* set including 983 bus frequency signals,
3) *Data set 3* with 164-speed deviation signals.

Figure 7.18 shows the speed deviations for data set 1.

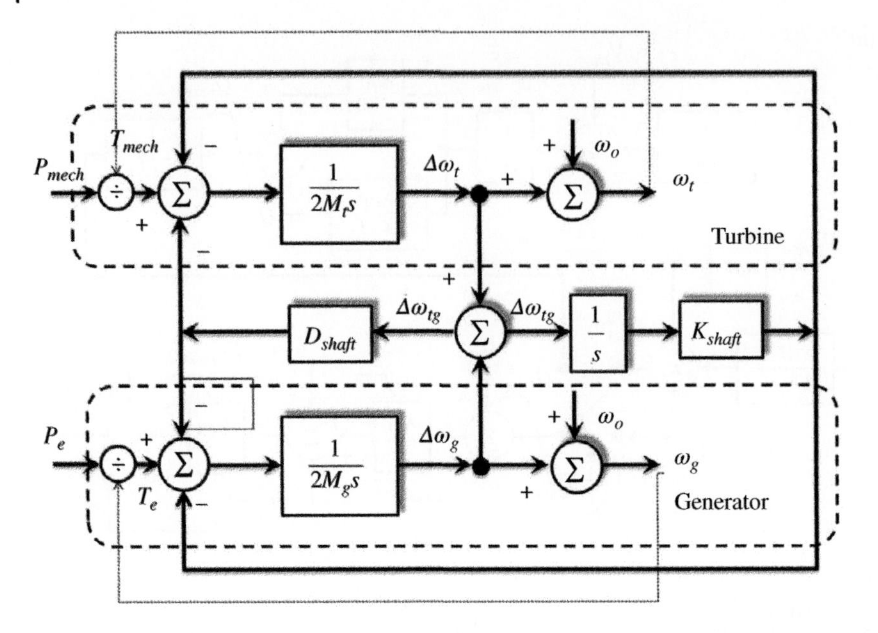

Figure 7.15 Two-mass wind mechanical model.

Table 7.18 compares the amplitude and frequency of the extracted modes using Prony analysis and DMD analysis for the 66-speed deviation signals (data set 1) with the eigensolution of the linear model $\dot{x} = Ax$. Results between Prony analysis and DMD analysis are found to be in good agreement although some discrepancies are found.

Table 7.15 Slowest interarea modes of the system.

Mode	Eigenvalue	Frequency (Hz)	Damping (%)	Swing pattern
1	$-0.0330 \pm j2.481$	0.394	1.33	Areas 1–3 vs. Areas 6 and 7
2	$-0.2415 \pm j3.430$	0.547	7.02	Area 1 vs. Areas 3, 6, and 7
3	$-0.3549 \pm j4.103$	0.653	8.62	Areas 1 vs. Areas 2, 3, and 6
3	$-0.3280 \pm j4.376$	0.696	7.47	Area 1 vs. Areas 2, 6, and 7
5	$-0.4944 \pm j5.1061$	0.812	9.64	Area 5 vs. Area 6

Base case condition.

Table 7.16 Contingency scenarios selected for the study.

Contingency scenario	Description
CS01	Loss of the largest unit in the system (generators 1 and 2, Table 8.1)
CS02	Three-phase fault at bus 153 in Area 3
CS03	Three-phase fault at the POIS[a] of WFs 1–19 in Area 6
CS04	Single-line outage, 8024–8052 in Area 6
CS05	The simultaneous outage of 400 kV lines 8024–8520 and 7415–7426 in Area 6
CS06	The outage of generator 1, unit 1 (750 MW) in Area 6

a Point of interconnection with the system.

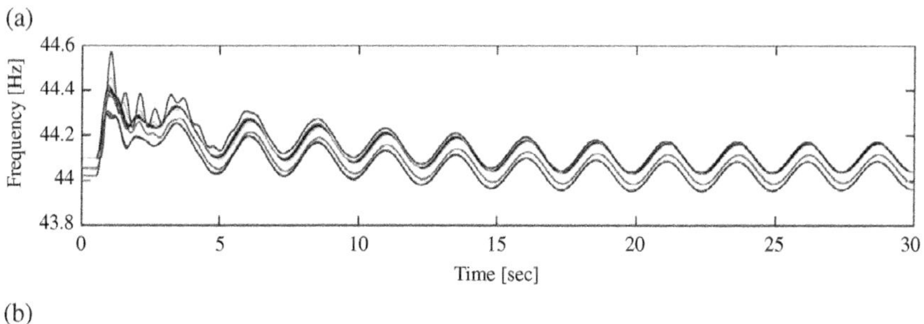

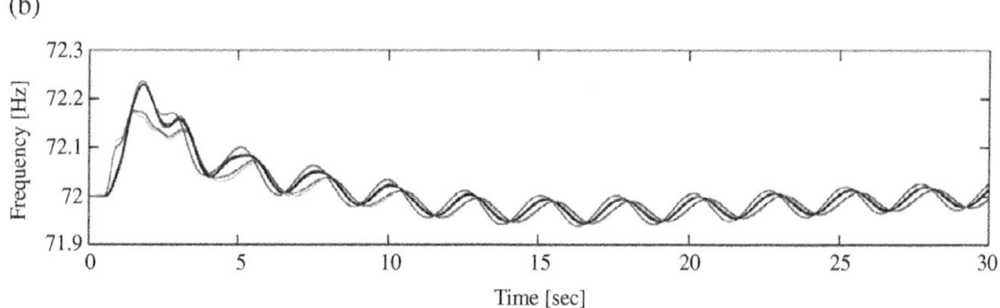

Figure 7.16 Frequency deviations for contingency scenario CS04; (a) WFs in Area 6, and (b) WFs in Areas 2 and 3.

Eigenvalues obtained using a small signal stability program exhibit good agreement with the time-domain studies. Study experience shows that conventional Prony analysis becomes prohibitive for high-dimensional data sets, and some expertise is needed to determine both the sampling interval and the study time window.

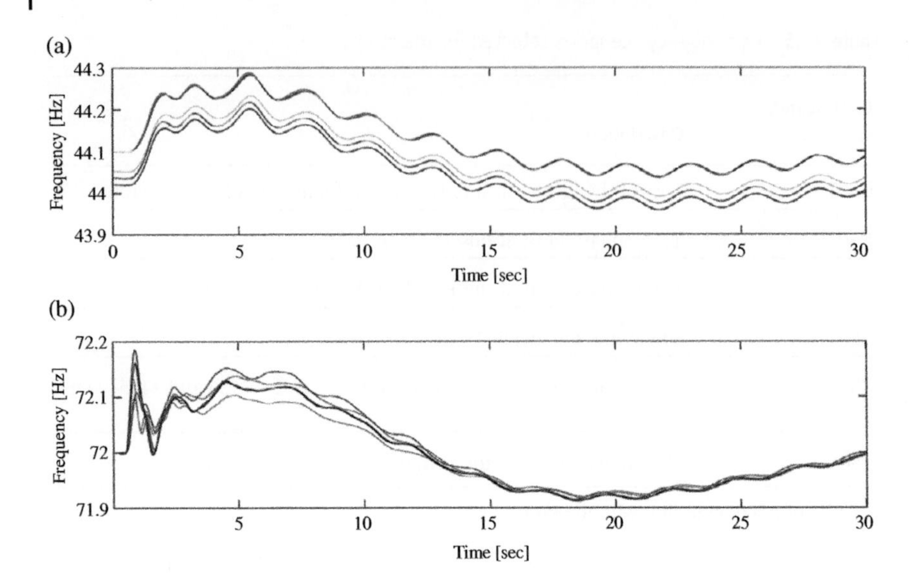

Figure 7.17 Frequency deviations for contingency scenario CS02: (a) WFs in Area 6 and (b) WFs in Areas 2 and 3.

Table 7.17 Frequencies and damping extracted from generator speed deviations using DMD for contingency scenario CS04.

Mode	Frequency	Damping
1	0.399	0.389
2	0.547	7.000
3	0.666	8.29
3	0.696	7.473
5	0.798	9.527

7.4.5.1 Reconstructed Flow Fields

To further verify the accuracy of the models, Koopman and DMD analyses were used to extract the dominant modes of data set 2 above, where $\boldsymbol{X}_f \in \Re^{2804 \times 983} = [\Delta\boldsymbol{f}_1 \quad \Delta\boldsymbol{f}_2 \dots \quad \Delta\boldsymbol{f}_{983}]$. Application of Koopman mode decomposition (KMD) and DMD in Table 7.19 to the data set identifies three dominant modes with frequencies of nearly 0.395, 0.555, and 0.697 Hz. Figure 7.19a shows the original signals, $\Delta\boldsymbol{f}_j$ while Figure 7.19b shows the reconstructed signals $\Delta\hat{\boldsymbol{f}}_j$ using the Koopman procedure in Section 7.2.3.

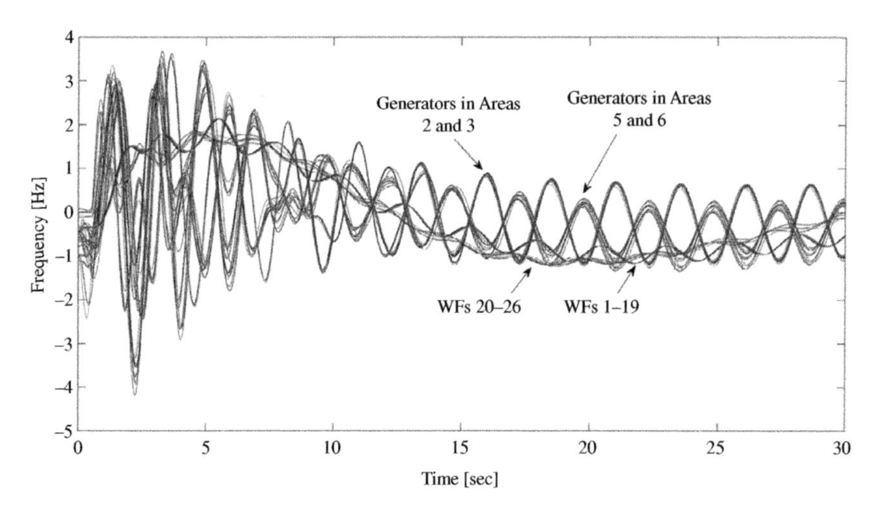

Figure 7.18 System response to a three-phase fault at the POIS (contingency scenario CS03) Signals detrended for clarity of presentation.

Table 7.18 Modal extraction analysis for simulation results in Figure 7.18.

	Multisignal Prony analysis		DMD analysis		Small-signal analysis	
Mode	Frequency (Hz)	Damping (%)	Frequency (Hz)	Damping (%)	Frequency (Hz)	Damping (%)
1	0.395	1.287	0.395	1.291	0.395	1.331
2	0.659	8.587	0.657	8.601	0.653	8.620
3	0.548	7.182	0.552	7.128	0.545	7.000

Contingency scenario CS04. Time window 20–30 seconds.

Similar results are obtained using DMD analysis. The overall reconstruction error, $\varepsilon = 1.277\times 10^{-3}$, is computed as

$$\varepsilon(\hat{x}_k, x_k) = \frac{\sum_{k=1}^{N} (\hat{x}_k)^2}{\sum_{k=1}^{N} (x_k)^2} \tag{7.52}$$

where $\hat{x}_k, k = 1, ..., N$ is the DMD approximation, and x_k is the simulated data.

The second problem of key importance is that of simultaneous analysis of measurements involving different physical units. To illustrate these ideas, assume that several data sets (such as speed deviations, frequency, bus voltages, etc.) are

Table 7.19 Comparison of Koopman and DMD results for data set 2.

	Koopman analysis		DMD analysis	
Mode	**Frequency (Hz)**	**Damping (%)**	**Frequency (Hz)**	**Damping (%)**
1	0.397	1.288	0.395	1.291
2	0.677	8.345	0.657	8.615
3	0.550	7.230	0.552	7.012

Time window 20–30 seconds.

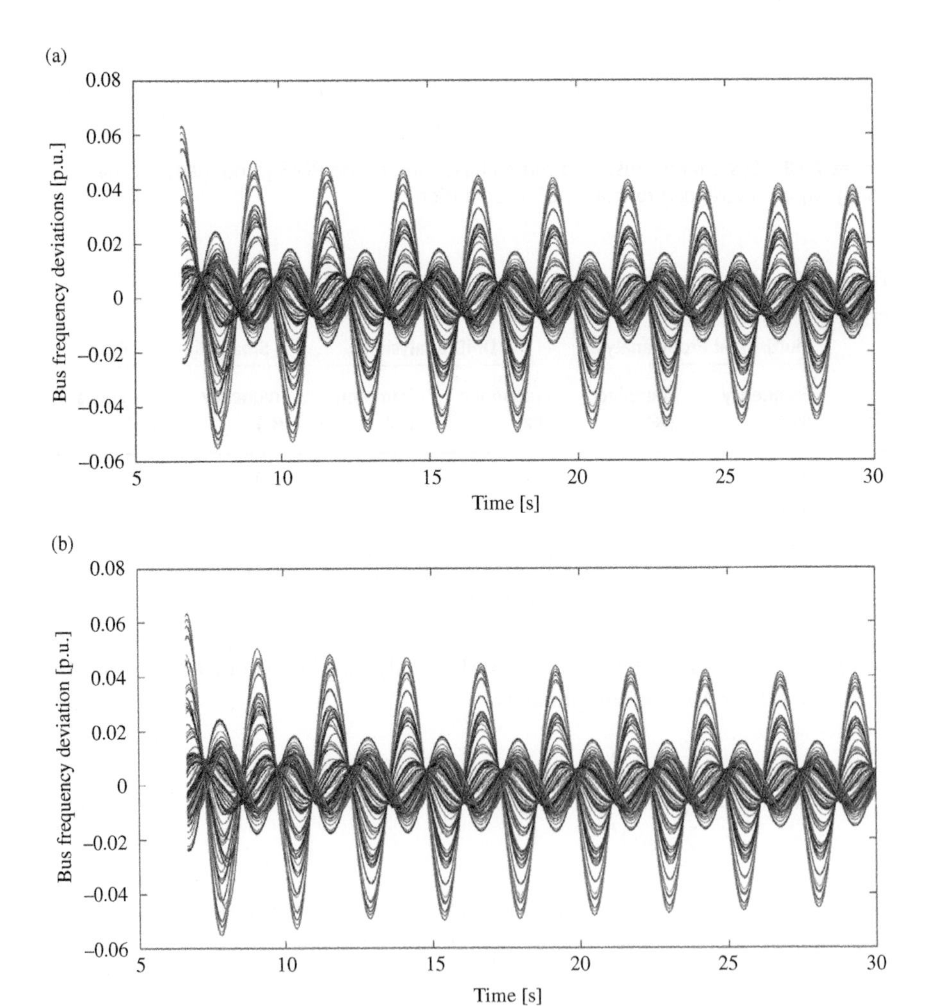

Figure 7.19 Bus frequency signals following a three-phase fault at the POIS (contingency scenario CS03): (a) original, simulated signals, and (b) reconstructed signals using DMD.

obtained from the same disturbance. They could represent measurements from multisignal records or represent historic information.

Let each data set be of the form $\boldsymbol{X}_{set\ 1} \in \Re^{N_1 \times m_1}, \boldsymbol{X}_{set\ 2} \in \Re^{N_2 \times m_2}, \ldots, \boldsymbol{X}_{set\ L} \in \Re^{N_L \times m_L}$. If each data matrix has the same length ($N_1 = N_2 = \ldots = N_L$), one can express the concatenated matrix as

$$\boldsymbol{X} = \begin{bmatrix} \boldsymbol{X}_{set\ 1} \\ \boldsymbol{X}_{set\ 2} \\ \vdots \\ \boldsymbol{X}_{set\ L} \end{bmatrix}$$

This motivates the need for a joint analysis of multiple data sets much as in the case of feature-based fusion of data. Three main issues are of interest here: (i) to study the complementary nature of data types, (ii) to assess the effect of a given data type of modal behavior and which combination of data modalities can be used for specific analysis or control purpose, and (iii) to find common structures and relationships between data types.

7.4.6 Identification of Mode Shapes Using DMD

As discussed earlier, identification of mode shapes in power systems with highly distributed energy resources is a critical objective in assessing the applicability of data-driven methods. Because of their inherent characteristics, mode decomposition methods (global multiscale methods) are well suited to extract mode shapes and other global characteristics.

Figure 7.20 shows the speed-based mode shape extracted from the speed deviations in Figure 7.16 (data set 1) using DMD analysis. To avoid cluttering, 40 generators and 26 WFs are selected for display. Figure 7.20 indicates that WFs in Area 6 swing (nearly) in phase with generators in Areas 6 and 7 and in opposition to generators and WFs in the south systems. It is also apparent in these plots that WFs 10, 17, and 19 in Area 6, and WFs 20 and 21 in Area 5 show the largest contribution to the slowest swing mode.

7.5 Analysis Results and Discussion

Multichannel modal extraction methods are rapidly becoming a powerful and ubiquitous tool for modal analysis of large sets of simulations or measurements. Effective modeling and characterization of global oscillatory phenomena

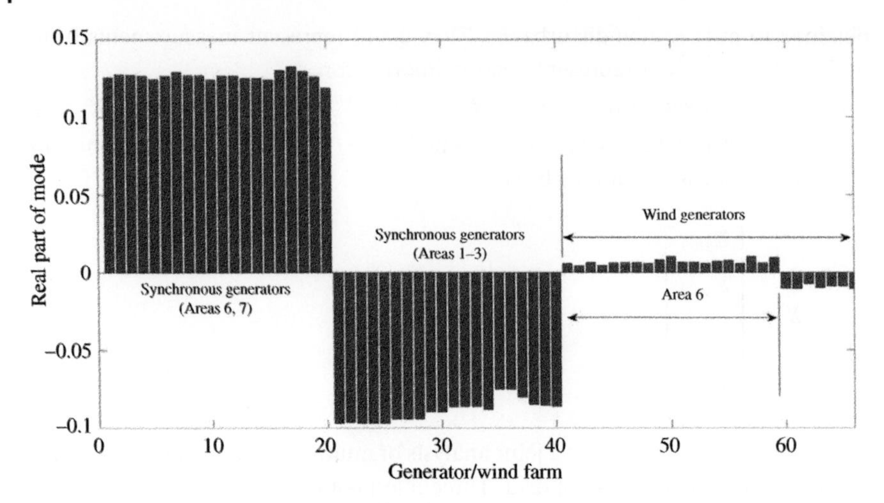

Figure 7.20 Speed-based mode shape of 0.395 Hz interarea mode 1.

involving large-scale penetration of RESs require many observational data often involving disparate spatial or temporal time scales. Extensive numerical experience shows that the effectivity of these methods depends on several interacting factors such as the presence of nonlinearities, the presence of noise, and the size of the observation window, to mention a few factors.

In this chapter, several multichannel modal extraction techniques have been benchmarked by comparison with multichannel Prony analysis and small-signal studies. These tools include techniques for extracting modal properties, determining spatiotemporal patterns, analyzing the system dynamics, and evaluating the system response. It is demonstrated that these methods can be efficiently applied to analyze large, complex data from selected system measurements and shed light on the impact of distributed generation on system oscillatory behavior. A crucial part of these applications concerns understanding the role of distributed energy resources in the observed oscillations. Methods for interpreting the structure and distribution of spatiotemporal patterns are given and techniques to visualize these patterns are pointed out.

Simulation experiences show that while WFs' speed signals can be used to extract spatial patterns, often other signals such as bus terminal frequencies offer a useful alternative to modal characterization. Other studies reveal that trends exhibited by WF signals may affect the interpretation of modal characteristics. Further analysis is needed to assess the performance of these methods when applied to measured, heterogeneous data.

References

1. Vittal, E., O'Malley, M., and Keane, A. (2012). Rotor angle stability with high penetrations of wind generation. *IEEE Transactions on Power Systems* **27** (1): 353–362.
2. Vittal, V. and Ayyanar, R. (2013). *Grid Integration and Dynamic Impact of Wind Energy*. New York, NY: Springer.
3. Gautam, D., Vittal, V., and Harbour, T. (2009). Impact of increased penetration of DFIG based wind turbine generators on transient and Small-signal stability of power systems. *IEEE Transactions on Power Systems* **24** (3): 1426–1434.
4. Power System Dynamic Performance Committee, Task Force on Identification of Electromechanical Modes, Sanchez-Gasca, J.J. Chair. (2012). Identification of electromechanical modes in power systems. IEEE/PES, Special Publication TP462 (June 2012).
5. Wilson, D., Bialek, J., and Gustavsson, N. (2008). Identifying sources of damping issues in the icelandic power system. Proceedings of the 16th Power Systems Computation Conference (PSCC'08).
6. Castellanos, R.B., Messina, A.R., Gonzalez, M.R., and Guizar, G.C. (2018). Assessment of frequency performance by grid integration in a large-scale power system. *Wind Energy* **21**: 1359–1371.
7. CIGRE Working Group C.6.08. (2011). Grid integration of wind generation (February 2011).
8. Browne, T.J., Vittal, V., Heydt, G.T., and Messina, A.R. (2008). A comparative assessment of two techniques for modal identification from power system measurements. *IEEE Transactions on Power Systems* **23** (3): 1408–1415.
9. Grebe, E., Kabouris, J., López Barba, S. et al. (2010). Low frequency oscillations in the interconnected system of Continental Europe. Proceedings of IEEE PES General Meeting.
10. Sanchez-Gasca, J.J., Miller, N.W., Price, W.W. (2004). A modal analysis of a two-area system with significant wind power penetration. Proceedings of the IEEE PES Power Systems Conference and Exposition.
11. Messina, A.R. and Vittal, V. (2007). Extraction of dynamic patterns from wide-area measurements using empirical orthogonal functions. *IEEE Transactions on Power Systems* **22** (2): 682–692.
12. Barocio, E., Pal, B.C., Thornhill, N.F., and Messina, A.R. (2015). A dynamic mode decomposition framework for global power system oscillation analysis. *IEEE Transactions on Power Systems* **30** (6): 2902–2912.
13. Feeny, B.F. and Kappagantu, R. (1998). On the physical interpretation of proper orthogonal modes in vibrations. *Journal of Sound and Vibration* **211** (4): 607–616.
14. Goebel, C.J. and Epstein, S.T. (1980). Motion of damped oscillators: expansion in normal modes. *American Journal of Physics* **48** (4): 289–291.

15. Hamdan, A.M.A. and Nayfeh, A.H. (1986). Coupling measures between modes and state variables. *International Journal of Control* **43** (3): 1029–1041.
16. Kundur, P. (1994). *Power System Stability and Control*. New York, NY: McGraw-Hill.
17. Hauer, J. (1991). Application of Prony analysis to the determination of modal content and equivalent models for measured power system response. *IEEE Transactions on Power Systems* **6** (3): 1062–1068.
18. Pierre, D.A., Trudnowski, D.J., and Hauer, J.F. (1990). Identifying reduced-order models for large nonlinear systems with arbitrary initial conditions and multiple outputs using prony signal analysis. Proceedings of the 1990 American Control Conference.
19. Trudnowski, D.J., Johnson, J.M., and Hauer, J.F. (1999). Making Prony analysis more accurate using multiple signals. *IEEE Transactions on Power Systems* **14** (1): 226–231.
20. Zhou, N., Pierre, J., and Trudnowski, D. (2013). Some considerations in using prony analysis to estimate electromechanical modes. Proceedings of the 2013 IEEE Power & Energy Society General Meeting.
21. Berti, E., Cardoso, V., González, J.A., and Sperhake, U. (2007). Mining information from binary black hole mergers: a comparison of estimation methods for complex exponentials in noise. *Physical Review D* **75** (124017): 1–17.
22. Borden, A. and Lesieutre, B. (2014). Variable projection and Prony analysis for power system modal identification. *IEEE Transactions on Power Systems* **29** (6): 2613–2620.
23. Hildebrand, F.B. (1956). *Introduction to Numerical Analysis*. New York, NY: Dover Publications, Inc.
24. Potts, D. and Tasche, M. (2013). Parameter estimation for nonincreasing exponential sums by Prony-like methods. *Linear Algebra and Applications* **439**: 1024–1039.
25. Koehl, P. (1999). Linear prediction spectral analysis of NMR data. *Progress in Nuclear Magnetic Resonance Spectroscopy* **34**: 257–299.
26. Kul'ment'ev, A.I. (2004). Analysis of positron lifetime spectra via a fast Prony algorithm. *The European Physical Journal of Applied Physics* **25**: 191–201.
27. Khazaei, J., Fang, L., Jiang, W., and Manjure, D. (2016). Distributed Prony analysis for real-world PMU data. *Electric Power Systems Research* **133**: 113–120.
28. Golpira, H., Atarodi, A., Amini, S. et al. (2020). Optimal energy storage system-based virtual inertia placement: a frequency stability point of view. *IEEE Transactions on Power Systems* **35** (6): 4824–4835.
29. Barkhuijsen, H., De Beer, R., and Van Ormondt, D. (1987). Improved algorithm for noniterative time-domain model fitting to exponentially damped magnetic resonance signals. *Journal of Magnetic Resonance* **73**: 553–557.

30. Kumaresan, R. and Tufts, D. (1982). Estimating the parameters of exponentially damped sinusoids and pole-zero modeling in noise. *IEEE Transactions on Acoustics, Speech and Signal Processing* **30** (6): 833–840.

31. Kumaresan, R., Tufts, D.W., and Scharf, L.L. (1984). A Prony method for noisy data: choosing the signal components and selecting the order in exponential signal models. *Proceedings of the IEEE* **72** (2): 230–233.

32. Almunif, A., Fan, L., and Miao, Z. (2020). A tutorial on data-driven eigenvalue identification: Prony analysis, matrix pencil, and eigensystem realization algorithm. *International Transactions on Electrical Energy Systems*: 1–17.

33. Susuki, Y. and Mezic, I. (Nov. 2011). Nonlinear Koopman modes and coherency identification of coupled swing dynamics. *IEEE Transactions on Power Systems* **26** (4): 1894–1904.

34. Susuki, Y. and Mezic, I. (2012). Nonlinear Koopman modes and a precursor to power system swing instabilities. *IEEE Transactions on Power Systems* **27** (3): 1182–1191.

35. Susuki, Y. and Mezic, I. (2014). Nonlinear Koopman modes and power stability assessment without models. *IEEE Transactions on Power Systems* **29** (2): 899–907.

36. Hernandez-Ortega, M.A. and Messina, A.R. (2018). Nonlinear power system analysis using Koopman mode decomposition and perturbation theory. *IEEE Transactions on Powers Systems* **33** (5): 5124–5134.

37. Alassaf, A. and Fan, L. (2020). Dynamic mode decomposition in various power system applications. Proceedings of the 2019 North American Power Symposium (NAPS).

38. Ruhe, A. (1984). Rational Krylov sequence methods for eigenvalue computation. *Linear Algebra and its Applications* **58**: 391–405.

39. Tu, J.H., Rowley, C.W., Dirk, M. et al. (2013). On dynamic mode decomposition: theory and applications. *Journal of Computational Dynamics* **1** (2): 31–421.

40. Bistrian, D.A. and Navon, I.M. (2015). An improved algorithm for the shallow water equations model reduction: dynamic mode decomposition vs POD. *International Journal for Numerical Methods in Fluids* **78** (9): 552–580.

41. Kutz, J.N., Brunton, S.L., Brunton, B.W., and Proctor, J.L. (2016). *Dynamic Mode Decomposition – Data-Driven Modeling of Complex Systems*. Phildelphia, PA: SIAM.

42. Sakata, I., Nagano, Y., Igarashi, I. et al. (2020). Normal mode analysis of a relaxation process with Bayesian inference. *Science and Technology of Advanced Materials* **21** (1): 67–78.

43. Tissot, G., Cordier, L., Benard, N., and Noack, B.R. (2014). Model reduction using dynamic mode decomposition. *Comptes Rendus Mécanique* **342** (6-7): 410–416.

44. A Study of Reactive Power Compensators for High-Voltage Power Systems. (1982). Advanced systems technology division and transmission and distribution systems engineering department, westinghouse electric corporation, contract 4-L60-6964P, Final Report (12 May 1981).

45. Roman-Messina, A. (2015). *Wide-Area Monitoring of Interconnected Power Systems, Power and Energy Series 77, Stevenage.* UK: IET, The Institution of Engineering and Technology.

46. WECC (2019). Converting REEC_B to REEC_A for solar PV generators WECC REMTF (13 June 2019). https://www.wecc.org/Reliability/Converting%20REEC_B%20to%20REEC_A%20for%20Solar%20PV%20Generators.pdf (accessed 2020).

47. Richwine, M.P., Sanchez-Gasca, J.J., and Miller, N.W. (2014). Validation of a second generation type 3 generic wind model. Proceedings of the IEEE PES General Meeting Conference & Exposition.

48. Byerly, R.T., Sherman, D.E., and McLain, D.K. (1975). Normal modes and mode shapes applied to dynamic stability analysis. *IEEE Transactions on Power Apparatus and Systems* **PAS-94** (2): 224–229.

49. Román-Messina, A. (2020). *Data Fusion and Data Mining for Power System Monitoring*. Boca Raton, FL: CRC Press.

8

Solar and Wind Integration Case Studies

In previous chapters, techniques to analyze and characterize transient behavior in power systems with high renewable power generation were introduced. This chapter describes the experience in the analysis of wind and solar integration in large-scale power grids with complex dynamics and operating characteristics.

Studies to evaluate the impact of geographically dispersed wind and solar generation on system dynamic behavior are conducted, and challenges related to the coordination of wind and hydro generation are described. The study assesses the impact of large clusters of wind and solar photovoltaic (PV) generation on system dynamic performance and identifies factors affecting small and large-signal stability and frequency regulation. A second objective is to evaluate transmission and generation limitations.

Methods for interpreting and understanding the impact of distributed renewable energy resources (RESs) on global oscillatory behavior are given, and tools for studying wind and hydro generation are developed and tested. The issues of network reactive control performance and tuning, inertia reduction, generation commitment, grid stability, and power system frequency response are discussed in detail. Steady-state eigenvalue analyses and detailed step-by-step simulations are used for assessing issues regarding wind and solar integration.

It is shown that dynamic performance considerations related to voltage collapse, oscillatory and frequency instability, and postdisturbance oscillations may limit high wind and solar PV integration.

8.1 General Context and Motivation

Large-scale penetration of inverter-based RESs presents some challenges and opportunities for the future power system. The use of emerging technologies such as inverter-based wind and PV resources and the increasing size and complexity of

Renewable Integrated Power System Stability and Control, First Edition.
Hêmin Golpîra, Arturo Román-Messina, and Hassan Bevrani.

modern power systems are expected to introduce new interactions and operational challenges [1–6].

Typically, factors that lead to integration issues include vulnerability to major perturbation, reduced system inertia, rotor angle, and voltage stability, transient stability and frequency performance, and postdisturbance oscillations [7–13]. In addition, the successful integration of distributed RESs requires an increased level of visibility (and controllability) of these devices [14, 15]. A minimum level of expected performance during power system disturbances is needed, which in turn requires knowledge of power system dynamic characteristics and the aggregated behavior of distributed RESs [16]. Sensitivity studies and data-based analysis techniques are of interest since they can supplement information on the system behavior, for instance, to understand the commitment and dynamic patterns, arising from the integration of new wind and PV generation.

Some recent investigations have also revealed a significant decline in the frequency performance of some major grids [10, 11, 17], prompting concerns on the effect of increased wind and solar penetration on the future inertial frequency performance. Understanding the nature of this trend is therefore critical to coordinating operational and control actions.

Recent advances in inertia emulation techniques have allowed modern wind turbines to provide inertial response functionalities under low-frequency conditions [18, 19]. Other developments include the use of advanced schemes for inertia emulation [20]. However, as discussed by several authors, the role of inverter-based inertial response and wind generation frequency response characteristics is not well understood [17], or knowledge of the real capabilities and limitations of inertial response from wind turbines is limited [21].

Other factors affecting system dynamic response include [22, 23]:

- The radial interconnection of wind or solar farms to the bulk transmission system,
- The lack of adequate dynamic voltage support near areas or zones with a high concentration of renewable generation,
- The intermittent nature and variability in power flow patterns and a more complex real and reactive power dispatch,
- Inertia reduction resulting from large increases in wind and solar generation and the associated displacement of thermal and hydro generation, along with high reactive power losses; and
- Control mode interactions.

Massive integration of (larger amounts of) wind-based renewable energy generation in weakly interconnected systems is influenced by many factors and constraints [3, 17, 24–26]. Modern generating plants such as wind and solar power plants consist of generating units that interface to the power system through frequency converters. These converters, however, are most often controlled in such a way that the operation of the generating plants is independent of the system frequency, resulting in reduced system inertia. A similar phenomenon is observed in loads and other resources that utilize power electronics [7].

Recent studies and operational experience suggest that the frequency response of major transmission systems in North America and Europe is deteriorating [7, 24]. In this sense, a recent study [27] showed that increased levels of wind penetration may lead to a reduction of the minimum short circuit levels and a reduction in system inertia. Similar observations are made in recent wind/PV integration reports [28].

Adding new solar PV and wind power generation might further degrade frequency response and voltage control. Moreover, the overall system response may be affected by the amount and type of generation committed and how it is dispatched. In Ref. [29], it has been found that as PV increases, the damping of dominant interarea modes decreases. A second observation is that the mode shape varies with the control strategy and new oscillation modes may emerge under inappropriate control settings.

Another trend is the connection of rotating motors to the power system through frequency converters [7]. In addition, high imports on high-voltage direct current (HVDC) connections to other synchronous systems are expected to displace traditional generation more often resulting in decreased system inertia. These factors will impact the system kinetic energy and inertia distribution [30].

Weak grids can pose challenges for connecting new resources and particularly for connecting inverter-based resources [24]. In particular, it has been observed in Ref. [27] that the trip of multiple hydropower plants on the system may result in poorly damped oscillatory behavior during the postcontingency conditions. While the frequency of large disturbances may not be increasing, their consequences seem to be growing with the size and complexity of the system involved [31]. Special protection and control actions are required to stop the power system degradation and minimize the impact of disturbances.

Potential mitigation measures include improved governor response and inertial control from wind plants, [32]. These issues are addressed in the following sections. First, the study system is briefly described.

8.2 Study System

Figure 8.1 is a simplified single-line diagram of the study system showing the main areas of concern for this study. Key buses and major transmission paths are indicated on the schematic map. The base system used in the studies described further

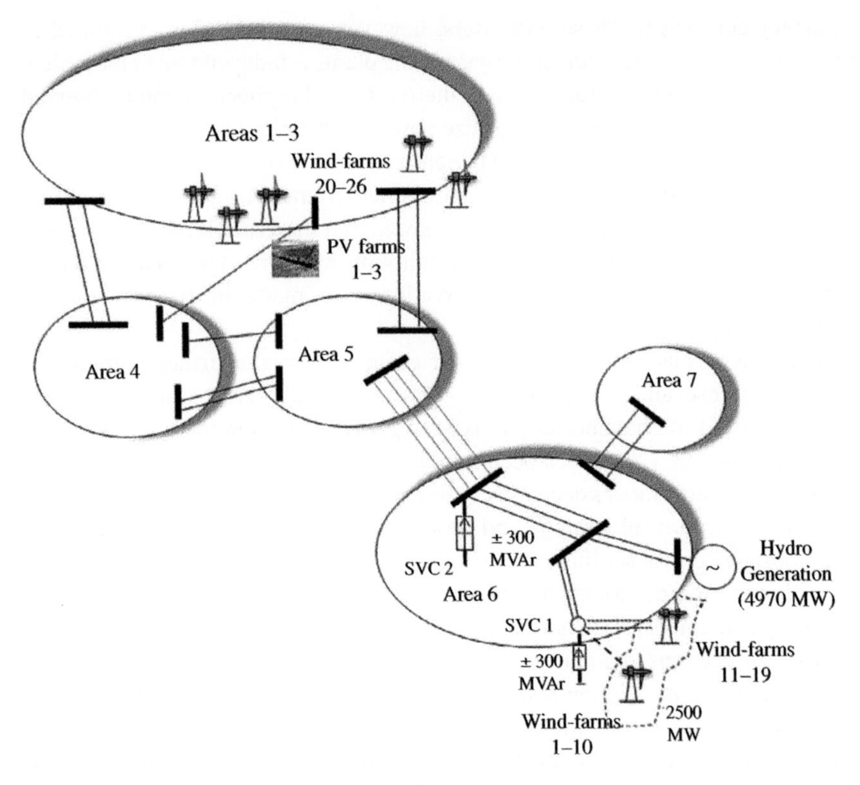

Figure 8.1 Simplified system representation.

is an expanded version of the given test system in Chapter 7 and includes the detailed representation of 635 generators, 26 explicitly modeled inverter-based wind farms (WFs), 3 solar PV farms, 5449 buses, and 5292 transmission lines. Turbine governors are represented on major machines. See Section 7.1 for detailed discussions of other modeling considerations. This model has been validated using measurement data from synchronized phasor measurement units (PMUs) [33].

Two wind and solar penetration scenarios are of interest here: (i) grid integration of wind energy in Area 6 and (ii) wind and solar integration in Area 4 and Area 5. The emphasis in the following analyses is placed on the scenario (i).

To evaluate the performance impact of wind and solar PV farms, appropriate models of wind and solar farms have been developed together with special remedial action schemes. Simulation studies include the detailed representation of the machines that provide governor response and discrete control devices such as underfrequency load shedding (UFLS) schemes.

The integration studies described further have two specific objectives: (i) define limits of integration of wind energy in the system, and (ii) evaluate actions to improve frequency regulation and voltage control on the 400 kV grid.

8.3 Wind Power Integration in the South Systems

Because of its geographical location and the sparse nature of the bulk transmission network, wind integration in Area 6 is of special concern. In this system, WFs 1–19 are located at the south extremity of Area 6; the current installed wind capacity in the study region is about 50% of the local hydropower generation. As wind penetration in the study area is expected to increase, hydro plants providing governor response will likely be displaced or be required to operate at reduced output (spinning reserve) to accommodate wind variations and provide primary frequency response as discussed in the following subsections.

Further, faults in this system resulting in loss-of-generation events will affect frequency performance and result in postdisturbance oscillations.

8.3.1 Study Region

The area under study represents portions of the south systems that include 19 WFs and a significant portion of hydro generation in the system. Figure 8.2 is a simplified diagram of this system around major WFs showing the interconnection to the 400 kV system. Three levels of transmission voltages are represented in detail: 115, 230, and 400 kV. For clarity of illustration, the low-voltage network around the WFs below 230 and 115 kV and small WFs are not represented in the diagram.

As shown in Figure 8.2, most of the existing hydropower generation is concentrated on four major hydro plants and amounts to about 4970 MW. By contrast, the wind generating capacity is over 2500 MW and constitutes about 50% of the local installed hydro generation capacity; the total area load is 5120 MW.

The transmission system at the WFs' point of interconnection with the system (POIS) consists of two radial 400 kV transmission lines and a large ±300 MVAr static volt-ampere reactive (VAR) compensator (static VAR compensator [SVC] 1 in Figure 8.2). SVC voltage support at the POIS is used to enhance the power transfer capability of this transmission system, prevent large postfault voltage swings, and increase reactive power reserves.

Fourteen generators (Gens 1–14) within the study region are selected to determine the synchronous generation (SG) to be displaced by wind generation. The main characteristics of the SG and WF plants within the study region are summarized in Tables 8.1 and 8.2. Since the primary frequency response of some generators is represented in detail, it can be inferred from Table 8.1 that the choice of the

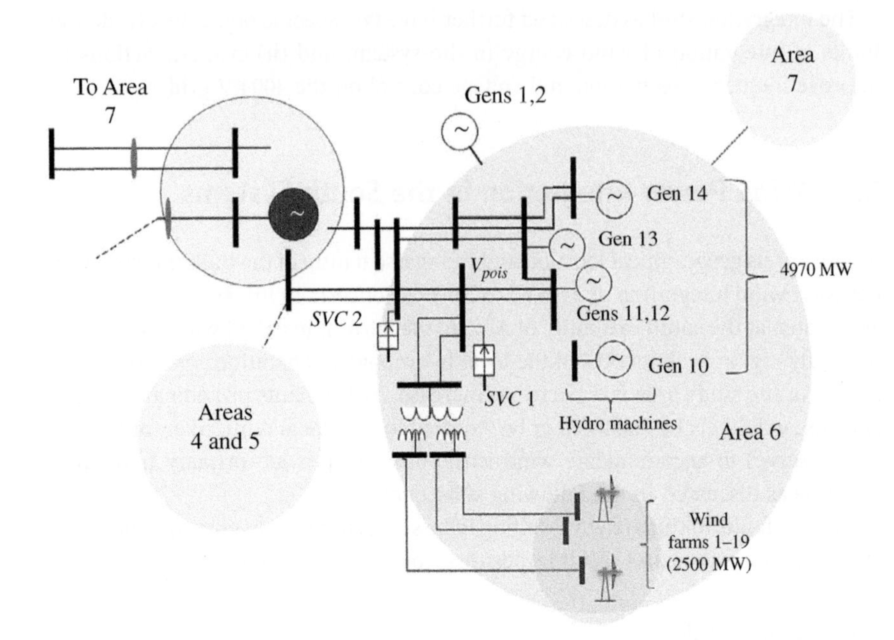

Figure 8.2 Simplified schematic of the study region within Area 6. Dashed lines represent interconnections to neighboring systems.

synchronous machines to be displaced by wind generation is likely to have a major impact on system dynamics. In addition to type and capacity, the use of power system stabilizer (PSS) and governor response for the main generators are determined in Table 8.1.

As shown in Table 8.2, the normal operating mode of doubly fed induction generators (DFIG) considered in these studies is voltage regulation and active power control. No inertial or governor control on wind plants is provided.

The procedure adopted to assess the impact of high wind penetration on system dynamic behavior is as follows:

- Wind generation was uniformly increased in proportion to its base case condition and generation at local hydro generators was decreased. The rationale behind the adopted approach is discussed in Section 8.5.2.
- Among the various generating sources, hydro generators 10–14 are used for balancing wind fluctuations and utilized as a spinning reserve for major contingencies.

Integration studies described further examine the impact of increased penetration of DFIG-based wind turbines on both transient and small-signal stability.

Table 8.1 Summary of the main generation in Area 6.

Generator	Type	Capacity (MW)	PSS	Governor response
1	Nuclear	750	No	No
2	Nuclear	750	No	No
3	Hydro	7.5	No	No
4	Hydro	66	No	No
5	Thermal	46	No	No
6	Thermal	46	No	No
7	Thermal	7	No	No
8	Hydro	122	No	No
9	Thermal	385	Yes	No
10	Hydro	191	No	Yes
11	Hydro	315	Yes	Yes
12	Hydro	315	Yes	Yes
13	Hydro	218	No	Yes
14	Hydro	115	No	Yes

Table 8.2 Main characteristics of DFIG WFs in Area 6.

WF	Type	Operating voltage (kV)	Voltage control	Active power control
1–2, 5–10, 14–15, 17–19	Type 3 DFIG	0.69	Yes	Yes
3, 11–13	Type 3 DFIG	12	Yes	Yes
4, 16	Type 3 DFIG	13.2	Yes	Yes

8.3.2 Existing System Limitations

Because of their location relative to the bulk 400 kV network and the weak transmission system, the integration of new wind generation in Area 6 presents multiple technical challenges to system security that need to be addressed [27]:

1) *Wind power concentration:* The concentration of wind generation and low load density within the study region. As discussed earlier, a large amount of wind power is concentrated on a small geographical region.
2) *Wind variability:* Wind power fluctuations ranging from 0 to 100 MW within an hour are not uncommon, especially early in the morning (during off-peak hours). Further, wind generation is out of phase with load demand thus adding to system variability.
3) *Relative size:* As noted earlier, the size of the WFs relative to hydro generation in the south systems is high (approximately 50%). As a result, the automatic generation control (AGC) system must compensate for fluctuations in wind power and load variations,
4) *Point of interconnection:* WFs 1–19 are connected radially to the 400 kV transmission system. A large ±300 MVar SVC is used to support local voltage. As a result, loss of SVC voltage support or loss of major transmission lines may result in system instability or poor postdisturbance system response.
5) *Impact on reserve margin:* Present levels of wind power penetration in the south systems may affect operating reserve margins.

In addition to these limitations, power transfers to the bulk 400 kV system are constrained by grid stability and voltage considerations.

8.4 Impact of Increased Wind Penetration on the System Performance

Detailed transient stability studies were conducted to evaluate the impact of high wind penetration on system stability and performance. The following discussion describes system response in the context of grid integration studies.

8.4.1 Study Considerations and Scenario Development

Table 8.3 summarizes the contingency scenarios selected for analysis. These contingencies represent various operating scenarios that result in poorly damped power oscillations.

Table 8.3 Contingency scenarios selected for the study.

Contingency scenario	Remarks
CS01	Outage without fault of transmission line 7415–8307 in Area 6
CS02	Three-phase stub fault at bus 153 in Area 3
CS03	Three-phase fault at the POIS of WFs 1–19 in Area 6
CS04	Outage without fault of transmission line 8024–8052 in Area 6
CS05	The simultaneous outage of two 400 kV lines 8024–8520 and 7415–7426 in Area 6
CS06	The outage of generator 1 unit 1 (750 MW) in Area 6
CS07	The outage of generator 10 units 1 and 2 (382 MW) in Area 6

In these analyses, three scenarios of wind penetration were selected for study:

Wind scenario 1 (WS1): This is the base operating case with about 400 MW wind generation in the south systems, representing about 4% wind penetration relative to the installed hydro generating capacity in generators 10–14 in Figure 8.2.
Wind scenario 2 (WS2): A low wind generation scenario.
Wind scenario 3 (WS3): A 20% wind scenario.

In all cases, flow patterns were adjusted to reflect current operating policies in the system. Note that, because some of the hydropower plants are equipped with PSSs, adding wind generation may affect the damping of major interarea modes associated with the study region.

8.4.2 Base Case Assessment

Figures 8.3–8.7 show the system time response of the base system to a three-phase short circuit at the POIS located close to the WFs (wind scenario CS03 in Table 8.3). Other contingency scenarios are discussed further.

As shown in these plots, critical faults in the neighborhood of the POIS can result in marginally damped power, frequency, and voltage oscillations in which wind generators participate in the (slowest) 0.394 Hz interarea mode 1; these oscillations are observed in the bus terminal frequency, the WFs' active (and reactive) power output and nearby SVCs. Refer to Table 7.15, in Chapter 7, for details about the main interarea modes in the system.

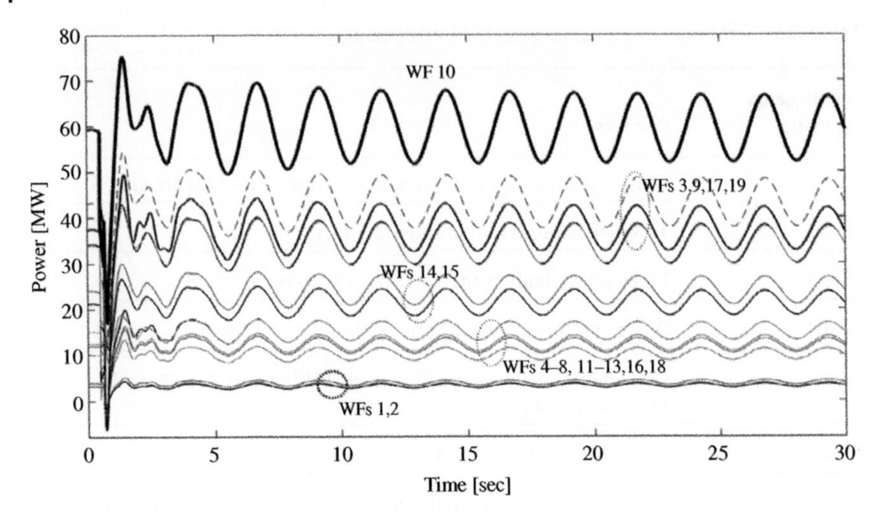

Figure 8.3 WFs' active power output following a stub three-phase fault at the POIS (contingency scenario CS03).

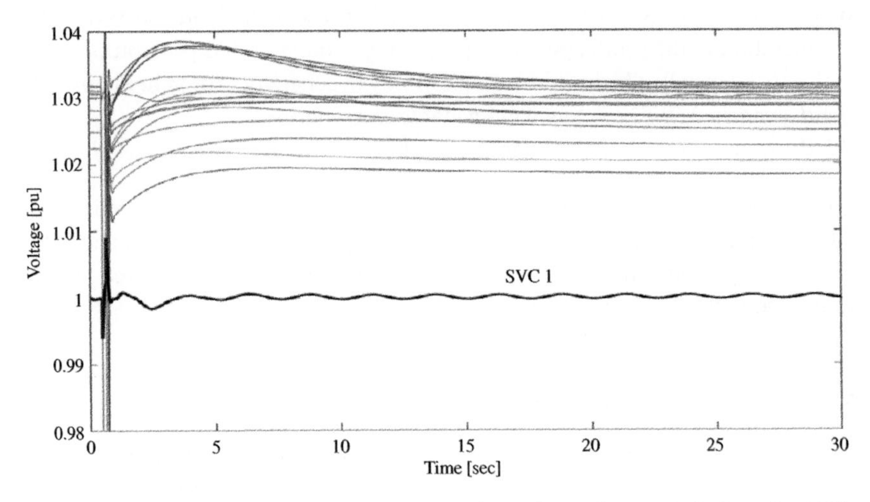

Figure 8.4 Wind farms terminal voltages following a three-phase stub fault at the POIS (contingency scenario CS03).

For reference, Table 8.4 presents a comparison of the frequency and damping of the dominant mode for the speed deviations of WFs for the contingency scenarios in Table 8.3. In all cases, WFs are found to have dominant participation in the

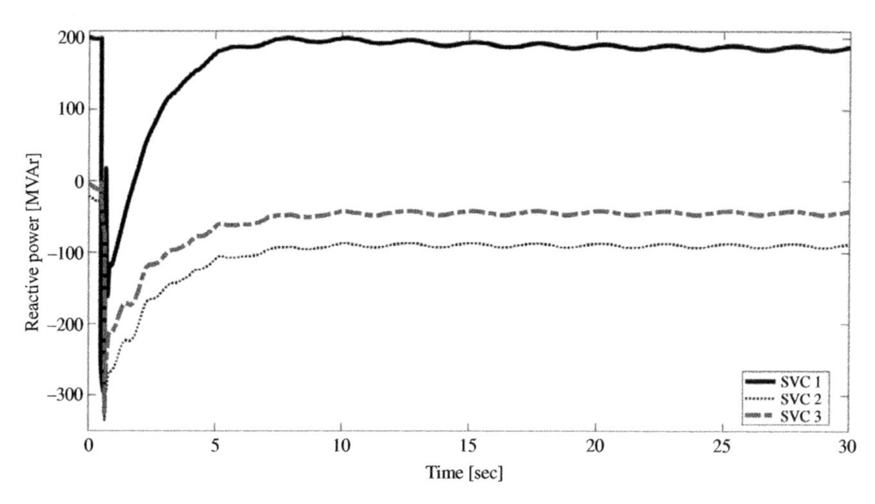

Figure 8.5 SVC reactive power output following a three-phase stub fault at the POIS (contingency scenario CS03).

0.394 Hz mode. From these results, contingency scenarios CS03, CS06, and CS07 are selected for analysis in the discussion that follows.

8.4.2.1 System Oscillatory Response

From Figure 8.6, the observed bus frequencies at the generator and WF locations exhibit a dominant mode at about 0.396 Hz showing that both WFs in Area 6 and

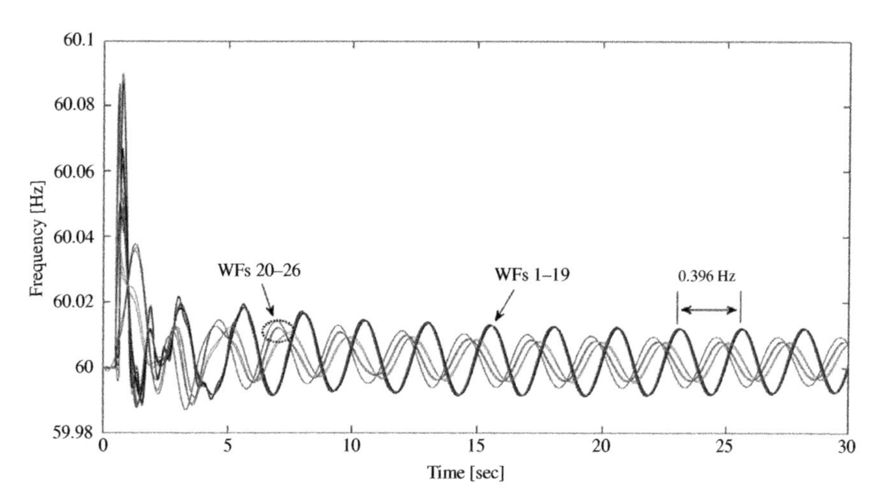

Figure 8.6 WFs' bus frequencies following a three-phase stub fault at the POIS (contingency scenario CS03).

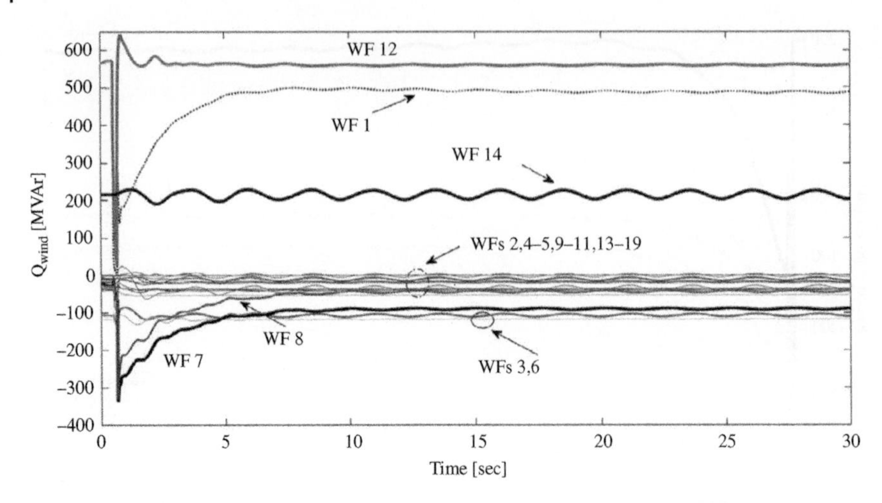

Figure 8.7 Wind farms reactive power output (contingency scenario CS03).

Table 8.4 Koopman analysis of rotor speed deviations for different contingency scenarios (wind scenario WS1). The bold entries indicate critical contingency scenarios with the lowest post-disturbance damping characteristics.

Contingency scenario	Frequency (Hz)	Damping (%)
CS01	0.397	0.555
CS02	0.395	0.475
CS03	0.396	**0.309**
CS04	0.395	0.445
CS05	0.395	0.687
CS06	0.395	**0.186**
CS07	0.396	**0.296**

WFs in areas 1–3 have an important participation in the slowest interarea mode in the system.

Insight into the nature of the WFs' contribution to these oscillations can be gleaned from the analysis of the mode phase for selected machines. Table 8.5 shows the amplitude and phase associated with the 0.394 Hz mode for WFs 1, 5, 24, and 25 and Gens 10–13 in Table 8.1, extracted using Koopman mode analysis. It is observed that WFs 1 and 5 swing in opposition to WFs 24 and 25 and Gens

Table 8.5 Phase relationships for dominant system generators extracted from generator speeds (contingency scenario CS03).

Generator	Area	Frequency (Hz)	Amplitude	Phase (°)
WF 1	6	0.396	0.0690	0.00
WF 5	6	0.395	0.0680	4.22
Gen 10	6	0.395	0.0130	104.75
Gen 11	6	0.395	0.0138	104.83
Gen 12	6	0.395	0.0136	104.86
Gen 13	6	0.395	0.0137	105.40
WF 24	3	0.397	0.0229	207.56
WF 25	3	0.397	0.0228	203.40

10–13 suggesting a local exchange of energy through the 400 kV transmission system in Figure 8.2. WFs 1 and 5 are found to have the largest participation in the 0.394 Hz mode.

8.4.3 High Wind Penetration Case

In this case, a worst-case scenario involving a critical contingency at the POIS was selected for analysis. Figures 8.8 and 8.9 show selected dynamic performance

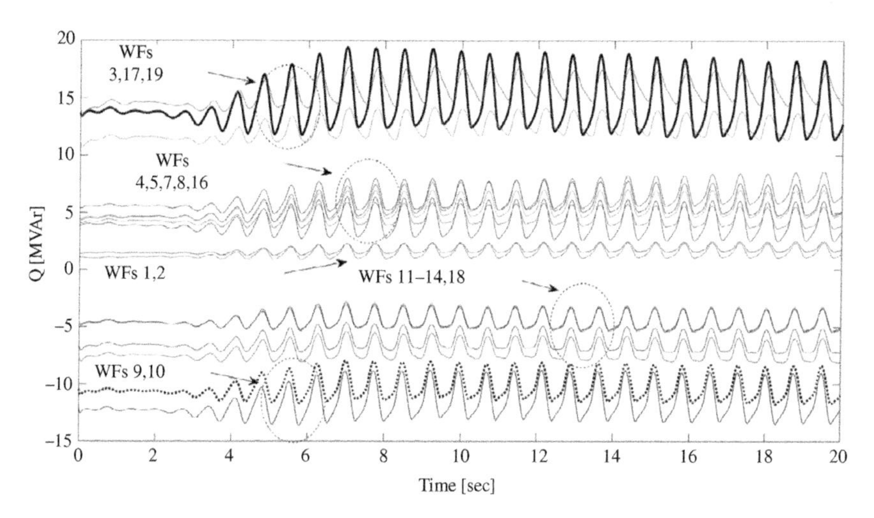

Figure 8.8 WFs reactive power output (contingency scenario CS03).

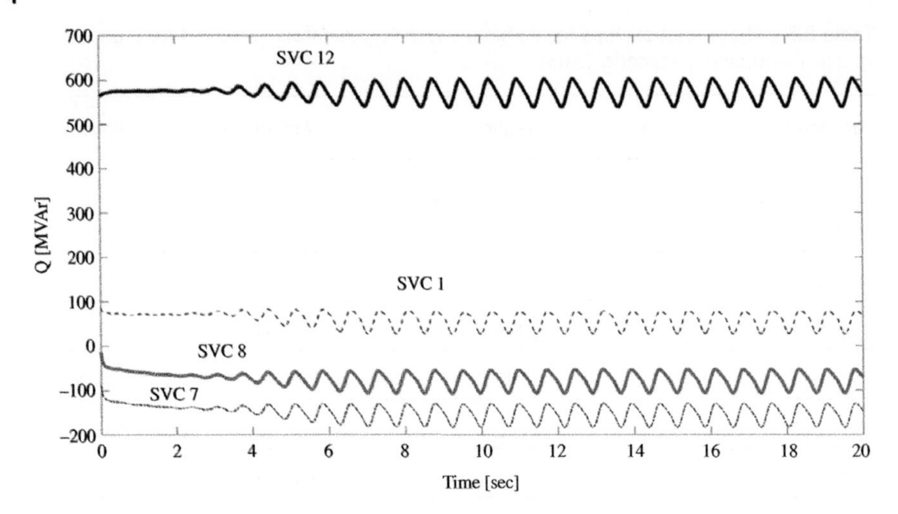

Figure 8.9 SVC reactive power output (contingency scenario CS03).

simulations for contingency scenario CS03. In this analysis, it is assumed that wind generation displaces hydroelectric generators equipped with PSSs.

From the results of Chapter 7, simulated data are approximated by a Koopman representation of the form

$$y_k = \sum_{j=1}^{m} \lambda_j^k \boldsymbol{\psi}_j(\boldsymbol{x}_o)\boldsymbol{v}_j \tag{8.1}$$

Close analysis of simulations for high wind generation cases in Figures 8.8 and 8.9 shows undamped reactive power oscillations involving two major electromechanical modes: a local mode at about 1.36 Hz and the 0.394 Hz interarea mode 1 ($m = 2$ in Eq. (8.1)). The 1.36 Hz mode is of special interest since it has no counterpart in linear (small signal) studies; similar findings are described in Ref. [29].

Simulation results highlight three key points: (i) careful analysis of Figures 8.8 and 8.9 shows that reactive power outputs at WFs swing in phase with nearby SVCs (SVCs 1,7, 8, and 12); (ii) while not explicitly addressed, the WFs swing out of phase with nearby hydro generators (Gens 10–14); and (iii) the oscillations manifest themselves as nonlinear oscillations with a dominant component at 1.36 Hz.

As a first step toward assessing the nature of energy exchange associated with these modes, phase relationships for several generators and wind generators were determined using dynamic mode decomposition (DMD) analysis in Chapter 7. In the interest of simplicity, Gens 1–14 in Table 8.1 and 19 WFs (Gens 1–19) were selected for the study.

Toward this end, the measurement matrix, $\boldsymbol{X}_{\omega}$, associated with wind and SG in Area 6 is represented as a row-concatenated matrix

$$\boldsymbol{X}_{\omega} = \left[\boldsymbol{X}_{\omega_{wind}}\ \boldsymbol{X}_{\omega_{gen}}\right] \tag{8.2}$$

where $\boldsymbol{X}_{\omega_{wind}} = \left[\Delta\omega_{WF1} \quad \Delta\omega_{WF2}\cdots \quad \Delta\omega_{WF19}\right]^T$, $\boldsymbol{X}_{\omega_{gen}} = \left[\Delta\omega_{g1} \quad \Delta\omega_{g2}\cdots\right.$ $\left.\Delta\omega_{g14}\right]^T$, and the $\Delta\omega\, i$ represent speed deviations. The accuracy of the model was improved by normalizing the individual data sets using the approach suggested in Section 5.4.

The idea is to expand the data matrix, $\boldsymbol{X}_{\omega}$, as a linear combination of functions of space and time as:

$$\hat{\boldsymbol{X}}_{\omega} = \left[\hat{\mathbf{x}}_1 \quad \hat{\mathbf{x}}_2\ldots \quad \hat{\mathbf{x}}_d\right] = \boldsymbol{\Phi}\boldsymbol{\Lambda}\boldsymbol{\Gamma}_m(t) \tag{8.3}$$

or

$$\hat{\boldsymbol{X}}_{\omega} = \left[\hat{\mathbf{x}}_1 \quad \hat{\mathbf{x}}_2\ldots \quad \hat{\mathbf{x}}_d\right] = \boldsymbol{\Phi}\boldsymbol{V}_{vand}\mathbf{c} \tag{8.4}$$

where, for clarity of illustration, the DMD method is used in the analysis, and the symbols in Eqs. (8.3) and (8.4) have the usual interpretation in Section 7.2.3.

In the studies presented further, generation was rescheduled to account for increases in wind generation. Starting from the base case, wind generation was increased, and active and reactive power was rescheduled.

For reference and validation, Table 8.6 shows the extracted modes using Koopman/DMD analysis for the high wind penetration scenario. Similar results are obtained for other operating conditions.

Further, Figure 8.10 shows the speed-based mode shape for the 1.36 Hz mode extracted using this approach.

Analysis of the mode shape in Figure 8.10 reveals several significant results. First, WFs 1–19 swing in opposition to machines 1–14 revealing the local nature

Table 8.6 The slowest modes of oscillation of the test system (contingency scenario CS03).

Eigenvalue	Frequency (Hz)	Damping in %
1	0.399	0.357
2	0.545	7.001
3	0.696	5.26
4	0.701	7.42
5	0.748	3.512

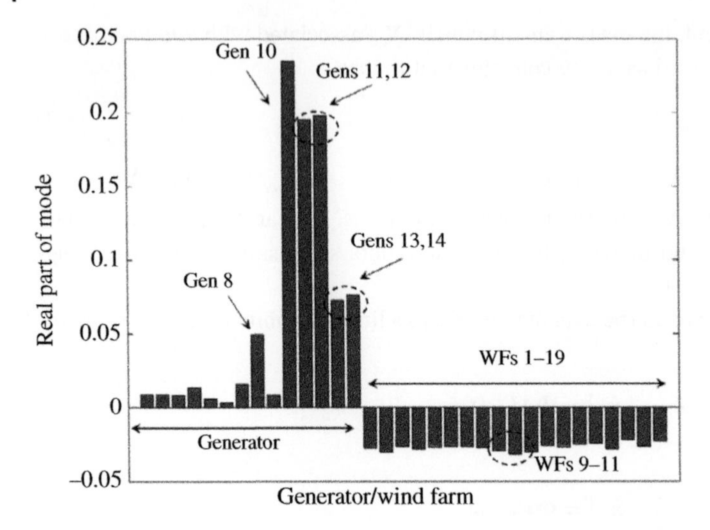

Figure 8.10 Speed-based mode shape (contingency scenario CS03).

of this mode; Gens 10–12 (hydro machines) and Gens 13–14 are seen to have the largest contribution to the mode. The WFs, on the other hand, show a similar contribution in terms of amplitudes and exhibit the largest participation in the mode. Physically, the form of the mode indicates the oscillatory mechanism involved in the observed time-domain responses.

8.5 Frequency Response

Disturbances that result in the loss of generation within the study area are of special concern. Three aspects of frequency dynamics are of interest in wind integration studies [24]: (i) inertial frequency response, (ii) the postdisturbance frequency fluctuations (primary frequency response) following major generator outages, and (iii) the amount and location of contingency reserves.

Inertial frequency response is inherent in the system and is associated with the rotating characteristics of loads and SG [6]. As a result, the amount and location of system inertia may affect it. Postdisturbance frequency fluctuations may be associated with inappropriate control settings and other operating and structural issues as discussed in the following section.

8.5.1 Frequency Variations

Figure 8.11 gives a conceptual representation of the system model used to assess inertial system response. A measured frequency excursion is also shown caused by a loss of generation event in the test system [33]. As discussed in Chapters 4 and 7, the location of contingency reserves plays a critical role in the ability of the system to limit frequency variations [34]. A simple criterion to estimate the effect of generator dropping on the frequency behavior at bus j, Δf_j, can be obtained from the sensitivity relationship [27]

$$\Delta f_j = \frac{\partial f_j}{\partial P_{g_{out}}} \frac{\partial P_{g_{out}}}{\partial P_{w_{gen}}} \Delta P_{w_{gen}} \tag{8.5}$$

where $P_{g_{out}}$ and $P_{w_{gen}}$ are the amount of generator outage and WF active power, respectively, and the sensitivities can be estimated from measurements or simulation studies involving frequency behavior to selected generator outages and wind generation scheduling strategies.

Another way of looking at frequency dynamics is by using two-area inertia-based dynamic equivalents derived from the center-of-gravity (COG) formulation in Chapter 6. Figure 8.11a gives a conceptual representation of this model, showing a recorded frequency disturbance event following the loss of a large hydro generating plant in Area 6. Figure 8.11b, in turn, shows the equivalent system for the analysis of wind and hydro coordination and frequency response analysis.

Using the same notation as in Section 6.2, the frequency deviation of the partial centers of inertias (COIs) can be approximated as [35, 36]

$$\Delta f_{COI_i} = \frac{\dfrac{M_{COG}}{M_i} \dfrac{\left[\Delta P_L - \sum_j P^{tie}_{COI_i,COG} - \Delta P^{MMG}_i\right]}{(\Delta P_{mech} - \Delta P_{elec})}}{1 + \dfrac{M_{COG}}{M_i} \dfrac{2\pi D}{(\Delta P_{mech} - \Delta P_{elec})} \Delta f_{COG}} \Delta f_{COG} \tag{8.6}$$

From (8.6), the relationship between the frequency of the COG and the motion of the local centers of angle is first determined, and expressions to compute local frequency deviations following major disturbances can be derived. Formally, the frequency deviation at bus j can then be approximated as

$$\Delta f_j = \frac{\dfrac{M_i}{M_j} \dfrac{\left[\Delta P_L - \sum_j P^{tie}_{COI_i,COG} - \Delta P^{MMG}_i\right]}{\left(\Delta P_j - P^{tie}_{ij} - \eta_j \Delta P^{MMG}_i\right)}}{1 + \dfrac{M_i}{M_j} \dfrac{2\pi D}{\left(\Delta P_j - P^{tie}_{ij} - \eta_j \Delta P^{MMG}_i\right)} \Delta f_{COI_i}} \Delta f_{COI_i} \tag{8.7}$$

where the symbols have been already described.

This approach provides alternate, complementary information concerning the COI frequency [10] and the local (cluster-based) frequency estimated using the c-means approach in Chapter 7.

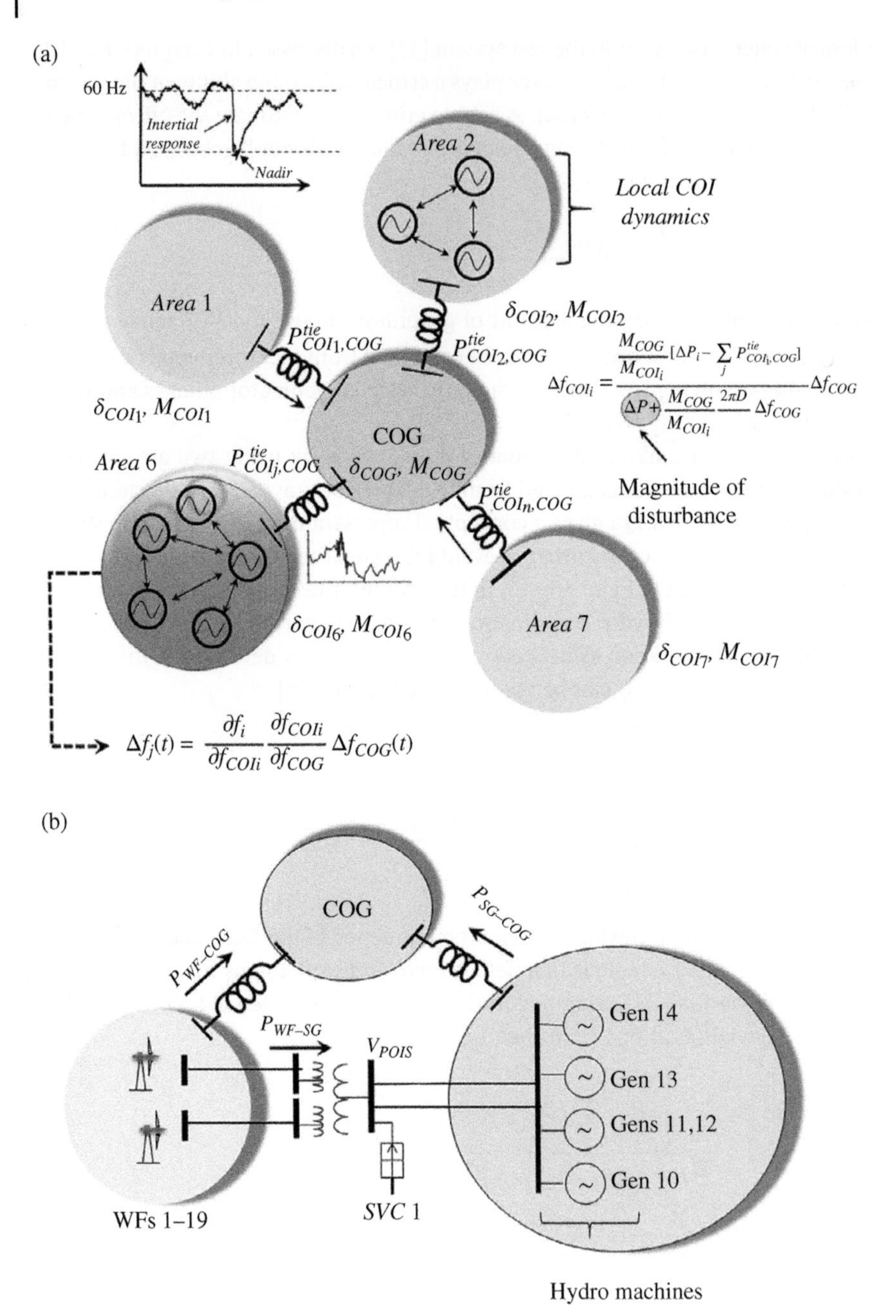

Figure 8.11 The COG dynamic equivalent adopted for assessing energy exchange and wind and hydro coordination; (a) multiarea representation and (b) two-area dynamic equivalent.

Key observations are:

- Frequency sensitivities can be obtained for various values of transmission and inertia in the system, which allows determining the maximum amount of wind energy penetration as a function of the frequency nadir [35].
- As a by-product, this approach allows determining the minimum amount of inertia required in the equivalent system (hydroelectric machines).

Once frequency sensitivities are estimated, the partial sensitivities in (8.5) can be determined, as suggested in Section 8.5.3.

8.5.2 Wind and Hydropower Coordination

A key problem of interest in assessing inertial frequency response is that of the coordinated operation of hydro and wind generation [28]. As noted previously and suggested in Figure 8.11, adding wind generation in Area 6 displaces local hydro generation (mainly Gens 10–14) making the development of commitment and dispatch of hydro generation procedures critical to assess frequency response and resulting in low system inertia and high rate of change of frequency (RoCoF) during system disturbances.

The first key issue is the correlation between wind generation and the hydro generation that needs to be displaced [28]. In this discussion, two aspects are of interest: the location of the machines where the spinning reserve is needed (inertial support) and the magnitude and distribution of the required spinning reserve. The first issue is addressed here using analytical models such as those in Eqs. (8.5)–(8.7) and the use of correlation analysis. The second requires extensive dynamic simulation studies.

Inspired by techniques that fuse or regress data, this section explores the use of correlation techniques to determine suitable machines to maintain spinning reserves. To introduce these ideas, consider two data sets $\boldsymbol{X} \in \Re^{N \times m_1}$ and $\boldsymbol{Y} \in \Re^{N \times m_2}$, where N is the number of snapshots, and m_1, m_2 denote the number of sensors or measuring locations, which in the more general case are assumed to represent distributed generation and conventional SG in (8.3), respectively [37].

In the present context, they could represent wind electric output power and hydro (thermal) active generation output and are assumed to have the same length N and have different measured signals or sensors, i.e. $m_1 \neq m_2$. The relationship between these two centered sets of variables is given by the covariance matrix $\boldsymbol{R} \in \Re^{N \times m_1} = \boldsymbol{Y}^T\boldsymbol{X}$.

Conceptually, these techniques decompose the data into a set of statistically independent and orthogonal components. Each component is defined in time by a unique series of weights assigned to each of the time samples; these weights are called component loadings. The extent to which each principal component

(PC) contributes to a particular oscillation in the original data set is reported as a component score [38].

The process can be summarized in three steps:

1) Create a concatenated data matrix from the individual models of distributed generation (i.e. wind and solar generation) and conventional SG (see Eq. (8.2)),
2) Perform correlation analysis on datasets $\boldsymbol{X}, \boldsymbol{Y}$,
3) Extract correlation measures.

To examine the applicability of this approach, active power responses were calculated for each wind generator and SG in the study system.

From partial least squares (PLS) regression analysis in Chapter 5, the measurement data are defined as

$$\begin{aligned} \boldsymbol{X} = \boldsymbol{X}_{sm} &= \left[\boldsymbol{P}_{g_{smj1}} \quad \boldsymbol{P}_{g_{smj2}} \cdots \quad \boldsymbol{P}_{g_{smj14}}\right]^T \\ \boldsymbol{Y} = \boldsymbol{X}_{wf} &= \left[\boldsymbol{P}_{g_{wfj1}} \quad \boldsymbol{P}_{g_{wfj2}} \cdots \quad \boldsymbol{P}_{g_{wfj19}}\right]^T \end{aligned} \tag{8.8}$$

where for the study system, $\boldsymbol{P}_{g_{wfj}} = \left[P_{g_j}(t_1) \quad P_{g_j}(t_2) \quad P_{g_j}(t_N)\right], j = 1, \ldots, 19$, $\boldsymbol{P}_{g_{smj}} = \left[P_{g_j}(t_1) \quad P_{g_j}(t_2) \quad P_{g_j}(t_N)\right], j = 1, \ldots, 14$ and $N = 3603$ samples.

For completeness, Figure 8.12 shows the time evolution of active power deviations for contingency scenario CS07.

Using these assumptions, the correlation between data sets $\boldsymbol{X}, \boldsymbol{Y}$ can be expressed as

$$\begin{aligned} \boldsymbol{X} &= \boldsymbol{t}_1\boldsymbol{p}_1^T + \boldsymbol{E} = \boldsymbol{T}\boldsymbol{P}^T + \boldsymbol{E} \\ \boldsymbol{Y} &= \boldsymbol{u}_1\boldsymbol{q}_1^T + \boldsymbol{F} = \boldsymbol{U}\boldsymbol{Q}^T + \boldsymbol{F} \end{aligned} \tag{8.9}$$

in which, as discussed in Chapter 5, the $\boldsymbol{T}, \boldsymbol{U}$ are matrices of the dominant extracted score vectors, matrices $\boldsymbol{P}, \boldsymbol{Q}$, represent the matrices of loadings, and $\boldsymbol{E}, \boldsymbol{F}$ are the matrices of residuals. The results for this scenario are shown in Figure 8.13, using the approach in Ref. [37].

In this case, the percentage of variation (PCTVAR), explained by the three dominant PCs, is

$$PCTVAR = \begin{bmatrix} \mathbf{0.8899} & 0.0899 & 0.0101 \\ \underbrace{\mathbf{0.1021}}_{PC1} & \underbrace{0.0190}_{PC2} & \underbrace{0.0289}_{PC3} \end{bmatrix} \tag{8.10}$$

It is apparent from column 1 in Eq. (8.10) that, overall, synchronous machines in Area 6 (first row) have the largest participation in mode 1 relative to WFs 1–19 (second row). In sharp contrast with this, PC modes 2 and 3 (columns 2 and 3 in Eq. (8.10)) are found to have a minor contribution to the observed oscillation.

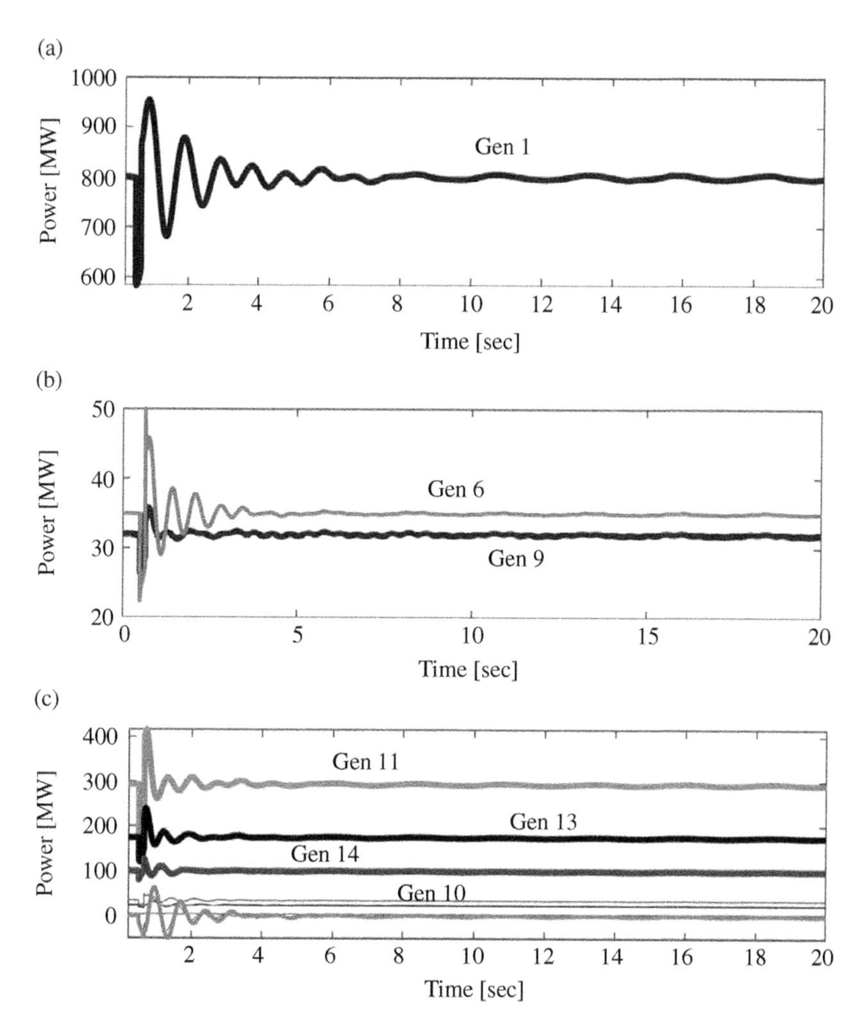

Figure 8.12 Generator active power response following the simultaneous outage of two generators (Contingency scenario CS07) for: (a) nuclear generation, (b) thermal generation, and (c) hydro generation.

In addition, one can infer from Figure 8.13 that WFs 3, 9–10, and 17, 19 have the largest contribution to the postdisturbance scenarios. In turn, nuclear generators 1 and 2 and hydro machines 9–13 show the largest contribution to the oscillation; the opposite is true for machines 3–8. Because nuclear generators are not expected to provide governor response, hydro generators 9 through 13 are selected for analysis.

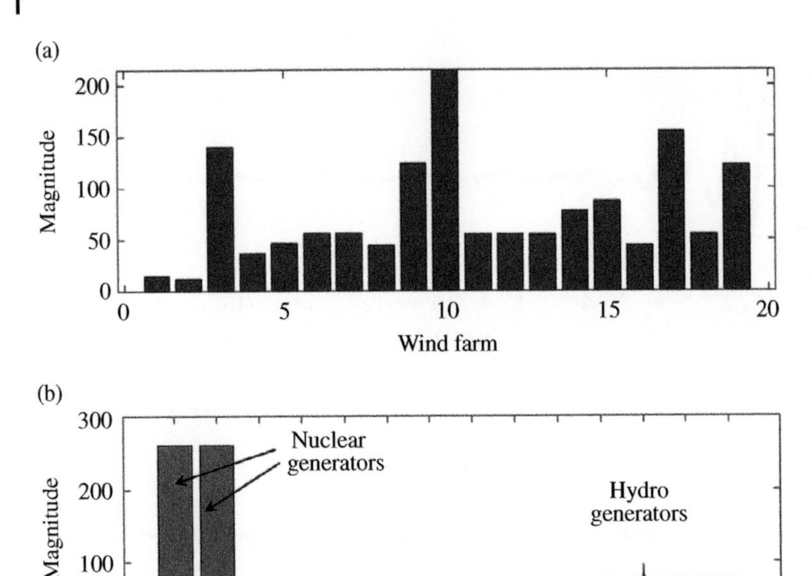

Figure 8.13 Partial least squares regression between WFs and generation sources in the south systems (contingency scenario CS07); (a) WFs 1–19 and (b) selected generators in Area 6.

From the system viewpoint, this information helps to easily identify the location and distribution of generation to be displaced; these findings can also be used to coordinate operational actions. A related issue is that of inertia support in hydro machines, as some of them could be operated as synchronous condensers.

8.5.3 Response to Loss-of-Generation Events

Large loss-of-generation events may result in complex postdisturbance oscillations and low frequency nadirs [27, 33]. To investigate inertial and primary frequency response, three-generation trip scenarios are considered: (i) loss of the largest generating unit in the system (750 MW), (ii) loss of a hydro generation without PSSs (315 MW), and (iii) loss of a large generation unit equipped with a PSS (315 MW).

For illustration, two wind penetration scenarios are considered for analysis: (i) a base case with 400 MW wind generation and (ii) a high wind generation case. In both cases, the increase in wind generation was offset by disconnecting main hydro generators located near the cluster of WFs 1–19 as suggested by analytical

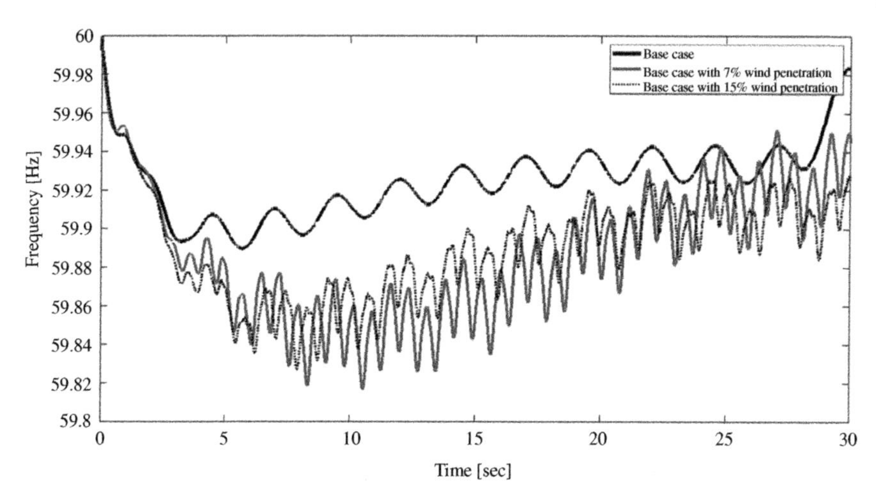

Figure 8.14 Frequency response at the POIS following the outage of a 750 MW nuclear unit (contingency scenario CS06).

results in Section 8.5.2. The metrics of interest are the RoCoF, the frequency nadir, and the settling frequency [10, 11].

Figure 8.14 is an example of postdisturbance frequency oscillations following a 750-MW generation loss (contingency scenario CS06). As shown in this plot, large loss-of-generation events result in negatively damped oscillations with a frequency of about 1.36 Hz (5–25 seconds). The mode at 1.36 Hz is identified as a local oscillation between the western portion (WFs 1–19) and the eastern portion (hydro generators) of Area 6.

For the base case, the frequency response reaches a nadir of about 59.9 Hz at approximately 3.8 seconds after the generation trip; while not shown in Figure 8.14, the frequency settles back to the prefault (nominal) condition at about 40 seconds. Several differences are obvious when comparing system performance. It can be seen that the slope of the inertial response increases as the amount of connected wind generation increases. Additional studies show that this effect is more pronounced in the case of stressed operating conditions.

Table 8.7 shows the eigenvalues computed using Koopman mode analysis for the contingency operating conditions in Figure 8.4. In this analysis, the study window is chosen to coincide with the frequency primary response period (5–15 seconds).

Analytical experience with other operating scenarios shows that frequency nadirs range from 59.92 Hz for low wind penetration levels to 59.65 Hz for high wind penetration levels for the cases with double- and triple-generation

Table 8.7 Prony results for frequency deviation signals in Figure 8.14 following the loss of critical system generators for contingency scenarios CS06 and CS07 in Table 8.1 (time interval 15–30 seconds).

	CS06-loss of unit 1 of Gen 1 (750 MW)		CS07-loss of two generating units (180 MW each)	
Operating scenario	**Frequency**	**Damping**	**Frequency**	**Damping**
Base	0.394	0.412	0.397	0.419
Base case + 20% increase in generation	0.393	−0.452	0.397	0.211
Maximum wind condition	0.393	−0.452	0.398	−0.397
	1.366	−0.010	1.378	0.010

contingency scenarios as shown in Figure 8.15. As expected, the critical mode is affected significantly depending on the amount of generation disconnected.

The analysis also suggests the need to coordinate turbine governor settings and other controllers to damp out the postdisturbance oscillations.

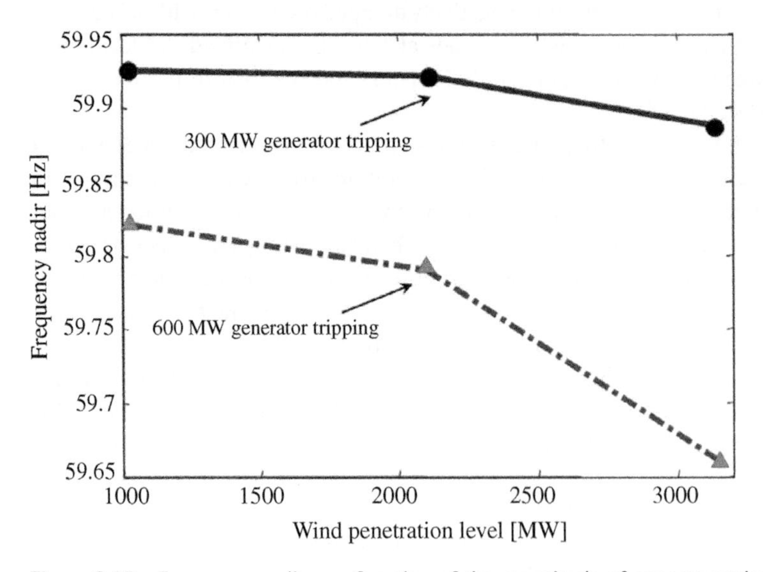

Figure 8.15 Frequency nadir as a function of the magnitude of generator tripping.

8.6 Effect of Voltage Control on System Dynamic Performance

The advent of modern, inverter-based wind and solar farms offers a direct means for controlling system voltage. Specifically, the WFs in the south systems are expected to provide local voltage support (Figure 8.16). Two main factors influence the effect of voltage control on system dynamic performance: (i) the location, size, and control characteristics of reactive support sources near the wind/solar PV plants; and (ii) the control characteristics of WFs.

8.6.1 Voltage Support and Reactive Power Dispatch

Because of the geographical location of wind generators, and the connection of large SVCs around the POIS, the need arises for coordinated voltage control. As illustrated in Figure 8.2, a large (400 kV, ±300 MVAr) SVC is connected near the WFs' POIS.

Major sources of reactive support in the area also include major hydro generating units in the area.

Three control strategies are suggested:

1) Individual voltage control of WFs
2) Hierarchical voltage control of the 19 WFs at the POIS
3) Hierarchical, voltage control involving the local SVC and WF control

The first issue is studied here. Based on these studies, the need for coordinated voltage control is established.

8.6.2 Effect of Voltage Control Characteristics

Assessing the impact of voltage control characteristics on system dynamic performance is also integral to the wind and solar PV integration studies. As discussed in the introductory section, voltage limitations in the south systems impose further constraints on wind generation and may be compounded by a weak connection grid at the POIS [24].

In the literature, a variety of different voltage control strategies are considered at WFs and dynamic reactive power compensators. Three operating modes are considered here [8]: (i) constant power factor control, (ii) constant Q control, and (iii) bus voltage control. Operating in voltage control mode is often considered to be the most efficient way to use their static and dynamic reactive power capability, which can reduce the burden on nearby SGs and SVCs on the power system [14].

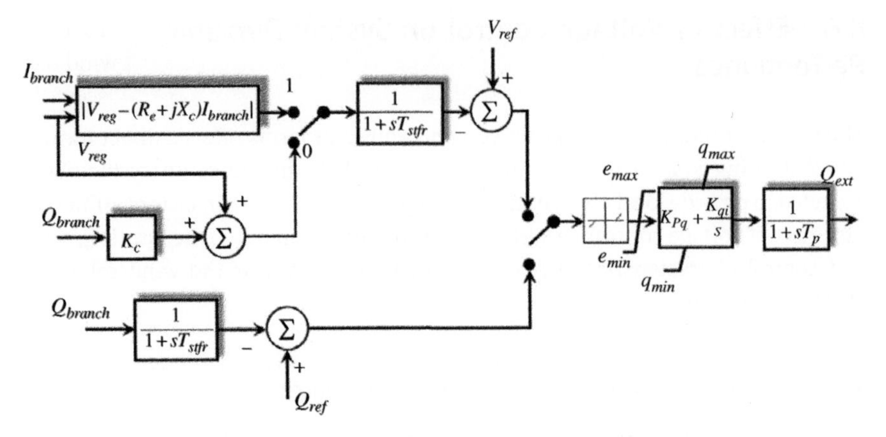

Figure 8.16 Simplified diagram of the WF reactive power control loop.

Figure 8.16 depicts the WF reactive power control loop used in the simulations. The primary voltage control loop includes a proportional–integral controller and droop control; the adjustable droop is obtained by the feedback of the WF reactive current and voltage and modulates the reference voltage, V_{ref}. The difference value is then processed through the PI controller to obtain the reactive command for the output, Q_{ext}.

Further, through the modification of the reactive power reference, Q_{ref}, the WFs can maintain constant reactive power output. A slow reactive control loop, automatic reactive power regulator (AQR) can also be utilized for the WFs. A suitable coordination control between the SVCs and the WFs [39] may be necessary as discussed in Chapter 5.

As a first step, extensive power flow analyses were conducted to determine the impact of transmission system limitations on the energy deliverability of the wind plant output. In these simulations, static compensation with constant reference voltage control is considered at SVCs 1 and 2 in Figure 8.2. Table 8.8 shows the SVC output reactive power as a function of the WFs' reactive power output

Table 8.8 SVC reactive power as a function of WFs' loading. The voltage control loop in Figure 8.16 is enabled.

SVC	WS1 (base case)[a]	WS2 (20%)[a]	WS3[a]
1	−200.78	−127.33	208.8
2	4.55	−6.05	11.44

a MVAr.

obtained from a steady-state power flow solution for the case when WFs 1–19 operate on voltage control mode. Three cases are considered, namely (i) wind scenario 1 (WS1) as the base case, (ii) wind scenario 2 (WS2), and (iii) wind scenario 3 (WS3).

By increasing the WFs' output, the reactive losses on the transmission system connecting the WFs with the 400 kV system (see Figure 8.11) increase, and the margin of dynamic reactive power to support voltage recovery decreases. Increased reactive losses, in turn, increase the possibility of voltage collapse as discussed further.

Based on these results, studies were conducted to assess voltage stability constraints. Emphasis was placed on the impact of solar and wind generation on power system performance. From the perspective of system modeling, three main aspects are of interest [1]: (i) optimizing voltage control, (ii) tuning voltage/VAR control loops [40], and (iii) providing sufficient static and dynamic reactive power capability [41]. These features are interrelated as discussed further.

To pursue these concepts, Figure 8.17 shows plots of terminal voltage at SVC 1 for various control strategies. The postdisturbance oscillatory response shows a marginally stable ($f = 1.366$ Hz, $\xi = 0.011\%$ oscillation and a negatively damped oscillation at 0.396 Hz ($\xi = -0.759\%$). Modal analysis discloses higher harmonics at 2.732 Hz ($\xi = -0.012\%$) revealing the nonlinear nature of the observed oscillations.

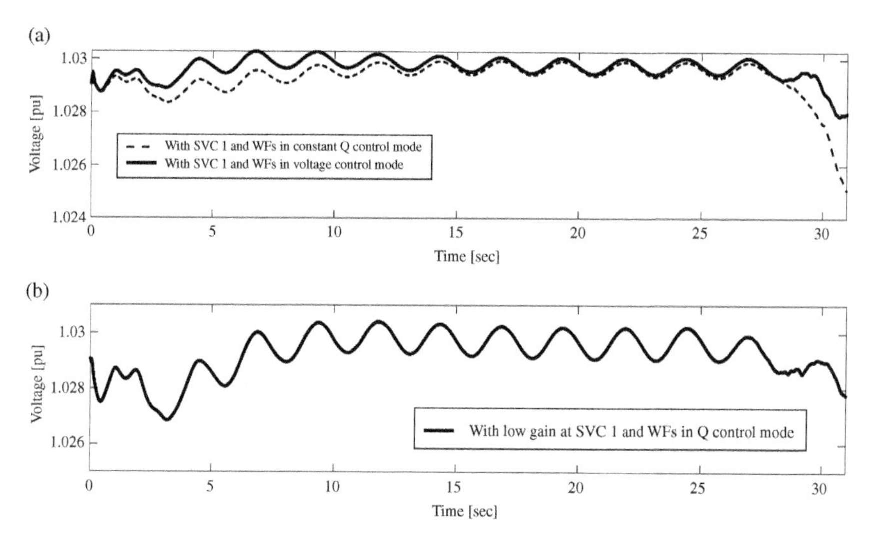

Figure 8.17 Bus voltage magnitudes at the POIS for various control and dynamic support characteristics (contingency scenario CS06); (a) SVC voltage support at the POIS and nominal SVC gain and (b) SVC voltage support at the POIS with low SVC gain.

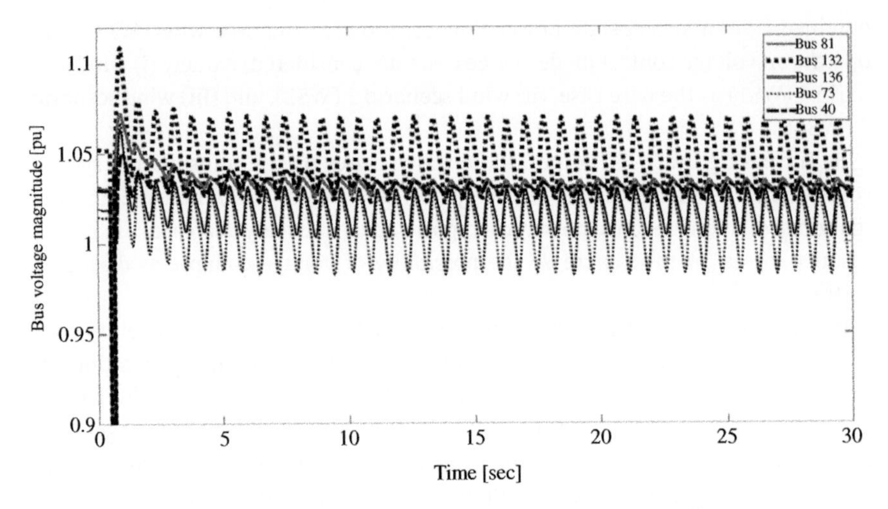

Figure 8.18 Bus voltage magnitudes at selected buses in Area 6 (contingency scenario CS06).

A crucial observation is that the study results in Figure 8.17a show that constant Q control may result in mid-term voltage instability. In turn, it can be seen in Figure 8.17b that improper SVC tuning can also result in degraded performance and severe voltage fluctuation or instability. Voltage instability is observed in Figure 8.17a at about 28 seconds into the simulation. The effect becomes more pronounced as the system is further stressed.

The need for additional (or more coordinated) dynamic reactive support is evident and deserves further investigation.

Based on the preceding results, system studies were conducted to identify modal characteristics in the observed oscillations. Figure 8.18 illustrates the results of transient stability simulations for contingency scenario CS06.

The nature of these oscillations becomes evident from the Fourier spectra of the observed oscillations in Figure 8.19 and Prony analysis of selected simulations in Table 8.9. From this Table, hydro plant generators and WFs in Area 6 (WFs 1 and 8) are seen to swing in opposition to WFs in Area 3 (WFs 23 and 26) and solar PV farm 1 in Area 4 as expected from physical considerations. It should be noted that the larger modal voltage swings for the dominant mode are located in zones or buses without voltage regulation (Bus 73) in agreement with Fourier spectral analysis in Figure 8.19.

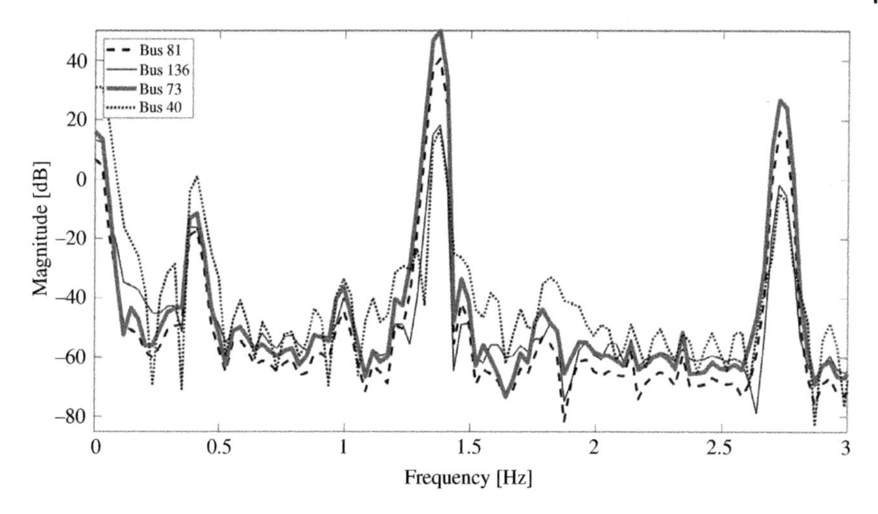

Figure 8.19 Fourier spectra of bus voltage magnitude deviations in Figure 8.18.

Table 8.9 Prony analysis of selected bus voltage magnitudes.

Generator/bus	Area	Amplitude	Phase (°)
Gen 10	6	0.0120	−59.71
Gen 11	6	0.0119	−59.70
Gen 12	6	0.0121	−58.49
WF 1	6	0.0105	−85.22
WF8	6	0.0105	−78.96
POIS	6	0.0106	−61.93
SVC 2	6	0.0102	−72.65
Bus 73	6	0.144	−57.71
WF 23	3	0.0059	32.12
WF 26	3	0.0061	39.04
PV 1	4	0.0032	43.05

8.7 Summary

In this chapter, studies to evaluate the impact of high wind penetration levels on the dynamic performance of a large, realistic power system model have been conducted. The studies examine the effect of geographically disperse wind and solar generation on transient stability and frequency regulation. The major emphasis was directed toward the problem of wind and hydro coordination and the study of the effect of coordinated voltage control on system dynamic performance. The applicability of correlation techniques to determine the conventional SG generation to be displaced by WF integration is also pointed out.

System stability studies have shown that dynamic performance considerations including voltage collapse, oscillatory and frequency instability, and postdisturbance oscillations may limit high wind and solar PV integration. The combination of these constraints makes the analysis of the high integration of wind and solar PV penetration difficult.

References

1. Miller, N.W., Larsen, E.V., and MacDowell, J.M. (2004) Advanced control of wind turbine-generators to improve power system dynamic performance. Proceedings of the IEEE International Conference on Harmonics and Quality of Power.
2. CIGRE Working Group C.6.08. (2011). Grid integration of wind generation (February 2011).
3. Alberta Electric System Operator, Wind Integration Impact Studies Phase 2. (2006). Assessing the impacts of increased wind power on AIES operations and mitigating measures (18 July 2006), Revision 2.
4. Miller, N.W., Guru, D., and Clark, K. (2008). Wind generation applications for the cement industry. Proceedings of the 2008 IEEE Cement Industry Technical Conference Record.
5. Winter, W. (ed.) (2010). European wind integration study (EWIS) – EWIS final report, Brussels. http://www.wind-integration.eu (accessed 2020).
6. Sharma, S., Huang, S.H., and Sarma, N., System inertial frequency response estimation and impact of renewable resources in ERCOT interconnection. Proceedings of the 2011 IEEE Power and Energy Society General Meeting.
7. European Network of Transmission System Operators for Electricity (ENTSO-E). Future system inertia – Report prepared by Energienet.dk, Fingrid, Stanett and Svenska kraftnät, Brussels. https://eepublicdownloads.blob.core.windows.net/public-cdn-container/clean-documents/Publications/SOC/Nordic/Nordic_report_Future_System_Inertia.pdf (accessed 2020).

8. Vittal, E., O'Malley, M., and Keane, A. (2010). A steady-state voltage stability analysis of power systems with high penetrations of wind. *IEEE Transactions on Power Systems* **25** (1): 433–442.

9. Vittal, V. and Ayyanar, R. (2013). *Grid Integration and Dynamic Impact of Wind Energy*. New York, NY: Springer.

10. Tan, J., Zhang, Y., You, S. et al. (2018). Frequency response study of a U.S. Western interconnection under extra-high photovoltaic generation penetrations. Proceedings of the IEEE Power and Energy Society General Meeting.

11. Tan, J., Zhang, Y., Veda, S. (2017). Developing high PV penetration cases for frequency response study of U.S. Western Interconnection. Proceedings of the IEEE Green Technologies Conference.

12. Quintero, J., Vittal, V., Heydt, G.T., and Zhang, H. (2014). The impact of increased penetration of converter control-based generators on power system modes of oscillation. *IEEE Transactions on Power Systems* **29** (5): 2248–2256.

13. Clark, K., Freeman, L.A., Jordan, G.A. et al. (2010). Impact of high levels of wind and other variable renewable generation on the grid operation: summary of major US studies. CIGRE 2010, C5_209_2010.

14. AEMO. (2019). Australian energy market operator, maintaining power system security with high penetrations of wind and solar generation. International Insights for Australia (October 2019).

15. MIGRATE – Massive InteGRATion of Power Electronic Devices. (2016). Deliverable D.1.1, Report on Systemic Issues. *Tech. Rep.* https://www.h2020-migrate (accessed 2020).

16. European Network of Transmission System Operators for Electricity (ENTSO-E). (2014). Dispersed generation impact on CE region security: dynamic study – 2014 report update – report of ENTSO-E SG SPD. http://www.entsoe.eu (accessed 2020).

17. Gevorgian, V., Zhang, Y., and Ela, E. (2015). Investigating the impacts of wind generation participation in interconnection frequency response. *IEEE Transactions on Sustainable Energy* **6** (3): 1004–1012.

18. Caldas, D., Fischer, M., and Engelken, S. (2015). Inertial response provided by full-converter wind turbines. Proceedings of the Windpower 2015 Conference and Exhibition.

19. Laudahn, S., Seidel, J., Engel, B. et al. (2016). Substitution of synchronous generator based instantaneous frequency control utilizing inverter-coupled DER. Proceedings of the 2016 IEEE Seventh International Symposium on Power Electronics for Distributed Generation Systems (PEDG).

20. Hydro-Québec TransÉnergie (2009). Transmission provider technical requirements for the connection of power plants to the hydro Québec transmission system. http://www.hydroquebec.com/transenergie/fr/commerce/pdf/exigence_raccordement_fev_09_en.pdf (accessed February 2009).

21. Fischer, M., Engelken, S., Mihov, N., and Mendonca, A. (2016). Operational experiences with inertial response provided by type 4 wind turbines. *IET Renewable Power Generation* **10** (1): 17–24.

22. Gautam, D., Vittal, V., and Harbour, T. (2009). Impact of increased penetration of DFIG-based wind turbine generators on transient and small signal stability of power systems. *IEEE Transactions on Power Systems* **24** (3): 1426–1434.

23. MIT Energy Initiative. (2011). Managing large-scale penetration of intermittent renewables. An MIT Energy Initiative Symposium (20 April 2011).

24. NERC. (2012). Frequency response initiative report - the reliability role of frequency response. Draft. http://www.nerc.com/docs/pc/FRI%20Report%209-30-12%20Clean.pdf (accessed September 2012).

25. North American Electric Reliability Corporation. (2017). Integrating inverter based resources into low short circuit strength system, reliability guideline (December 2017).

26. Milligan, M., Lew, D., Corbus, D. et al. (2009). Large-scale wind integration studies in the United States: preliminary results. Proceedings of the Eighth International Workshop on Large Scale Integration of Wind Power and on Transmission Networks for Offshore Wind Farms.

27. Bustamante, R.C., Roman Messina, A., Gonzalez, M.R., and Guizar, G.C. (2018). Assessment of frequency performance by grid integration in a large-scale power system. *Wind Energy* **21**: 1359–1371.

28. Acker, T. (2011). National Renewable Energy Laboratory, IEA Wind Task 24, Integration of Wind and Hydropower Systems, Volume 1: Issues, Impacts, and Economics of Wind and Hydropower Integration. *Tech. Rep. NREL/TP-5000-50181* (December 2011).

29. You, S., Kou, G., Liu, Y. et al. (2017). Impact of high PV penetration on the inter-area oscillations in the U.S. Eastern Interconnection. *IEEE Access* **5**: 4361–4436.

30. Tuttelberg, K., Kilter, J., Wilson, D., and Uhlen, K. (2019). Estimation of power system inertia from ambient wide area measurements. Proceedings of the 2019 IEEE Power & Energy Society General Meeting.

31. Begovic, M.M. and Messina, A.R. (2010). Wide area monitoring, protection and control, special issue on wide-area monitoring, protection and control. *IET Generation, Transmission & Distribution*: 1083–1085.

32. Miller, N.W., Shao, M., D'aquila, R. et al. Frequency response of the US Eastern Interconnection under conditions of high wind and solar generation. Proceedings of the 2015 Seventh Annual IEEE Green Technologies Conference.

33. Martinez, E. and Messina, A.R. (2011). Modal analysis of measured inter-area oscillations in the Mexican Interconnected System: The July 31, 2008 event. Proceedings of the 2011 IEEE Power and Energy Society General Meeting.

34. Muljadi, E., Gevorgian, V., Singh, M., and Santoso, S. (2012). Understanding Inertial and frequency response of wind power plants. Proceedings of the IEEE Symposium on Power Electronics and Machines in Wind Applications.

35. Golpira, H. and Messina, A.R. (2018). A center-of-gravity-based approach to estimate slow power and frequency variations. *IEEE Transaction on Power Systems* **33** (1): 1026–1035.
36. Golpîra, H., Seifi, H., Messina, A.R., and Haghifam, M.-R. (2016). Maximum penetration level of micro-grids in large-scale power systems: frequency stability viewpoint. *IEEE Transactions on Power Systems* **31** (6): 5163–5171.
37. Roman-Messina, A. (2020). *Data Fusion and Data Mining for Power System Monitoring*. Boca Raton, FL: CRC Press.
38. Rosipal, R. and Kramer, N. (2006). Overview and recent advances in partial least squares. In: *Subspace, Latent Structure and Feature Selection: Statistical and Optimization Perspectives Workshop (SLSFS 2005), Revised Selected Papers (Lecture Notes in Computer Science 3940)*, 34–51. Berlin: Springer-Verlag.
39. Irokawa, S., Andersen, L., Pritchard, D., and Buckley, N. (2008). A coordination control between SVC and shunt capacitor for windfarm. *CIGRE* **B4-307**.
40. Bevrani, H., Watanabe, M., and Mitani, Y. (2014). *Power System Monitoring and Control*. New York, NY: IEEE-Wiley Press.
41. Order 661. (2005). Interconnection for wind energy, issued by Federal Energy Regulatory Commission (FERC) of United States (2 June 2005).

Index

Renewable Integrated Power System Stability and Control, First Edition.
Hêmin Golpîra, Arturo Román-Messina, and Hassan Bevrani.